BIOJET FUEL IN AVIATION APPLICATIONS

BIOJET FUEL IN AVIATION APPLICATIONS

Production, Usage and Impact of Biofuels

CHENG TUNG CHONG

JO-HAN NG

ELSEVIER

Elsevier
Radarweg 29, PO Box 211, 1000 AE Amsterdam, Netherlands
The Boulevard, Langford Lane, Kidlington, Oxford OX5 1GB, United Kingdom
50 Hampshire Street, 5th Floor, Cambridge, MA 02139, United States

Library of Congress Cataloging-in-Publication Data
A catalog record for this book is available from the Library of Congress

British Library Cataloguing-in-Publication Data
A catalogue record for this book is available from the British Library

ISBN: 978-0-12-822854-8

Publisher: Matthew Deans
Acquisitions Editor: Carrie Bolger
Editorial Project Manager: Aleksandra Packowska
Production Project Manager: Prasanna Kalyanaraman
Cover Designer: Greg Harris

Typeset by TNQ Technologies

Contents

Preface

Biojet fuel is an emerging renewable energy for aviation applications that will soon become an essential part of the aeronautical sector. This paradigm shift meant that the study of biojet fuel is increasingly becoming part of mainstream elective courses for undergraduate students pursuing degrees in chemical engineering, mechanical engineering, and sustainable energy engineering. This book is intended for use by the aforementioned *undergraduate* students, with emphasis placed to give students a holistic view in terms of the technical, economical, political, and social aspects of biojet fuel. The text is also intended as a gateway for *postgraduate* degree studies or as supplementary text for introductory courses into alternative fuels.

The philosophy behind this book is for it to be the definitive "first" book for readers wanting to know about the basic fundamental and practical issues on biojet fuels. This supports the authors' main goals in writing the book, which is to provide a comprehensive book for use in classrooms and also for self-study. Thus, the book is written in an accessible manner to encourage readers to develop deep understanding on the subject matter, by linking up scientific knowledge, established facts, latest real-world data, and viewpoints on biojet fuels.

In addition to students and researchers, the authors are expecting this book, ***Biojet Fuel in Aviation Applications: Production, Usage and Impact of Biofuels***, will also appeal academics preparing for new courses to usher in the age of sustainable fuels, government officials in charge of energy and environmental policies, industrial players desiring the keep-up with the key knowledge about the future of aviation fuels, and general public with an inquisitive mind.

Book organization by chapter

The authors arranged the chapters in a logical manner to bring readers through a journey of understanding the rationale behind the rise of biojet fuel around the world, followed by the bulk technoeconomical concerns, and culminating in its sustainability impacts on planet Earth. The following paragraphs provide insights on the ensuing chapters:

Chapter 1 addresses the biojet fuel policies, legislations, initiatives, and roadmaps for global aviation. In this chapter, readers will learn about the simultaneous efforts by individual governments around the world to decarbonise their domestic aviation sector and how they combined their efforts for international flights through the Carbon Offset and Reduction Scheme for International Aviation (CORSIA). The market-based measures, mandates, fuel standards, initiatives, reporting tools, and legally binding commitments all synergistically help to support the top-down development of the biojet fuel industry.

The primary goal of Chapter 2 is to provide readers firm grasp on the production methods, primarily categorized as oil-to-jet, alcohol-to-jet, gas-to-jet, and sugar-to-jet methods. Each of the broader categories contains production pathways, many of which pertaining to the ASTM D7566 approved pathways. The chapter also discusses how the current biojet production processes have developed to improve their yields and where they are in the technology maturity curve.

Chapter 3 highlights the characteristics of biojet fuel that distinguish it from conventional jet aviation fuel. This covers the typical chemical composition, physicochemical properties, and their compatibility with present-day aviation sector infrastructure and usage in jet engines. Readers will understand the significance of the "drop-in" requirement of biojet fuel in blends with fossil jet fuel.

This ties in with Chapter 4 where the neat and blended biojet fuel performances under combustion are the key focuses. The mechanisms of biojet fuel spray, combustion, and emissions formation are fundamentally discussed and validated by research data. This is complemented by the myriad of flight tests conducted around the world using the various biojet fuels.

Chapter 5 emphasizes on the economics of biojet fuel and identifies the practical factors affecting the supply—demand scenario such as crude oil prices, biojet fuel production costs, feedstock prices, taxation, and subsidies. In addition to economic concerns, the availability of feedstocks and barriers to commercialization are also highlighted. The chapter also placed importance on the postpandemic cost issues and the recent development of price discovery for biojet fuel.

The final chapter, Chapter 6, provides an overview of pertinent issues pertaining sustainability and energy balance via a life cycle assessment (LCA) methodology. This is augmented with a holistic view using an energy—water—food (EWF) nexus approach to resource management. The true impacts of biojet fuel are fully elucidated in this chapter.

Consistent chapter organization

While the book is intended to be read in the arranged order, the authors purposefully wrote each chapter in a self-contained manner. This allows readers to approach the chapters in any order and will still gain the same insights as those faithfully following the chapters as intended. Within each chapters, the structure order starts with a general introduction, followed by the main contents which cover the most salient information, and ending with a chapter summary to provide readers with the take-home messages. Each chapter uses numerous tables and figures interspersed with text to provide data for comparison, reveal trends, summarize concepts, illustrate concepts, and support conclusions.

Acknowledgments

The authors would like to thank and acknowledge the contributions of Elsevier and its staff for the professional support provided in the preparation of this book. In particular, we would want due recognition to be given to Carrie Bolger, Acquisition Editor, who provided guidance during the book proposal stage leading to the project being approved; Aleksandra Packowska, Editorial Project Manager, who provided top-notch professional support and encouragement throughout the writing process; Rajaganapathy Essaki Pandyan, Payee Information Manager, and Kavitha Balasundram, Copyrights Coordinator, for shedding light on publishing-related matters.

We would also like to express our appreciation to the book proposal and manuscript reviewers. Their remarks and comments help us to gain focus on the topics to write and also improve the quality of the book.

Finally, we would also want to convey heartfelt thanks to our family, Stella, Hoe Jay and Chen Xi (Cheng Tung Chong) and Wong Minh Chjiat Isabelle and Einstein Ng Gi Neer (Jo-Han Ng) for their continued patience, abundance in support, and unconditional love throughout this project.

CHAPTER 1

Global Aviation and Biojet Fuel Policies, Legislations, Initiatives, and Roadmaps

1.1 Introduction

Emissions from aviation contribute to 2.0% of the total global CO_2 emissions. While the proportion is relatively small compared with other forms of transport, air travel per capita emissions is among the highest with aviation contributing to 12% of CO_2 from all transport sources. Also worrying is the release of emissions at higher altitudes as compared with other pollution methods, leading to greater global warming effects.

Policies and legislations regarding biojet fuel will play key roles in shaping the industry and steer the market adoption of the alternative aviation fuel to supplant its fossil counterparts. Favorable policies could be introduced to provide financial incentives to attract investment into the nascent market, while legislations will provide mandates for legally binding commitments. They are frequently combined when governments need to encourage and regulate a new sector of national importance.

Comprehensive regulatory framework for biojet fuels at international and national levels is crucial to improve energy security, improve environmental sustainability, grow the sector for economic well-being, linking up stakeholders and resolve technical difficulties. It will improve the chances of breaking status quo and provide a smooth path toward the mass adoption of biojet fuel for the aviation industry.

1.2 Global—International Civil Agency Organization

1.2.1 Carbon Offset and Reduction Scheme for International Aviation

From the 2% of total global CO_2 emissions, international aviation emissions account for 1.3% of the global CO_2 emissions, while domestic aviation contributes to the other 0.7% (Deane and Pye, 2018). The former falls

Biojet Fuel in Aviation Applications
ISBN 978-0-12-822854-8
https://doi.org/10.1016/B978-0-12-822854-8.00004-4

under the responsibility of the International Civil Agency Organization (ICAO) as flights cross international boundaries, while the latter is reported under the United Nations Framework Convention on Climate Change (UNFCCC) with the responsibilities held by the countries covered under the framework. As such, emissions produced from the international aviation category are not included under the Paris Agreement's Nationally Determined Contributions (NDCs).

ICAO is influential on the global stage since its inception in 1944 under the Chicago Convention, it has grown to have 193 contracting states agreeing to multilateral conventions. In the 1970s, ICAO tackled aviation-related environmental issues through the Committee on Aircraft Noise (CAN) and Committee on Aircraft Engine Emissions (CAEE), which were formed in 1970 and 1977, respectively (ICAO, 2019f). These technical committees of the ICAO council then developed Standards and Recommended Practices (SARPs) to deal with aircraft noise and control of aircraft engine emissions, which were parked under SARPs Annex 16. In 1983, the Committee on Aviation Environmental Protection (CAEP) was formed to merge and supersede both CAN and CAEE. The CAEP focuses on both the original aims of CAN (for aircraft noise) and CAEE (for aircraft emissions), which are then combined for a more general coverage of aviation environmental impacts.

Fuel requirements are specified in the SARPs Annex 6, of which the various sovereign national aviation authorities or regulating authorities could adjust to better match the needs and characteristics of their airspace. States are expected to undertake measures to comply to the standard portion of the SARPs or immediately file a difference if they implement any deviation, while being recommended on the best practices for the Recommended Practice of the SARPs. The focus of the SARPs with respect to fuel covers primarily on matters such as sufficiency to complete flights, fuel contingency requirements, in-flight fuel checks, and fuel emergency situation. The SARPs do not specify biojet fuels per se. Ultimately, the SARPs only concern themselves with flight operating-related State Safety Programmes (SSP) and Safety Management Systems (SMS) by service providers.

However, to address the annual increase in total global CO_2 emissions, ICAO adopted a global carbon-offset scheme in October 2016 for nondomestic aviation under the Carbon Offset and Reduction Scheme for International Aviation (CORSIA). CORSIA is formed under Working Group 4 (WG4) of CAEP. Under the scheme, aircraft operators operating within signee countries are encouraged to offset their emissions against the average level of international aviation CO_2 emissions for the years 2019 and 2020. Aircraft operators are required to monitor emissions on all international flight routes and offset emissions by purchasing eligible emissions units. The eligible emissions units need to be generated by emissions reduction projects in other sectors such as the renewable energy sector. This meant that biojet aviation fuel could generate eligible emissions units used for carbon offsetting.

This represents the basis for carbon neutral growth from 2020 onward, where the baseline is set for comparison against future years (ICAO, 2020a). The difference between the international aviation CO_2 emissions as covered by the scheme and the average baseline emissions of years 2019 and 2020 will represent the required sector offset in any year from 2021 onward. The carbon offsets can be obtained from either emissions trading scheme or the Clean Development Mechanism (CDM) as defined in Article 12 of the Kyoto Protocol.

Sixty-nine states (as of May 24, 2017) have stated their intention to voluntarily participate in the scheme from the outset. While they represent more than 87% of international aviation activities (Deane and Pye, 2018), notable countries such as India and Russia are not covered under CORSIA. This pilot phase will apply from 2021 through 2023. The subsequent first phase and second phase will apply from 2024 through 2026 and from 2027 through 2035, respectively. Alongside states volunteering in the pilot phase, additional states may also opt in to participate in the first phase. All European Union (EU) countries will join the scheme from the onset. The second phase is made mandatory for states having an individual share of international aviation activities on the basis of revenue ton-kilometers (RTK) above 0.5% of total RTKs in 2018 or is listed under the cumulative share (from highest to lowest) of RTK up to 90% of total RTK. Exceptions are given to least developed countries (LDCs), small island developing states (SIDSs), and landlocked developing countries (LLDCs), although they are allowed to voluntarily participate in the second phase. Fig. 1.1 shows the states implementing CORSIA (Openairlines, 2018). During the 15-year period of 2021–35, CORSIA is envisioned to offset about 80% of total emissions above 2020 levels.

1.2.2 Sustainable Aviation Fuels

ICAO recognizes sustainable aviation fuels (SAFs) as an important element to reduce aviation emissions and also to eventually ensure the success of

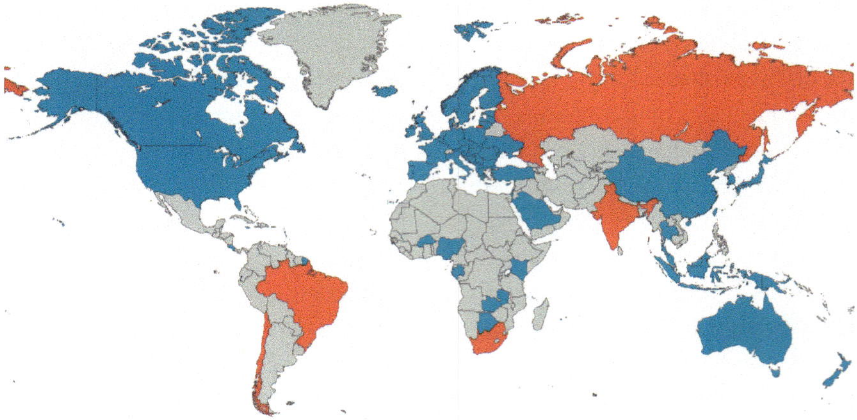

Figure 1.1 States implementing CORSIA. *CORSIA*, Carbon Offset and Reduction Scheme for International Aviation. *(Adapted from Openairlines, 2018. CORSIA: Who Needs to Be Participating in the Scheme?. https://blog.openairlines.com/corsia-who-needs-to-be-participating.)*

CORSIA. This includes appreciating the importance of biojet fuel (under the general umbrella of alternative fuels) and urges member states to take due account of ICAO policies and guidance on emissions related to environmental protection and climate change under ICAO Resolution A38-18 (ICAO, 2013). A further resolution by ICAO under Resolution A40-18 by the ICAO Assembly also acknowledges the need to develop SAF in an economically, socially, and environmentally sustainable manner. States are requested by ICAO to assess the sustainability of all alternative fuels for use in aviation, where they should achieve net greenhouse gas (GHG) emissions reduction on a life cycle basis and work together through ICAO and other relevant international bodies to exchange information and best practices on the sustainability of alternative fuels for aviation. ICAO also pursues three key programs with regard to SAF, namely the ICAO Global Framework for Aviation Alternative Fuels (GFAAF), the 2050 ICAO Vision, and the ICAO Stocktaking Process (ICAO, 2020b).

The ICAO GFAAF was formulated as the then tangible product of the 2009 ICAO Conference on Aviation Alternative Fuels. The GFAAF is an online database containing information, projects, and news announcements of aviation fuels dating back to 2005. While states and stakeholders can share relevant information with ICAO through this portal, it also serves a secondary function of being able to keep tabs of the progress of alternative fuels in aviation through State's action plans and work with financial institutions to facilitate financing of alternative fuel projects to overcome initial market hurdles. A live feed of the ongoing alternative fuel purchase agreements inclusive of batch delivery and ongoing deliveries through offtake agreement is shown on the GFAAF portal, although it does not necessarily equate to the quantity of alternative fuel used on flights. This is due to the gap in information regarding the airports' fuel blending procedures. Nonetheless, it is a good proxy of the SAF activities for airlines and airports.

ICAO also initiated the 2050 ICAO Vision for Sustainable Aviation Fuel during the 2017 edition of the ICAO Conference on Aviation and Alternative Fuels (CAAF/2) in Mexico (ICAO, 2018). The vision is to have stakeholders within the international aviation sector to operate flights using a significant proportion of SAF by 2050. The uptake of SAF is established to be a key contributor to meet ICAO's climate objectives and also allow the aviation sector to contribute in 13 out of the 17 United Nations Sustainable Development Goals (SDGs). As a corollary to the increase in SAF usage, international civil aviation should also reduce carbon emissions significantly. The vision also ties in with the GFAAF where stakeholders are expected under the vision to continuously update the portal. The 2050 ICAO Vision also identifies key steps to meet the vision which include the

- role of ICAO as a facilitator to support states to develop and deploy SAF;
- development of guidance materials describing the drop-in nature of SAFs;
- support from states to approve new conversion processes;
- support from states to develop and implement stable policies to facilitate deployment of SAF;
- evaluation of policy effectiveness through qualitative metrics by states;
- evaluation and facilitation of funding sources to implement SAFs; and
- collaborative initiatives among states alongside industries to reduce the price gap between SAF and conventional aviation fuels.

It should, however, be noted that the 2050 ICAO Vision will not set a precedent or prejudge the periodic review of CORSIA as stated under paragraph 18 of Assembly Resolution A39-3.

The third major initiative on SAF is the ICAO Stocktaking Process which stemmed from a decision made during CAAF/2. The stocktaking exercise has the objective of assessing the SAF development and deployment progress.

During the first ICAO Stocktaking Process held from April 30, 2019 to May 1, 2019 in Canada, the stocktaking process was conducted through the means of a simple questionnaire, which requires information on conducted projects, project partners, project duration, feedstock used, feedstock origin, amount of aviation fuel produced, and if the SAF has been certified by any Sustainable Certification Scheme (SCS).

The self-reported stocktaking data will complement environmental trends analysis to provide an overall picture of the impacts of SAF on the aviation industry and also environment at large. In addition to assessing the progress of SAF development and deployment, the aggregated data can also be used to steer political updates for member states, provide confidence for financial institutions to support SAF projects, match providers and requestors of assistance, and compile the data for outreach purposes to dispel the notion of SAF competition with food and water.

1.2.3 CORSIA Eligible Fuels

The CAEP through Fuels Task Group (FTG), which is one of the 11 groups with CAEP membership, is tasked to develop the processes and methodologies to define what qualifies as SAF under CORSIA, or more precisely CORSIA eligible fuel (CEF). This is requested under ICAO Assembly Resolution A39-3 and defined in the context of CORSIA, Annex 16, Volume IV. Both renewable and fossil-based aviation fuels have the potential to be a CEF. The CORSIA sustainable aviation fuel refers to a renewable or waste-derived aviation fuel that meets the CORSIA Sustainability Criteria, while the CORSIA lower carbon aviation fuel is the counterpart for fossil-based aviation fuels (ICAO Secretariat, 2019). The focus is on sustainability criteria and life cycle methodologies.

To ensure that the CEF meets the CORSIA Sustainability Criteria, Sustainability Certification Schemes (SCSs) are developed by ICAO to conduct the sustainability certification process. The current CORSIA Sustainability Criteria specifying the sustainability criteria required to be certified as a CEF is valid through the end of the CORSIA pilot phase in 2023. Once a fuel is deemed to be a CEF, its life cycle emissions value (LS_f) is evaluated, and their default values are listed in the "CORSIA Default Life Cycle Emissions Values for CORSIA Eligible Fuels" document.

Table 1.1 shows the assigned CORSIA default life cycle emissions values for the 16 feedstocks evaluated to have the potential to be a CEF (ICAO, 2019c). The LS_f indicates the expected CO_2-equivalent reduction from the

Table 1.1 Feedstocks with potential to be CORSIA eligible fuel with their CORSIA default life cycle emissions value.

Fuel conversion process	Region	Feedstock	Type	Core LCA value (gCO_2e/MJ)	ILUC LCA value (gCO_2e/MJ)	LS_f (gCO_2e/MJ)
Fischer-Tropsch	Global	Agricultural residues	Waste	7.7	0	7.7
		Forestry residues	Waste	8.3	0	8.3
		Municipal solid waste (MSW) with 0% nonbiogenic carbon (NBC)	Waste	5.2	0	5.2
		MSW with NBC given as a %	Waste	NBC*170.5 + 5.2	0	NBC*170.5 + 5.2
	United States	Miscanthus	Energy crop	10.4	−32.9	−22.5
		Poplar	Woody crop	12.2	−5.2	7.0
		Switchgrass	Energy crop	10.4	−3.8	6.6
	European Union	Miscanthus	Energy crop	10.4	−22.0	−11.6

Continued

Table 1.1 Feedstocks with potential to be CORSIA eligible fuel with their CORSIA default life cycle emissions value.—cont'd

Fuel conversion process	Region	Feedstock	Type	Core LCA value (gCO_2e/MJ)	ILUC LCA value (gCO_2e/MJ)	LS_f (gCO_2e/MJ)
Hydroprocessed esters and fatty acids	Global	Corn oil (extracted from dry mill ethanol plants)	By-product	17.2	0	17.2
		Palm fatty acid distillate	Waste	20.7	0	20.7
		Tallow	Waste	22.5	0	22.5
		Used cooking oil	Waste	13.9	0	13.9
	United States	Soybean oil	Oil crop	40.4	24.5	64.9
	European Union	Rapeseed oil	Oil crop	47.4	24.1	71.5
	Brazil	Soybean oil	Oil crop	40.4	27.0	67.4
	Malaysia and Indonesia	Palm oil—closed pond	Oil crop	37.4	39.1	76.5
		Palm oil—open pond	Oil crop	60.0	39.1	99.1
Alcohol (ethanol) to jet	United States	Corn grain	Food crop	65.7	25.1	90.8
	Brazil	Sugarcane	Food crop	24.1	8.7	32.8

Alcohol (isobutanol) to jet	Global	Agricultural residues	Waste	29.3	0	29.3
		Forestry residues	Waste	23.8	0	23.8
	United States	Corn grain	Food crop	55.8	22.1	77.9
		Miscanthus	Energy crop	43.4	−54.1	−10.7
		Switchgrass	Energy crop	43.4	−14.5	28.9
	European Union	Miscanthus	Energy crop	43.4	−31.0	12.4
	Brazil	Sugarcane	Food crop	24.0	7.3	31.3
Synthesized isoparaffins)	European Union	Sugar beet	Food crop	32.4	20.2	52.6
	Brazil	Sugarcane	Food crop	32.8	11.3	44.1

CORSIA, Carbon Offset and Reduction Scheme for International Aviation; *ILUC*, indirect land usage change; *LCA*, life cycle assessment.

use of a CEF due to the sustainable fuel conversion pathway, production region, feedstock type, land size usage, and the type of land used. The general guiding principle is for waste, residue, or by-product to only consider the core life cycle assessment (LCA) value as the LS_f, whereas the other cases will require the additional indirect land usage change (ILUC) to be factored in the LS_f value.

Some of the ILUC values are negative, which means that additional carbon sequestration will be larger overall than the associated carbon emissions from land use changes. The use of negative ILUC values is a point of contention as it is said to introduce possible optimism bias. This is compounded by the CAEP's decision to base ILUC values using the lower value of the competing models of GTAP-BIO and GLOBIOM models favored by the US and EU delegations, respectively. It should be noted that the final values used are obtained through the process of model reconciliation instead of taking any particular model as the base. It is still an open point of debate if the values used by CAEP for CORSIA are due to modeling improvement or the stance to adopt a more biojet fuel-positive approach. Using this point of contention as the backdrop, a review will be conducted at the end of the pilot phase to determine if negative ILUC values should be allowed.

It should also be noted that the default LS_f values can be challenged by fuel producers by using the calculation methodologies stated under the "CORSIA Methodology for Calculating Actual Life Cycle Emissions Values" document approved in November 2019 (ICAO, 2019d). Using the specified methodology and proof of technical information, fuel producers are allowed the liberty to define a lower LS_f values than that of the default value. This is also particularly useful for fuel producers if their fuel production pathway does not yet have a default core life cycle value.

As per the methodology set by the CORSIA policy, fuel producers need to determine the CEF emissions reductions (ER_y) using Eq. (1.1) (ICAO, 2019b):

$$ER_y = FCF\left[\sum_f MS_{f,y} \times \left(1 - \frac{LS_f}{LC}\right)\right] \tag{1.1}$$

where the subscript y denotes the year, subscript f refers to the fuel type, *FCF* is the fixed value fuel conversion factor, *MS* denotes the mass of *CEF* claimed, and *LC* refers to the fixed value baseline life cycle emissions.

The fixed values for FCF are 3.16 kg CO_2/kg fuel for Jet-A/Jet-A1 and 3.10 kg CO_2/kg fuel for AvGas/Jet-B. The LCs for jet fuel and AvGas are 89 and 95 gCO_{2e}/MJ, respectively.

1.2.4 CORSIA Central Registry

The CORSIA Central Registry (CCR) keeps three key sets of information nested under the "CORSIA Central Registry: Information and Data for the Implementation of CORSIA." As stated from the title, the documents serve to provide important information to support the implementation of CORSIA. The first among the three documents is the "CORSIA 2020 Emissions" where the total CO_2 emissions arising from international aviation in 2020 will be published. This is relevant to biojet fuel as its efficacy in reducing CO_2 equivalent emissions under the CORSIA program can be gauged. However, it will be interesting to see by how much the CO_2 emissions will decrease as the global COVID-19 pandemic, which has its first wave peaked in 2020. The pandemic has all but decimated the passenger subsector of the aviation industry. The expected low CO_2 emissions in 2020 will skew the data for future comparison. It will be prudent to chalk off emissions reduction from 2020 as a one-off outlier rather than being part of the underlying trend. As CORSIA will only start its pilot phase in 2021, there is no report to be made available until the second half of 2021.

The second key document is the "CORSIA Aeroplane Operator to State Attributions." This is not directly linked to biojet fuels as it only states the airplane operator name, the attribution method, and the identifier for each carrier. The current third edition of the document published in December 2019 provides information on 690 airplane operators from 122 states.

The third document is the "CORSIA Annual Sector's Growth Factor (SGF)." The first edition of the document is envisioned to be published in 2022, midway through the pilot stage of CORSIA. Prior to the global pandemic, the February 2019 estimated SGF from 2021 to 2035 was expected to rise from 6% in 2021 to 38% in 2035 as shown in Fig. 1.2 (ICAO, 2019a).

The prepandemic sector outlook was optimistic at a compounded annual growth rate of 4.3% in terms of revenue passenger-kilometers with Africa having the most potential for growth due to its emerging industrial sector and large developing population. The South East Asian (SEA) region was also expected to see rapid growth due to the boom in low-cost carriers,

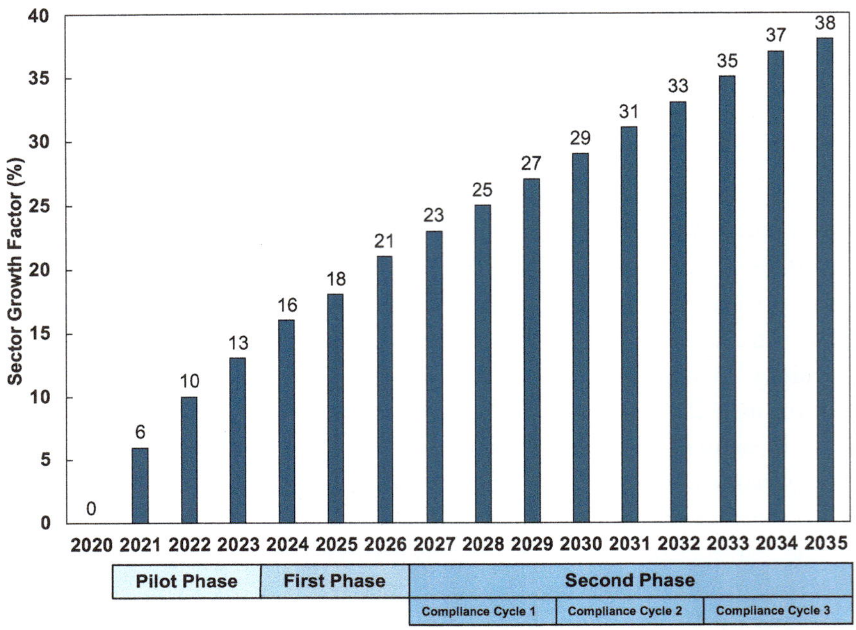

Figure 1.2 Estimated sector growth factor (SGF) from 2021 to 2035. *(Adapted from ICAO, 2019a. Committee on Aviation Environmental Protection (CAEP). https://www.icao.int/environmental-protection/CORSIA/Documents/CAEP_Analysis%20on%20the%20-estimation%20of%20CO2%20emissions%20reductions%20and%20costs%20from%20CORSIA.pdf.)*

growing middle class, and efforts to liberalize air traffic regulations. It is also in the SEA region where biojet fuel could potentially take a stronghold as these countries such as Malaysia and Indonesia are rich with palm oil as feedstock and the Philippines with coconut oil as potential feedstock. The sudden and rapid contraction of the aviation industry will impact growth of the overall sector and also the biojet fuel industry.

ICAO also frequently updates the "CORSIA Central Registry (CCR): Information and Data for Transparency" listing of verification bodies accredited in member states. In the latest April 2020 sixth edition, 40 verification bodies from 17 states are accredited and listed. The United States and China lead the list with eight and six verification bodies, respectively. These verification bodies can conduct verifications for carbon offsetting and GHG inventory reports under the CORSIA scheme. The number of verification bodies is expected to grow substantially as not all voluntary member states of CORSIA have accredited verification bodies. In fact, in the first edition of the document in May 2019, there were only 22 accredited verification bodies from seven member states. The addition of accredited verification bodies will make it more efficient for stakeholders of CORSIA to submit verified reports and for ICAO to determine if the biojet fuels entering the aviation industry actually help to achieve carbon neutral growth. A summary of CORSIA-related reporting required for the CCR can be found in Table 1.2. CO_2 emissions and CEF will be reported.

1.2.5 CORSIA CO_2 Estimation and Reporting Tool

Under CORSIA, airplane operators within voluntary member states are required to report the CO_2 emissions or estimated CO_2 generated. The CORSIA CO_2 Estimation and Reporting Tool (CERT) exists to help airplane operators to generate a summary assessment for airline operators with relatively "lower" levels of activities. This is for aircraft operators to fulfill the monitoring and reporting requirements in accordance with ICAO Annex 16, Volume IV, Part II, Chapter 2, 2.2.1 and Appendix 3.

CERT contains a set of equations for the estimation of CO_2 emissions based on the Great Circle Distance or Block Time for a given aircraft type (ICAO, 2019e). CERT uses its standardized emissions monitoring plan and emissions report to assess eligibility to utilise the fuel use monitoring methods, assess the scope applicability of monitoring, reporting, and verification (MRV) requirements, and help to fill in any CO_2 emissions data gaps. The last of the three is the most pertinent as data are often difficult to come by and to estimate. CERT itself refers to ICAO's aircraft database, location indicators, aircraft type designators, and fuel formula. It also refers to the European Union Aviation Safety Agency (EASA)—approved noise level by maximum take-off mass (MTOM).

There is no specific category for biojet fuels under CERT, but biojet fuels that meet the standards (for example, Jet-A1) can be included as equivalent fuel. The undifferentiated categorization of biojet fuel possibly stemmed from the present drop-in fuel mechanism where biojet fuel meeting the standards cannot be differentiated postblending with fossil-based aviation fuel. Since they have the same or similar properties, the emissions arising from the combustion of either fuel will also be comparable. Furthermore, the categorization of biojet fuel as equivalent fuel is practical as the CERT is meant to be a simplified tool to easily estimate CO_2 emissions levels.

Table 1.2 CORSIA-related reporting required for the CORSIA Central Registry for baseline, pilot phase, and first phase of CORSIA.

	Baseline		Pilot phase				First phase	
Information	**2019**	**2020**	**2021**	**2022**	**2023**	**2024**	**2025**	**2026**
Airplane operators	Yes	Yes	Yes	Yes	Yes	Yes	Yes	Yes
Verification bodies	Yes	Yes	Yes	Yes	Yes	Yes	Yes	Yes
CO_2 emissions		Yes 2019 data	Yes 2020 data	Yes 2021 data	Yes 2022 data	Yes 2023 data	Yes 2024 data	Yes 2025 data
CORSIA eligible fuels		Optional 2019 data	Optional 2020 data	Yes 2021 data	Yes 2022 data	Yes 2023 data	Yes 2024 data	Yes 2025 data
Canceled emissions units							Yes 2021–23 data	

CORSIA, Carbon Offset and Reduction Scheme for International Aviation.

1.2.6 Impact of COVID-19 on CORSIA

The COVID-19 pandemic has caused unprecedented disruption to the aviation sector, especially international air travel. Air travel has greatly reduced in 2020 with airlines around the world still assessing the impact. An ICAO report on the state of the industry in September 2020 stated a 50% reduction in passenger seats offered by airlines. This will invariably reduce the CO_2 emissions in 2020 and beyond, with great level of uncertainty. In fact, it represents a reversal of expectations where the International Air Transport Association (IATA)'s 2019 end-year report estimated a 2.3% increase in CO_2 emissions over the 2019 levels (IATA, 2019).

As such, it greatly impacts the implementation of CORSIA as the baseline is set upon a 2-year emissions average for 2019 and 2020. This decreases the CORSIA's sectoral baseline sharply when compared with a non-COVID-19 projection. The greatly reduced baseline will unfairly burden the airline industry groups as that will mean greater offsetting costs.

ICAO reacted by invoking ICAO Assembly Resolution A40-19 to provide a safeguard by adjusting CORSIA, as allowed in the unforeseen circumstances where the sustainability of the scheme is affected. The extraordinary disruption in the form of COVID-19 has led to ICAO agreeing to use only the 2019 emissions level to determine the baseline levels. This means that airlines will be allowed to discharge 30% more emissions as compared with 2019's level, which amounts to 81 million metric tons during CORSIA's pilot phase.

Future implications beyond the pilot phase (2021—23) cannot be known as it is unclear if the industry will undergo a "V" (full and fast recovery), "U" (slow recovery with dampened long-term growth), or "L" (emissions fall then level off) shaped recovery. As such, ICAO's move to shift the baseline determination rules amid the uncertainty is welcomed by airline operators. Airline operators will now get reprieve from an economic standpoint.

However, critics are of the opinion that the change in CORSIA baseline for the pilot phase will all but practically eliminate offset requirements. It is expected that this adjustment will also delay the implementation of aviation carbon offset by up to 5 years. Furthermore, it has the potential to dampen the green energy market as the rules are inconsistent. This shines a negative light on CORSIA's credibility and long-term stability.

The first review of CORSIA is not due until 2022, with offsetting targets for the first phase (2024—26) not being finalized until the end of 2023. It will be prudent to adopt a practical route of observing the rebound level of air travel before making any further adjustments.

1.3 European Union

1.3.1 European Union Emissions Trading Scheme

The European Union has a head start in policies regarding to aviation emissions. In fact, in year 2012, the EU preceded ICAO's initiatives by effecting aviation market-based measures (MBMs) through the inclusion of the sector under the European Union Emissions Trading System (EU ETS) (Deane and Pye, 2018).

In this scheme, the onus is on the airlines to reduce aviation-related emissions. Airlines operating in the European Union irrespective of being European or non-European are required to monitor, report, and verify their emissions level. From it, tradable allowances can be received depending on the flight emissions level per year.

The scheme in its original form had the ambition of covering all of European Union's aviation emission, although it was contested by the industry. Geographically, the legislation was also initially designed to apply to emissions from flights from, to, and within the European Economic Area (EEA). EEA covers all of the EU member states plus three other countries, namely, Iceland, Liechtenstein, and Norway (European Commission, 2020).

However, in 2013, the European Union decided to limit the reach of EU ETS to just internal flights within the confines of EEA until 2016. This "stop the clock" measure on the implementation of international aviation law was taken to support the efforts of ICAO in developing a global system to combat aviation emissions, which eventually came in the form of CORSIA in 2016. "Stopping the clock" was widely regarded as being the crucial component in the provision of political negotiation space to ICAO for the formation of an international framework in tackling aviation carbon emissions.

With the introduction of CORSIA, the European Union retained the geographic scope of EU ETS from year 2017 with a view to review the EU ETS for aviation subjected to the codevelopment of CORSIA. It should be noted that the EU ETS would relapse to its original scope covering extra-EEA flights from 2024.

1.3.2 Renewable Energy Directives

The European Commission (EC) recognizes the need to set out the path to climate neutrality by 2050. This can be achieved through the decarbonization of economic activities in all sectors and reduce GHG emissions. As the energy sector contributes over three quarters of European Union's GHG emissions, tackling the emissions arising from the energy sector will be a keystone in achieving climate neutrality (European Commission, 2018).

The European Union has had various directives and indicative targets related to renewables in the past, namely the 1997 indicative EU target of 12% renewables by 2010, the 2001 directive on electricity production from renewables, and the 2003 directive on biofuels and renewable fuels for transport. In 2009, the EC started to set nationally binding targets of 20% renewables by 2020 through the original Renewable Energy Directive (2009/28/EC). The Directive allocates individual national targets to member states, ranging from the lowest at 10% for Malta to 49% for Sweden (Deane et al., 2017). Member states must also have 10% transport fuels originating from renewable sources by 2020 under the 10% RES-T target. In principle, as long as the biofuel meets the specific sustainability criteria, it can be included as meeting the quota. However, for RED (2009/28/EC), biojet fuel contributed to a negligible amount to this target as road transport biofuels such as bioethanol and biodiesel remained as the lower hanging fruits for member states to meet their quota. Also, road transport accounts for the bulk of the total EU-28 transport emission in 2012 at 72%. Revisions to RED were made to also specify aviation sector-specific aspects. In fact, the renewables for the aviation sector (alongside the maritime sector) were given a boost where they are weighted 20% more.

RED was revised in 2018 as Renewable Energy Directive 2018/2001/EU with a more ambitious binding target of at least 32% for 2030. The revised RED is also widely known as RED II. The revision was made by the EC to keep the EU at the front of the pack for renewables, while also bringing the EU one step closer to meeting its Paris Agreement commitments. In terms of the transport sector, the share of renewable fuel target was increased to 14% by 2030. The criteria for bioenergy sustainability were also specified and strengthened. The main elements and key provisions of the revised RED are summarized in Table 1.3 (Chiaramonti and Goumas, 2019).

Advanced biofuel sources are also explicitly defined into two separate groups. The groupings are tabulated in Table 1.4 (USDA GAIN Report, 2019). Feedstocks as listed in part A must form 0.2%, 1.0%, and at least 3.5% of transport energy in 2022, 2025, and 2030, respectively. The feedstocks consist of only nonfood sources with algae, municipal waste, agricultural waste, glycerine, and forestry residue forming the bulk of the list. On the other hand, part B consists of used cooking oil and some categories of animal fat. Part B sources will be capped at 1.7% in 2030. Due to advanced biofuels category being a subset of the overall RTF, it can be double-counted by member states toward both the RTF (14%) and advanced biofuel (3.5%) shares. It should be noted that the mandates by member states as of 2019 are

Table 1.3 Main elements and key provisions for the revised Renewable Energy Directive 2018/2001/EU.

Main elements	Key provisions
Share of renewable transport fuels by 2030	Minimum 14%, of which 3.5% must be advanced biofuels
GHG savings	65% advanced biofuels from 2021, 70% renewable fuels of nonbiological origin
Advanced biofuels that may be double-counted by member states	Capped at 1.7%
Multipliers in specific end-use sectors	Biofuels in aviation: 1.2 Biofuels in maritime: 1.2 Electricity in road: 4 Electricity in rail: 1.5
Advanced biofuel growth pathway	0.2% in 2022, 1% in 2025, 3.5% by 2030
Food-feed crop-based biofuels	Maximum of 7%
High-ILUC risk biofuels	Below 2019 consumption level to gradually being phased out totally by 2030
Low-ILUC risk biofuels	Exempted from phasing out

GHG, greenhouse gas; *ILUC*, indirect land usage change.

Table 1.4 Advanced biofuel feedstocks.

Category	Feedstock
Part A	Algae Animal manure Bacteria Bagasse Biomass fraction of industrial waste not fit for use in the food or feed chain Biowaste from private households subject to separate collection Cobs cleaned of kernels of corn Crude glycerine Husks Forest residues Grape marc Nut shells Organic fraction of municipal waste Palm oil mill effluent and empty palm fruit bunches Sewage sludge Straw Tall oil pitch Wine lees Other nonfood cellulosic material Other lignocellulosic material except saw logs and veneer logs
Part B	Used cooking oil Some categories of animal fats

still heavily weighted upon biodiesel and bioethanol. In fact, only the United Kingdom has set a requirement for blending development fuel, which must be aviation fuel, hydrogen, or substitute natural gas.

The aviation sector is unique within the implementation of RED due to the nature of international aviation, where emissions and environmental effects are not bounded to just the national borders of the EU member states. In reality, the modes of implementation must be compliant to both the RED and CORSIA frameworks. This requires greater coordination than just regulations and policies being issued in silo. The risk of carbon accounting discrepancies is possible especially in how emissions accounting methods are defined. Furthermore, the different frameworks also have differing GHG savings requirements where CORSIA requires 10% instead of the more stringent 65% (after January 2026) imposed under the RED II.

The RED scope as defined under Article 3(4) calculates the overall renewable energy in transport as in Eq. (1.2):

$$RED(\%) = \frac{\textit{All types of energy from renewable sources consumed in all forms of transport}}{\textit{Petrol, diesel, biofuels consumed in road and rail transport, and electricity (in transport) but excluding off} - \textit{road}} \tag{1.2}$$

Multipliers for the denominator can be used in specific end sector, for example, a 1.2 value is assigned to biofuels in aviation. This increased the desirability for member states to encourage the use of biojet fuel.

The EU Directive 2015/1513 amended and harmonized both the Fuel Quality Directive, (FQD) (98/70/EC) and RED (2009/28/EC). Key changes pertinent to the aviation sector include the ability of EU member states to voluntarily opt in for the RED aviation opt-in mechanism. When the amendment was made, all the 28 EU member states were categorized according to the potential to implement the voluntary aviation opt-in. From the exercise, Germany, Ireland, Italy, Portugal, Spain, and the United Kingdom were deemed to have high potential in implementing the voluntary aviation opt-in. The six member states were identified in addition to the Netherlands, which already implemented voluntary sustainable aviation fuel for their RED in 2013. Clearly, it was the Netherlands' move to include the aviation opt-in in their RED that triggered the EU-wide reform that led to the ability of other member states to do the same.

The member states were evaluated for certificate system, policy incentives, and local sustainable aviation fuel development opportunities. Among the three factors, the certificate system is of utmost importance as the existence of a certification system will allow quick transition to adopt the voluntary aviation opt-in. Other policy incentives such as existing tax exempts on road biofuels that could be modified to cover sustainable aviation fuels also provide positive pointers for a higher categorization of member states. Local fuel development opportunities may include a set of criteria such as current biojet fuel production levels, domestic jet fuel demand, sustainability scores of local airliners, and explicit support from member state governmental organizations. The member state categorization for the voluntary aviation opt-in for RED in a study executed by SkyNRG in collaboration with Boeing is tabulated in Table 1.5 (Meijerink, 2016). It is apparent from the list that Western Europe member states are generally in a position to implement the measures as compared with their Eastern European counterparts.

The point of trading the certificates independently from the physical biofuels is important for category 2 as the premise of the voluntary aviation opt-in works on the concept of price premium. It cannot work economically if the certificates produced from the biojet fuels cannot be traded to obligated parties from the road transport sector. The policy, production, demand, and governmental bodies support aspects for European countries with high potential for voluntary aviation opt-in are tabulated in Table 1.6 (Meijerink, 2016).

Member states are also required to submit their forecast of the expected renewables as contained in the Directive. This allows coordination of the "cooperation mechanism" where member states can agree to statistically exchange a given quantity of renewable energy produced (European Commission, 2009). This will allow member states to meet their RED target in a cost-effective manner. Table 1.7 summarizes the intended use of the cooperation mechanisms under RED as per the respective EU member state's National Renewable Energy Action Plans (NREAP). Negative surplus values are denoted in parenthesis. It should be noted that the forecast is not only for the aviation sector but also for all sectors under the scope of RED.

From the member state forecasts, at least 10 member states are expected to generate a surplus in 2020 as compared with their binding target share of renewable energy. The bulk of the 5.5 Mtoe surplus will come from Spain and Germany with 2.7 and 1.4 Mtoe, respectively. On the other hand, five member states are expected to face a deficit in 2020 and would require transfers from another member state or a third country. In absolute terms, Italy is expected to have the largest deficit with 1.2 Mtoe. By percentage, Denmark is forecasted to have the largest deficit at 2%, and Luxembourg

Table 1.5 Member state categorization for the voluntary aviation opt-in for RED by SkyNRG in collaboration with Boeing.

Category	Description	Member states
Category 1	Member states already have aviation opt-in included in the legislation.	The Netherlands
Category 2	Member states with a tradable certificate system in place for road biofuels. The tradable certificates can be traded independently from the physical biofuel.	Germany Ireland Italy Portugal Spain The United Kingdom
Category 3	Member states have a mix of policies. Their certificate system may be for power generation, or biofuel certificates without the tradable elements in them. Member states may also have large local demand, existing second-generation biofuel company or biofuel policies.	Belgium Croatia Denmark Finland France Sweden
Category 4	Member states have no specific biofuel policy and score poorly on the other criteria.	Austria Bulgaria Cyprus Czech Republic Estonia Greece Hungary Latvia Lithuania Luxembourg Malta Poland Romania Slovakia Slovenia

faces an uncertainty of 1%–6% deficit as compared with the binding renewable energy share. Overall, member states should collectively exceed its 20% target by 0.3%.

The latest available achieved targets in 2016 against the RES targets for each EU member state are shown in Table 1.8 (JRC EU, 2020). From the forecast, the EU as a whole were expected to have interim renewable energy surplus in 2016. Instead, there are 18 member states with an overall RES share deficit as opposed to just nine surpluses in 2016. For RES-T for

Table 1.6 The policy, production, demand, and governmental bodies support aspects for European countries with high potential for voluntary aviation opt-in.

Member state	Policy	Local production	Fuel demand	Governmental bodies support
Germany	Has a GHG quota system where trading between obligated and nonobligated parties are permissible. Biofuels profit from reduced taxes on production.	No large commercial facilities but has many advanced biofuels research facilities. Average feedstock opportunity.	In 2014, domestic fuel demand is 8,793,847 metric tons/year. National carrier Lufthansa has a clear sustainability strategy with CO_2 reduction and reporting with biojet fuel-powered flights conducted.	None.
Ireland	Has a tradable certificate scheme to put biofuels on the market under the Biofuel Obligation Scheme (BOS) under the administration of the National Oil Reserves Agency (NORA).	No advanced biofuel producers and below average feedstock opportunity.	In 2014, domestic fuel demand is 633,109 metric tons/year. National carrier Air Lingus has no sustainability goals.	None.
Italy	Has a tradable certificate system for making biofuels available for consumption issued by the Ministry for Agriculture, Food and Forestry Policies (MiPAAF).	Two advanced biofuel producers, namely ENI and Beta Renewable. Above average feedstock opportunity.	In 2014, domestic fuel demand is 3,708,840 metric tons/year. National carrier Alitalia has a clear sustainability strategy with CO_2 reduction and reporting without having conducted biojet fuel-powered flights.	Military body "Flotta Verde" is developing biofuels for military purposes in collaboration with "Green Fleet" of the US Navy.

Portugal	The "Titulo de Biocombustiel" (TdB's) system includes biofuel entitlements as tradable units. Partial and total tax exemptions available for biofuels.	One advanced biofuel producer, IncBio. Above average feedstock opportunity.	In 2014, domestic fuel demand is 1,049,937 metric tons/year. National carrier TAP Air Portugal has a clear sustainability strategy with CO_2 reduction and reporting but without including biojet fuel.	None.
Spain	Has a certificate trading system between obligated parties managed by the National Energy Commission (NEC). Has an investment system to subsidise existing and future renewable fuel facilities.	One advanced biofuel producer, Abengoa. Above average feedstock opportunity.	In 2014, domestic fuel demand is 5,149,217 metric tons/year. National carrier Iberia has a clear sustainability strategy with CO_2 reduction and reporting without having conducted biojet fuel-powered flights.	None.
The United Kingdom	Has a tradable certificate system which can be traded independently from the biofuel between obligated and nonobligated parties under the Renewable Transport Fuel Obligation (RTFO).	No large commercial facilities.	Domestic fuel demand is 11,364,889 metric tons/year, making it the highest domestic fuel demand among all member states. National carrier British Airways has a clear sustainability strategy with CO_2 reduction and reporting with biojet fuel-powered flights conducted.	The UK Department for Transport launched the Advanced Biofuel Demonstration Competition which provides GBP 25 million in grant funding to support the production of the UK-based advanced biofuels.

Table 1.7 EU member state targets, forecasts, and expected cooperation mechanism actions in 2020.

	Surplus energy (ktoe)					2020 National Renewable Energy Action Plans (NREAP)		Cooperation mechanism actions following the 2020 target
Member state	2011–12	2013–14	2015–16	2017–18	2020	2020 forecast (%)	2020 target (%)	
Austria	0	0	0	0	0	34	34	Not expected to produce a surplus or require a transfer to meet its target
Belgium	675	875	812	521	(−279)	12.3	13	Deficit
Bulgaria	1–144	186–346	231–481	53–375	(−140) to 289	18.7	16	Surplus
Cyprus	0	0	0	0	0	13	13	Not expected to produce a surplus or require a transfer to meet its target
Czech Republic	0	0	0	0	0	13	13	Not expected to produce a surplus or require a transfer to meet its target
Denmark	613–809	769–784	473–657	333–366	(−337)	28	30	Deficit

Estonia	47–69	78–96	79–88	52–67	3	25.1	25	Surplus
Finland	0	0	0	0	0	38	38	Not expected to produce a surplus or require a transfer to meet its target
France	0	0	0	0	0	23	23	Not expected to produce a surplus or require a transfer to meet its target
Germany	5930–7058	5866–6977	4657–5917	3842–5088	1387	18.7	18	Surplus
Greece	—	—	70.9	239.4	488	20	18	Surplus
Hungary	0	0	0	0	0	13	13	Not expected to produce a surplus or require a transfer to meet its target
Ireland	251–259	255–272	403–430	138–148	0	16	16	Not expected to produce a surplus or require a transfer to meet its target
Italy	—	(−86)	(−860)	(−1170)	(−1170)	16	17	Deficit
Latvia	0	0	0	0	0	40	40	Not expected to produce a surplus or require a transfer to meet its target

Continued

Table 1.7 EU member state targets, forecasts, and expected cooperation mechanism actions in 2020.—cont'd

Member state	Surplus energy (ktoe)					2020 National Renewable Energy Action Plans (NREAP)		Cooperation mechanism actions following the 2020 target
	2011–12	2013–14	2015–16	2017–18	2020	2020 forecast (%)	2020 target (%)	
Lithuania	96.3	93.9	79.7	52.9	18.3	23.3	23	Surplus
Luxemburg	—	—	—	—	(−43) to (−300)	5–10	11	Deficit
Malta	2.8	6.2	7.1	14.1	(−43.5)	9.2	10	Deficit
Netherlands	0	0	0	0	0	14	14	Not expected to produce a surplus or require a transfer to meet its target
Poland	519–866	705–1032	647–1162	613–1129	333	15.5	15	Surplus
Portugal	0	0	0	0	>0	31	31	Surplus
Romania	0	0	0	0	0	24	24	Not expected to produce a surplus or require a transfer to meet its target
Slovakia	56	112	134	167	143	15.2	14	Surplus
Slovenia	0	0	0	0	0	25	25	Not expected to produce a surplus or require a transfer to meet its target

Spain	4200	–	4791	–	2700	22.7	20	Surplus
Sweden	1074	1273	1286	1105	486	50.2	49	Surplus
UK	−119	−210	−254	40	–	15	15	Not expected to produce a surplus or require a transfer to meet its target
Net surplus	13,346 −15,190	9,905 −11,573	12,557 −14,802	6,270 −8,102	3,546 −3,718	20.3%	20	0.3% surplus available for transfer to other member states

Table 1.8 Achieved targets in 2016 against the NREAP RES share targets for each EU member state.

	2020 National Renewable Energy Action Plans (NREAPs) targets		Current trends (2016)		Gap to reach the 2020 RES share target for member states in 2016 (%)	
Member state	Overall RES share (%)	RES-T share (%)	Overall RES share (%)	RES-T share (%)	Overall RES share (%)	RES-T share (%)
Austria	34.2	11.6	33.5	10.6	−0.7	−1.0
Belgium	13.0	10.1	8.7	5.9	−4.3	−4.2
Bulgaria	16.0	10.8	18.8	7.3	+2.8	−3.5
Cyprus	13.0	4.9	9.3	2.7	−3.7	−2.2
Czech Republic	14.0	10.8	14.9	6.4	+0.9	−4.4
Denmark	30.4	10.1	32.2	6.6	+1.8	−3.5
Estonia	25.0	10.0	28.8	0.4	+3.8	−9.6
Finland	38.0	20.0	38.7	8.4	+0.7	−11.6
France	23.0	10.5	16.0	8.9	−7.0	−1.6
Germany	19.6	13.2	14.8	6.9	−4.8	−6.3
Greece	18.0	10.1	15.2	1.4	−2.8	−8.7
Hungary	14.7	10.0	14.2	7.4	−0.5	−2.6
Ireland	16.0	10.0	9.5	5.0	−6.5	−5.0
Italy	17.0	10.1	17.3	7.2	+0.3	−2.9
Latvia	40.0	10.0	37.2	2.8	−2.8	−7.2
Lithuania	24.0	10.0	25.6	3.6	+1.6	−6.4
Luxemburg	11.0	10.0	5.4	5.9	−5.6	−4.1
Malta	10.0	10.1	6.0	5.4	−4.0	−4.7
Netherlands	14.5	10.3	6.0	4.6	−8.5	−5.7
Poland	15.9	11.4	11.3	3.9	−4.6	−7.5
Portugal	34.5	34.5	28.5	7.5	−6.0	−27.0
Romania	24.0	10.0	25.0	6.2	+1.0	−3.8
Slovakia	14.0	10.0	12.0	7.5	−2.0	−2.5
Slovenia	25.3	10.5	21.3	1.6	−4.0	−8.9
Spain	20.8	11.3	17.3	5.3	−3.5	−6.0
Sweden	50.2	13.8	53.8	30.3	+3.6	+16.5
The United Kingdom	15.0	10.3	9.3	4.9	−5.7	−5.4

the transportation sector, the picture is bleaker with 26 of the 27 member states not yet meeting its 2020 target. The exception is Sweden, which exceeded its target by a large margin.

It is clear that meeting the targets in its original form looks bleak. The RED II recasting of energy policy will provide EU member states with another chance to meet the new overall targets of 32% and 14% for RES and RES-T, respectively. This time the European policymakers have decided that for the transport target, member states are no longer obliged to use crop-based biofuels. Also, there is a push toward advanced biofuels. This can only benefit the biojet fuel industry in its attempt to gain a market share for renewables in transport, which is currently dominated by biodiesel and bioethanol.

1.3.3 European Advanced Biofuels FlightPath

In June 2011, the European Advanced Biofuels FlightPath initiative was launched by the EC (DG Energy). This was done in close coordination with stakeholders including leading European airlines such as Lufthansa, Air France, KLM, and British Airways; major European biofuel producers such as Neste Oils, Biomass Technology Group, UPM, Chemtex Italia, and UOP; and aircraft maker Airbus (ICAO, 2011a). The initiative is introduced with the objectives to define a roadmap with defined milestones to achieve an ambitious target of 2 million tons (Mt) of sustainable biofuels in European civil sector aviation by 2020 and get sustainably produced biofuels to the market through the construction of advanced biofuels production plants in Europe. The aim is to get the two sets of plants to be operational by 2015 or 2016 and by 2020. This will speed up the commercialization of Sustainable Aviation Fuels (SAF) in Europe. The target of 2 Mt represents approximately 1% of the total world jet fuel consumption or 4% of EU jet fuel consumption projected for the year 2020 (Deane et al., 2017). The trends are certainly close to the cited values but will no longer be valid for the year 2020 since the COVID-19 pandemic has disrupted the global aviation sector greatly.

The volunteering members have a shared commitment to support and promote the production, storage, and distribution of sustainably produced drop-in aviation biofuels. Drop-in fuel refers to interchangeable substitute for conventionally derived petroleum fuel, which does not require adaptation of the engines, fuel systems, or fuel distribution network. To achieve the targets, the FlightPath needs to host workshop with financial institutions to find funding and facilitates the signing of purchase agreements between the stakeholders.

To facilitate a possible FlightPath, key activities have to be achieved by a trifecta of stakeholders, which would require substantial investment of resources, time, and money. They are summarized in Table 1.9 (European Commission, 2013).

Ideally, the FlightPath would have its implementation plan validated in 2014. Subsequently, it would have 300,000, 800,000, and 2,000,000 tons of biofuel produced for use in the aviation sector in 2016, 2018, and 2020, respectively. The amount of biofuel in 2020 would have been produced from a total of nine plants. However, progress in the first 6 years was insufficient to meet the 2 Mt aviation biofuels usages in 2020, despite the availability of various production technologies, which are ready for commercial deployment. This also meant that economic concerns, policies, and

Table 1.9 Key activities for stakeholders of the European Advanced Biofuels FlightPath.

Stakeholders		
Policymakers	**Biofuel supply chain**	**Aviation sector**
- Ensure the availability of supporting policies, including stable sustainability criteria - Availability of financial support mechanisms for research, demonstration, and commercial application for second generation biofuels - Safeguard an international level playing field	- Ensure clear understanding and use of effective financial mechanisms for technology developers and investors to construct novel plants - Development of quality standards and certified use of biofuels - Ensure sufficient supply of sustainably produced feedstock - Develop mechanisms for a real aviation biofuel market through policies and specific financial support instruments	- Ensure an operational off-take agreement with biokerosene supply chains stakeholders - Enable the validation of biofuels with on-flight testing - Facilitate and promote the policy dialogue with EU national governments, European Parliament, and EC

market uptake are the barriers instead of technical constraints. The FlightPath initiative has not solved the issues of financial investment into the sector and regulatory landscape sufficiently. Additionally, biofuels are embroiled in a perception battle on sustainability such as the food versus fuel debate and effects of ILUC. As such, fresh impulses are needed by the FlightPath beyond just the addition of new aviation sector members such as BiojetMap, IAG, IATA, and SkyNRG, alongside biofuel producers such as Honeywell-UOP, Mossi Ghisolfi, Swedish Biofuels, and Total.

In November 2019, a conference was organized by the EC, SENASA, FlightPath, ARTFuels, and Airport Regions Conference. Focus was given to the development of local production capabilities for feedstock and aviation biofuels conversion through the creation of regional bioports in Europe to support the FlightPath.

Another initiative to meet the objectives set by the FlightPath is the Initiative Towards Sustainable Kerosene for Aviation (ITAKA), which exists to develop a full value chain in Europe for the production of sustainable Synthetic Paraffinic Kerosene (SPK) (ICAO, 2016). The EC-funded research project (EU contribution is Euro 9,378,083.40 out of total budget of Euro 15,955,672.35) sets the ambitions of producing sufficient SPK at a scale capable of testing the existing European logistic infrastructure and normal flight operations. ITAKA homes in on camelina oil and used cooking oil (UCO) due to their availability in Europe. The lipid will then be converted into biojet fuels through the HEFA pathway. Key achievements of the 4 year project, which ended on October 31, 2016, include large-scale Roundtable of Sustainable Biomaterials (RSB)—certified camelina plantations implemented, UCO-based biojet fuel flights on A330-200 and Embraer E190, and EU RED-compliant HEFA biojet fuel produced from camelina oil in Europe.

1.3.4 FlightPath 2050

FlightPath 2050 refers to the European Union's vision for aviation in which global leadership in sustainable aviation products and services is maintained and society's needs are served. The vision, which is set in 2011, has highly ambitious goals divided into five categories (European Commission, 2011a), namely:

(1) Meeting societal and market needs (5 goals)
(2) Maintaining and extending industrial leadership (3 goals)
(3) Protecting the environment and the energy supply (5 goals)

(4) Ensuring safety and security (6 goals)
(5) Prioritizing research, testing capabilities, and education (4 goals)

Particular to biojet fuels will be goal no. 3 under "Protecting the environment and the energy supply" which states "Europe is established as a center of excellence on sustainable alternative fuels." Here, the idea would be to reduce dependency on the more polluting crude oil through substitution with drop-in liquid fuels from renewable sources at a competitive cost. Furthermore, the technologies (inclusive of the developed biojet fuel) and procedures in 2050 will allow a 75% and 90% reduction in CO_2 and NO_x emissions per passenger kilometer.

There is a push to move toward fuel cells, electrification, and batteries in the aviation sector, but they are limited to ground operations. At best, they could only be used to power ancillary systems. Liquid-based biojet fuel is the only viable alternative for aircrafts in the foreseeable future due to the energy density of the fuel. As such, fuel innovation research will be pursued aggressively and be funded through revenues from the Emissions Trading Scheme (ETS).

For Vision 2050 of the FlightPath to be realized, the EU aviation industry will need to be underpinned by simple and effective policy and regulatory framework. The policy and regulatory framework must also resolve the all-important funding and financing issues. This is because an estimated EUR 100 billion is required for research funding to meet this vision. Pertaining to biojet fuels, the FlightPath also specified that the success of alternative fuel research requires the governance, funding, and financing framework to coordinated oversight of a comprehensive research program.

1.3.5 EU Fuel Quality Directive 98/70/EC

The EU Fuel Quality Directive 98/70/EC of October 13, 1998, was first launched with the objectives of ensuring the quality of petrol and diesel fuels. The strict quality requirements are imposed within the European Union for petrol and diesel fuels used in cars, trucks, and other vehicles to protect human health and the environment. Among the key pushes of the 1998 FQD was to rule out the marketing of leaded petrol for all member states and for diesel fuels complying to key environmental specifications such as cetane number, density, distillation, polycyclic aromatic hydrocarbons (PAHs), and sulfur content. In this iteration, aviation jet fuel was not part of its considerations. In terms of biofuels, only low blends of biofuels below 30% biofuel content are within the scope of FQD. From Article 7a of the FQD, the original iteration calculated the overall GHG emissions reduction in transport, aviation, and electricity used in rails, where Eq. (1.3) is

$$FQD(-\%) = \frac{\textit{Fossil transport fuel GHG intensity } 2010 - \textit{All transport fuels GHG intensity } 2020}{\textit{Fossil transport fuels GHG intensity in } 2010} \tag{1.3}$$

The FQD contributed to the reduction of pollutants from the transportation sector (European Commission, 2017). Over the 1995–2013 period, sulfur oxides (SO_x), lead, nitrogen oxides (NO_x), PM_{10}, and PAHs reduced by 98%, 95%, 51%, 42%, and 62%, respectively. While it is widely agreed that the transport fuel covered by FQD contributed to the largest share of the reduction, other forms of transport such as international aviation, shipping, and railways could have also contributed. The inability to separate the effects of the non-FQD transport fuel shows the weaknesses of this directive. The FQD also represented a possible loss of opportunity in greening the aviation sector as compliance for petrol fuel samples range from 74% to 100% with a median value of 99%, while the range for diesel is 89%–100% with a median compliance value of 100%. Had FQD been extended to jet aviation fuels perhaps through the introduction of low-level biojet fuel blends, the contribution of the aviation sector to reduce emissions could have been greater.

The FQD has always had strong interactions with the RED. In fact, a new EU Directive 2015/1513 amended and harmonized both the FQD (98/70/EC) and RED (2009/28/EC) in terms of sustainability criteria and ILUC emissions requirements (Scarlat and Dallemand, 2019). The harmonization of the RED and FQD made sense as they focused on the renewable energy replacement and GHG emissions savings, respectively. The two key aims need to be harmonized to prevent having dual narratives which are conflicting.

The amendment also involves the permissions by member states to suppliers of aviation biofuel in becoming contributors to the reduction of obligations provided that those biofuels comply with the sustainability criteria. This amendment meant that all member states will have the same opportunity in adopting the voluntary aviation biojet fuel opt-in, as compared with the preamendment rulings where the ability to opt in is dependent on the implementation of the related directives. The voluntary aviation opt-in will allow biojet fuels to play a part in meeting the overall FQD aims of protecting human health and the environment through strict fuel requirements for the transport sector.

1.3.6 White Paper on Transport

While not a policy or a regulation, the EC adopted a White Paper on Transport in 2011 (European Commission, 2011b). The strategies established within the White Paper defined 10 challenging goals to guide policy actions and accountably measure progress. The 10 goals as outlined by roadmap in the White Paper will create a competitive and resource-efficient transport system to reduce 60% in CO_2 emissions and comparable reduction in petroleum dependency by 2050. Goal No. 2 which is particular to low-carbon sustainable fuels in aviation refers to an ambitious goal to increase the share of SAF to 40% by 2050. The other goals related to the biojet fuel within the aviation sector are Goals No. 5, No. 7, and No. 10, which mention a fully functional EU-wide multimodal Trans-European Transport Networks (TEN-T) "core network" by 2030, about the modernization of air traffic management infrastructure (SESAR) and adoption of a "user pays and polluter pays" principle, respectively.

The White Paper is also underpinned by 40 initiatives to be developed over the decade. Key initiatives unique or relevant to the aviation sector are tabulated in Table 1.10 (European Commission, 2011b).

1.4 United Kingdom

1.4.1 Renewable Transport Fuel Obligation

The Renewable Transport Fuel Obligation (RTFO) was introduced in November 2005 and came into effect in April 2008 to mandate that 5% of all road vehicle fuel must be sourced from renewables by 2010. The obligation made possible under the Energy Act 2004 was initially expected to meet the target through bioethanol, biomethanol, and biodiesel.

The RTFO ties in well with the EU biofuels directives which required all EU member states to meet the 2% and 5.75% targets of biofuel blends by the end of 2005 and 2010, respectively. The sustainability criteria of RED have also been implemented in RTFO through the December 2011 RTFO amendment. In 2013, RTFO was again amended to transpose the requirements of EU FQD 2009/30/EC.

The amendment to RTFO in September 2017 brought biojet fuel into the market trading mechanism (MTM) of RTFO for the first time (DfT, 2017). The emphasis on biojet fuel came from the realization that domestic transport is the largest GHG emitting sector, and liquid fuel cannot be decoupled from the aviation industry even in the longer term.

Table 1.10 Key initiatives of the white paper on transport related to the aviation sector.

Wider strategy	Narrow strategy	No	Initiative	Directly related to biojet fuel
An efficient and integrated mobility system	A single European transport area	2	Completion of the single European sky	No
		3	Capacity and quality of airports	No
	Promoting quality jobs and working conditions	10	A socially responsible aviation sector	No
	Secure transport	12	Cargo security	No
		13	High levels of passenger security with minimum hassle	No
		15	"End-to-end" security	No
	Acting on transport safety: saving thousands of lives	17	A European strategy for civil aviation safety	No
		20	Transport of dangerous goods	No
Innovating for the future: technology and behavior	A European transport research and innovation policy	24	A technology roadmap	Yes, to include a sustainable alternative fuels strategy with appropriate infrastructure
		25	An innovation and deployment strategy	Yes, to include measure for promoting the replacement rate of inefficient and polluting vehicles
		26	A regulatory framework for innovative transport	Yes, to include appropriate standards for CO_2 emissions, guidelines for refueling infrastructures and better implementation of existing rules and standards
	Promoting more sustainable behavior	29	Carbon footprint calculators	Yes, to estimate carbon footprint of each passengers to allow easier marketing of cleaner transport solutions

Continued

Table 1.10 Key initiatives of the white paper on transport related to the aviation sector.—cont'd

Wider strategy	Narrow strategy	No	Initiative	Directly related to biojet fuel
Modern infrastructure and smart funding	Transport infrastructure: territorial cohesion and economic growth	35	Multimodal freight corridors for sustainable transport networks	No
	Getting prices right and avoiding distortions	39	Smart pricing and taxation	No
	The external dimension	40	Transport in the world: the external dimension	Yes, to build established research and innovation partnerships to find common answers for sustainable low-carbon fuels

The amendment was made to accelerate the delivery of biojet fuel, allowing the United Kingdom to lead in the development and deployment of the fuel. This benefits the United Kingdom in terms of decarbonization of air travel, meeting the climate change commitments and industrial opportunities.

Aviation fuel can now be eligible for the Renewable Transport Fuel Certificates (RTFCs) as long as they are made from an eligible feedstock. One certificate can be claimed for every liter of renewable fuels produced, although renewable fuels from nonbiological origins (RFNBOs) and dedicated energy crops are incentivized with double the RTFC amounts. The MDM nature of RTFCs meant that it can be traded openly in the market.

The RTFO also defines "development fuels," which can be a renewable aviation fuel that must be made from waste and residues. Development fuels now form the subtarget in addition to the overall biofuel targets. The target levels are summarized in Table 1.11 (DfT, 2017; DfT, 2018). The RTFO policy is expected to reduce a total of 52 $MtCO_{2e}$ for the 2018–32 period additional to the baseline. The maximum allowable share of crop-derived fuels by volume will also be reduced from 4% in 2018 to 2% in 2032. The rationale to cap crop-derived fuels is to reduce the additional carbon emissions from ILUC through the planting of crop-based biofuels.

RTFO's contributions to RED including fuels that are double-rewarded are 3.24%, 4.12%, and 4.01% in 2012/13, 2013/14, and 2014/15, respectively. It should be noted that RTFO only requires that UK fuel suppliers provide 4.75% by volume of road transport from renewables. The present renewable fuel supply under RTFO is 3.3% by volume, increasing to 4.75% when double rewarding is taken into consideration. Similarly, it is 2.6% by energy, rising to 4.0% when double-rewarding is factored in. This is substantially lower than the carbon budgets and RED's transport subtarget, which mandates 10% of road transport fuel by energy to come from renewable sources in 2020. The inclusion of biojet fuel from the 2017 amendment will not substantially bring RTFO closer to meeting RED's subtarget for transport fuel.

It is still not entirely clear on how the Brexit invoked through Article 50 of the Treaty of the European Union will affect the United Kingdom's current EU requirements. Regardless, leaving the European Union means that the United Kingdom will get another opportunity to look at how the carbon budget reductions of the transport sector (inclusive of the aviation sector) can be met. On this front, the Brexit is unlikely to have a material effect on the direction of the RTFO policy.

Table 1.11 RTFO targets for biofuels and development fuels.

RTFO year	Obligation period	All biofuels			Development fuels		
		Specified amount as share of fossil fuel, by volume (%)	Specified amount as share of total fuel, by volume (%)[a]	Expected total volume (million liters)	Subtarget (obligation) level, including double rewarding (%)	Resultant "development fuel" supply as proportion of total fuel, by volume (%)[b]	Expected fuel supplied under the development fuel subtarget (million liters)
11	2018[c]	7.82	7.25	361	—	—	—
12	2019	9.29	8.50	719	0.1	0.05	3
13	2020	10.80	9.75	1071	0.15	0.075	15
14	2021	11.24	10.10	1414	0.5	0.25	101
15	2022	11.61	10.40	1489	0.8	0.4	173
16	2023	11.86	10.60	1553	1	0.5	220
17	2024	12.11	10.80	1594	1.2	0.6	267
18	2025	12.36	11.00	1635	1.4	0.7	313
19	2026	12.61	11.20	1673	1.6	0.8	359
20	2027	12.87	11.40	1716	1.8	0.9	404
21	2028	13.12	11.60	1757	2	1	449
22	2029	13.38	11.80	1797	2.2	1.1	494
23	2030	13.64	12.00	1842	2.4	1.2	539
24	2031	13.90	12.20	1887	2.6	1.3	584
25	2032	14.16	12.40	1931	2.8	1.4	630

[a]Excluding the effects of double counting and carryover.
[b]The resultant development fuel supply is calculated as half of the subtarget as development fuels are eligible for double reward.
[c]The obligation period is from April 15, 2018 to December 31, 2018 instead of the full year.

1.4.2 Fuels for Flight and Freight Competition (F4C)

The UK Department for Transport (DfT) launched the Advanced Biofuels Demonstration Competition (ABDC) in December 2014 to support the development of a domestic advanced biofuel industry. The need for this governmental initiative is supported by an independent feasibility study, which pinpointed potential financial gains from converting low-value waste into high-value transport fuels. The competition targets all biofuels and not particular to biojet fuels. Two projects have received funds for new biofuel plants, namely from Nova Pangaea Technologies with its production of bioethanol from wood waste, which is then blended with petrol and Advanced Biofuels Solutions with biomethane from municipal solid waste (MSW) and forestry waste. The competition supported the RTFO policy of that time, as the focus was still very much on the road transport sector.

The UK DfT then launched the Future Fuels for Flight and Freight Competition (F4C) in April 2017 to promote the development of advanced low carbon fuels to tackle the emissions problem from the hard-to-decarbonized aviation and heavy goods vehicle (HGV) sectors. Like ABDC, F4C is not a policy per se, but it is conceptualized to support the addition of aviation and HGV sectors in the RTFO amendment in 2017, which in turn will contribute to the UK's RED II target for the transport sector. The F4C organized jointly by DfT, Ricardo, and E4tech will provide capital grant funding to improve the supplier capabilities and skills in relevant technologies, while also maximizing the outcome for taxpayers.

The F4C covers two stages, namely stage 1 (project development) and stage 2 (capital funding) (RICARDO, 2017). The former was from June to November 2018 where GBP 2 million was awarded for the development of proposals. In the latter, F4C shortlisted four projects and are expected to grant GBP 20 million over a 3-year period of 2019–21 for the major demonstration projects.

The standout project for potential biojet fuel in the future is from Kew Projects Limited with the "Integrated ATC & F-T Demonstration Plant." The project was initially awarded GBP 312,300 in funding to produce the diesel substitute in stage 1 and was awarded another GBP 1.5 million in stage 2. Other two projects are currently in the shortlist, and they are related directly to biojet fuels. They include Altalto Immingham Ltd's "Altalto (Velocys Waste to Jet Fuel Project)" and LanzaJet UK Limited's "Sustainable Aviation Fuel from Waste-Based Ethanol." The Altalto project received GBP 434,000 in stage 1 to develop a proposal for kerosene and petrol substitutes and is being supported financially and technically by Shell and British Airways. The Lanza project obtained GBP 410,000 for stage 1 to develop a proposal for kerosene and diesel substitutes through a large-scale alcohol-to-jet (ATJ) facility. The proposal has partners from various sectors such as aviation, steel mills, research, and sustainability.

The competition is expected to deliver meaningful technical, route-to-market, and/or supply chain innovation. It is also expected to have projects to produce quality "development fuel" for testing or sale, while having a distinct commercial potential and plan, which is of value to the United Kingdom. The produced fuel also needs to show significant GHG reduction against their baseline fossil counterparts and be held to the highest of sustainability standards. The outcome of the competition might produce a localized solution, which will benefit the local biojet fuel industry and reduce imports of renewable fuels.

1.5 Scandinavia

1.5.1 Nordic Initiative for Sustainable Aviation

The Nordic Initiative for Sustainable Aviation (NISA) is neither a policy nor regulation. Instead, it is an initiative consisting of stakeholders through the entire value-chain steps including airlines, authorities, airports, and manufacturers in Denmark, Finland, Norway, and Sweden (ICAO, 2014b). The initiative aims to facilitate and strengthen the conditions for a conducive sustainable aviation industry. It is aligned to the EU FlightPath initiative. The Scandinavian region—focused NISA also bears the ambition of deploying new sustainable aviation fuels, spurring new green jobs, and attracting investments in the sector, which will combine to allow the Nordic bloc to be a global leader in green development.

1.5.2 Legislations in Nordic Countries

Within the Nordic region, Norway and Iceland are not official EU member states. However, they are both part of the European Civil Aviation Conference (ECAC), which is fully committed to combat climate change. As such, the Norwegian Ministry of Climate and Environment decreed in 2018 that airlines operating in Norway will need to have 0.5% biojet fuel blend in aviation jet fuel by 2020. The target will be increased to 30% of sustainable aviation jet fuel a decade later in 2030.

Iceland also has a Climate Action Plan for a 5-year period of 2018–23 to phase out transport sector's petroleum-based fuels. The plan includes the aviation sector as the industry is a large economic driving force, contributing to about 6.6% of GDP and employs around 9200 people (ICAO, 2012). Iceland shares the view that the reduction of aviation emissions is necessary. To achieve this, Iceland will use a two-pronged approach with greater fuel efficiencies from flights and less dependency on fossil fuels.

EU members such as Denmark, Finland, and Sweden also have their own legislations and priorities. Denmark has an ambitious target of making the transport sector free of fossil fuel by 2050. Airline industry players have taken up the task to plan for meeting the national targets. Denmark also takes an encouragement approach instead of a punitive approach by proposing for a climate fund instead of carbon tax. This is expected to boost the development and production of biojet fuel in the country. Finland targets 30% sustainable biofuels by blend for aviation by 2030 (NER, 2020).

Sweden recognizes the possible contribution of biojet fuel to meet its 2035 and 2050 climate goals. The 2017 aviation strategy includes support from the government for research to reduce costs for biofuel production. A sum of Euro 9.5 million was awarded to the Swedish Energy Agency to support initiatives that will lead to profitable biofuels. This is a smart strategy as the largest barrier to the success of biojet fuels is really the cost as opposed to the price of its fossil counterpart in the market. A more punitive obligation was proposed by the Swedish authorities, where penalty is applied to airlines not meeting the GHG intensity reduction targets.

1.6 United States of America

1.6.1 Renewable Fuel Standard

In the United States, the Renewable Fuel Standard (RFS) was established in 2005 through the Energy Policy Act of 2005. This federal program by the Congress mandated that renewable fuels must consist of a minimum volumetric amount for transportation fuel sold within the country. The US Environmental Protection Agency (EPA) administers the RFS. Annual target of 4 billion gallons of biofuels in used were set for year 2006, with it rising by 87.5% of the original target in 2012.

The scope and ambitions of RFS were further expanded with the amended form through RFS2 in 2007. The new RFS was passed through the Energy Independence and Security Act of 2007, which renewed the biofuels usage targets. Under RFS2, 9 billion gallons of biofuels should be in use for the year 2008, with the annual target facing a scheduled rise to 36 billion gallons in 2022. In both iterations of RFS, the defined renewable fuel must emit lower levels of GHGs as compared with the fossil-based fuels that it displaces. To improve the sustainable measures of biofuel sources, corn starch ethanol and cellulosic biofuels have a quota of not more than 15 billion gallons and not less than 16 billion gallons from cellulosic biofuels, respectively. This practically and symbolically shows the transition to the new generation biofuels.

Although the initial RFS is very ethanol and biodiesel-centric, sustainable fuels for the aviation sector have also entered the picture as biojet fuels from camelina oil, sugar cane, napier grass, distillers corn oil, distillers sorghum oil, and cellulosic components have since completed the pathway assessments as specified by the RFS Regulations at Title 40, Chapter 1, Subchapter C, Part 80, Subpart M (US EPA, 2020a). There are currently 20 generally applicable pathways under RFS, of which five of them are applicable for biojet fuels. The remaining pathways are for ethanol and biodiesel.

Fuel types meeting the pathway requirements will be allowed to generate Renewable Identification Numbers (RINs). Biojet fuels with RIN demonstrate compliance under the RFS program. The RIN credits are used by EPA to track and ensure compliance toward meeting the mandates from the RFS. The generated RINs (which will be attached to the renewable fuel) by the renewable fuel producers or importers can be traded with an obligated party in the form of a refiner or importer of fuel. Once the transaction is completed, the RIN is separated from the renewable fuel and may be independently traded until it is retired to meet the volume obligation. This provides an open market approach with the prices and transactions tracked under the EPA Moderated Transaction System (EMTS).

The fuels will also be assigned a RIN D-code depending on the types of renewable fuel produced. The classification for the D-code is summarized under Table 1.12. Qualified fuel pathways may be assigned more than one D-code as long as they meet the qualifying standards (Celignis Analytical, 2020). The more advanced renewable fuels count against compliance for the less advanced counterparts.

Table 1.12 RIN D-codes and the specific fuel pathway requirements.

RIN D-Codes	Category of renewable fuel	Specific fuel pathway requirements	Compliance
3	Cellulosic biofuel	Produced from cellulose, hemicellulose, or lignin. Must reduce life cycle GHG emissions by at least 60% as compared with the petroleum baseline.	Also counts against compliance for D5 and D6
4	Biomass-based diesel	Must reduce lifecycle GHG emissions by at least 50% as compared with the diesel baseline. Examples include biodiesel and renewable diesel.	Also counts against compliance for D5 and D6
5	Advanced biofuel	Produced from any type of renewable biomass except corn starch ethanol. Must reduce life cycle GHG emissions by at least 50% as compared with the petroleum baseline.	Also counts against compliance for D5
6	Renewable fuel	Fuel produced (in facilities or extended capacity after December 19, 2007) must reduce life cycle GHG emissions by at least 20% as compared with the average 2005 petroleum baseline. Examples include corn starch-derived ethanol or any other qualifying renewable fuel.	
7	Cellulosic biofuel (cellulosic diesel)	Same as RIN D-code 3 with the additional condition that fuel must be cellulosic diesel.	Same as D3

GHG, greenhouse gas; *RIN*, Renewable Identification Number.

The RIN codes follow the format of KYYYYCCCCFFFFBBBB RRDSSSSSSSSEEEEEEEE (Celignis Analytical, 2020), where

- K: Identifies if the RIN is attached to gallon
- YYYY: Year of production
- CCCC: Company ID

- FFFF: Plant facility ID
- BBBB: Batch number
- RR: Biofuels equivalence value
- D: Renewable fuel category
- SSSSSSSS: Start number for the biofuel batch
- EEEEEEEE: End number for the biofuel batch

For the biofuels equivalence value, RR, it is determined from the energy and renewable content of the RFS-compliant fuel when compared with denatured ethanol. In this case, ethanol will have an equivalence value (EV) of 1. The EV of a typical biojet drop-in fuel is 1.6. As decimal value is assigned to the EV, the RR code multiplies the EV by a factor of 10. Thus, the RR for a typical biojet fuel is 16. This also has the implication where 1.6 credits are generated for every gallon of biojet fuel produced, leading to 10 gallons of biojet fuel being able to replace 16 gallons of ethanol under the RFS program.

Companies producing biojet fuels or finished fuels with potential aviation sector application passing the completed pathway assessments are compiled in Table 1.13 (US EPA, 2020b). This excludes ethanol as an additional chemical process is still required to convert ethanol to ethanol-derived jet fuel. In contrast with the 12 completed pathway assessments for potential aviation usage, ethanol has 105 with the earliest since January 2013. The first alternative jet fuel which completed the pathway assessment is from Sustainable Oils in March 2013 using camelina oil.

In the event that feedstocks are mixed during the renewable biofuel conversion process, EPA evaluates the life cycle GHGs separately as per the individual component's weightage. The life cycle is evaluated for all of the process energy and materials used during the fuel production process until the point where it is a finished fuel. A fuel type is deemed to be a finished fuel if no further chemical alteration is required prior to its usage. Fuel blending is not considered under the life cycle calculations as it is a physical process.

The price of RIN has been tracked since 2010 when the transition from RFS to RFS2 was made. The weekly prices of RIN D3-D6 fuel types for the past decade until October 2020 is shown in Fig. 1.3 (US EPA, 2020c). The trading prices of D3 have always been the highest, followed by D4, D5, and D6, although the gap between D4 and D5 has closed since 2013. The price convergence of D4 and D5 is because obligated parties wanting to fulfill its D5 volume obligations can either purchase D3, D4, or D5 RINs. While D3 could serve the purpose of D4 or D5, the cellulosic

Table 1.13 Biojet fuels or finished fuels with potential aviation sector application passing the completed pathway assessment.

Organization	Fuel	Feedstock	D-code	Date
East Kansas Agri-Energy	Naphtha	Distillers corn oil, distillers sorghum oil	5	December 19, 2019
REG Geismar	Naphtha, LPG	Commingled distillers corn oil and sorghum oil	5	December 19, 2019
Texmark-Neste	Renewable jet fuel	Renewable diesel	4	September 23, 2019
National Sorghum Producers	Biodiesel, renewable diesel, jet fuel, and heating oil	Distillers sorghum oil	4/5	August 2, 2018
Butamax Advanced Biofuels	Biodiesel, renewable diesel, jet fuel, and heating oil	Distillers corn oil	4	August 2, 2018
REG Geismar	Naphtha. LPG	Biogenic waste oils/fats/greases	5	February 23, 2018
REG Geismar	Naphtha, LPG	Nonfood grade corn oil	5	April 13, 2017
ENVIA energy, LLC	Diesel, naphtha	Landfill biogas	3 or 7	May 8, 2015
Diamond Green Diesel, LLC	Naphtha, LPG	Nonfood grade corn oil	5	October 28, 2013
Chemtex Group	Cellulosic biofuel	Giant reed (Arundo donax)	3 or 7	July 11, 2013
BP Biofuels North America, LLC	Ethanol, cellulosic diesel, jet fuel, and heating oil; naphtha	Energy cane, napier grass	3 or 7	March 5, 2013
Sustainable Oils	Biodiesel, renewable diesel, jet fuel, heating oil, naphtha, LPG	Camelina sativa oil	4 or 5	March 5, 2013

biofuel criteria for D3 also provide cellulosic waiver credit (CWC), which is valuable on its own. Thus, the price of D3 should logically be the price of D4 or D5 summed with the price of a CWC. The CWC price ranges from USD 0.49 to USD 2.00, where the lowest price occurred in 2013 while the

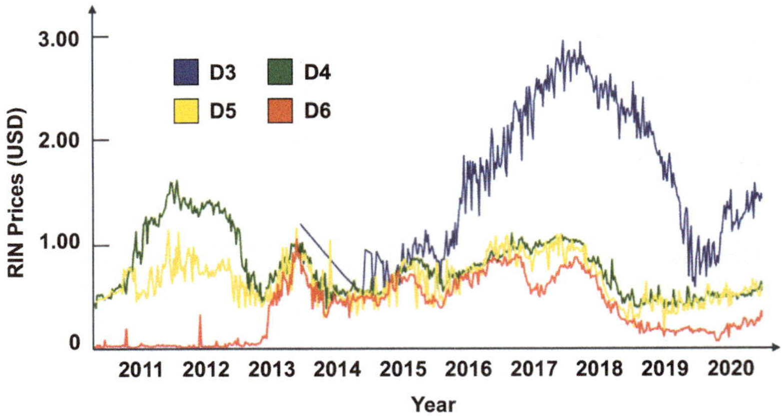

Figure 1.3 Weekly RINs prices for D3, D4, D5, and D6 from January 2010 to October 2020. *RINs*, Renewable Identification Numbers. *(Adapted from US EPA, 2020c. RIN Trades and Price Information, https://www.epa.gov/fuels-registration-reporting-and-compliance-help/rin-trades-and-price-information.)*

peak was attained in 2017. The peak number of CWC traded was in 2016 where 33,155,196 CWCs changed hands. The number of CWCs traded has since fallen to 1,474,354 in 2019 although price remained above the 2-year average at USD 1.77. The price of a CWC is set at USD 3 minus the 12-month average of the wholesale gasoline price, with a minimum price fixed as USD 0.25. The price mechanism and interchangeable usage (or nesting) of the RINs add to the flexibility of the RFS program. This makes cellulosic-based biojet fuel, which will carry the D7 code potentially lucrative for biojet fuel producers.

The total available RINs generated, retired, and available as of October 29, 2020, are tabulated in Table 1.14. Currently, cellulosic diesel RINs have not been generated despite three D7 codes awarded, with the first in 2013. It is also not a surprise for D6 to dominate at 71.2% of the RINs generated as it is the usual code used for sustainable ethanol.

Great progress is made for renewable jet fuel with the absolute amount and proportion of biojet fuel as a share of total renewable fuels increasing throughout the decade. This is illustrated in Fig. 1.4 where the RINs generated, volume of biojet fuels, and their proportions against that of all sustainable fuels from 2010 to present day are shown. As of October 2020, US domestic producers generated 6,459,392 RINs for a total of 4,037,120 gallons of biojet fuel. This is still miniscule in terms of overall proportions with just 0.0483% and 0.0333% of RINs generated and volume, respectively.

Table 1.14 Total available RINs generated, retired, and available as of October 29, 2020 (US EPA, 2020d).

D-code	Assignments	Total generated	Total retired	Total available (locked)	Total available (unlocked)
3	Assigned	322,219,110	218,820	0	324,501
3	Separated	0	2,268,078	6,465,479	312,942,232
4	Assigned	3,310,192,342	70,987,329	1,173,877	304,261,316
4	Separated	0	437,464,737	13,584,482	2,482,720,601
5	Assigned	218,011,830	0	0	5,359,877
5	Separated	0	14,151,546	0	198,500,407
6	Assigned	9,521,195,664	105,814,644	128,311	611,216,914
6	Separated	0	179,529,639	416,402,027	8,208,104,129
7	Assigned	0	0	0	0
7	Separated	0	0	0	0

RINs, Renewable Identification Numbers.

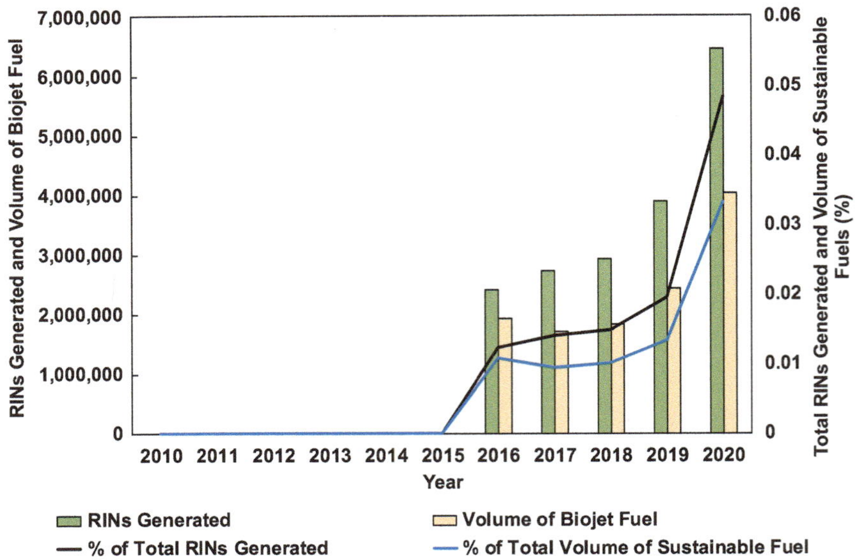

Figure 1.4 RINs generated, volume of biojet fuels, and their proportions against that of all sustainable fuels from 2010 to October 2020. *RINs*, Renewable Identification Numbers.

Nonetheless, compared with the all-time proportion for a decade since 2010, the proportions are 0.0102% and 0.0071% of total RINs generated and volume of fuel, respectively.

Small refineries may be exempted from RFS, provided petitions submitted to EPA on an annual basis to be relieved from their renewable volume obligations (RVOs). Exemptions are given on the basis of demonstrated disproportionate economic hardship. Compared with the overall sustainable fuel volume, the exemptions given are just a blip with only less than 50 petitions submitted in any given year. The COVID-19 pandemic might have greatly affected the production as the number of petitions fell from 31 in 2019 to just 4 in October 2020. To date, no exemptions were given for biojet fuels.

While the RFS is an ambitious program to ensure the sustainability of the transport sector and also to emphasize that biofuels remain an important piece of the overall US strategy to address climate change and enhance energy security, its efficacy is up for debates. Table 1.15 shows the end-of-year compliance deficit for the RFS from 2010 to 2019. Compliance deficit refers to the amount accumulated when obligated party or exporter does not meet the annual RVO target through the retirement of RINs.

Table 1.15 The end-of-year compliance deficit for the RFS from 2010 to 2019.

Compliance year	Cellulosic biofuel (D3 or D7)	Biomass-based diesel (D4)	Advanced biofuel (D5)	Renewable fuel (D6)
2010	31,543	227,120,812	229,693,190	163,353,609
2011	0	45,802,515	37,394,607	68,763,484
2012	0	28,773,252	19,690,744	75,521,355
2013	31	23,535,850	26,605,634	68,855,617
2014	7615	20,276,804	21,654,525	141,478,021
2015	22,823	5,555,088	4,897,514	10,419,624
2016	4,454,758	40,944,372	66,462,857	390,514,451
2017	19,976,793	75,247,335	137,768,294	681,328,963
2018	2,962,896	19,737,837	28,957,783	129,279,042
2019	7,986,724	105,622,268	169,887,781	458,886,372

RFS, Renewable Fuel Standard.

Compliance deficits are carried over to the next compliance year, hence possibly creating a scenario where the deficit could be higher than any particular annual RVO target.

Since the commencement of the program, there has been compliance deficit year-on-year. The zero deficits for cellulosic biofuel in 2011 and 2012 were due to their removal arising from a successful legal challenge mounted to remove them. Despite the deficit, the May 2020 data show that the deficits are relatively low compared with the total RVO (which includes present year reported RVO and reported prior year deficit). The 2019 compliance deficit volumes of D3 (7,986,724 gallons), D4 (105,622,268 gallons), D5 (169,887,781 gallons), and D6 (458,886,372 gallons) came against the target of 2018s total RVO of 421,498,816 gallons, 3,526,218,417 gallons, 4,960,402,759 gallons, and 20,413,716,106 gallons, respectively. They represent deficits of 1.89% (D3 or D7), 3.00% (D4), 3.42% (D5), and 2.25% (D6). Biojet fuel could typically contribute to the D3, D4, D5, and D7 categories to bridge the deficit gap.

1.6.2 Farm to Fly

The Farm to Fly initiative was launched by the US Department of Agriculture (USDA), Airlines for America (A4A), and the Boeing Company in July 2010. The initiative aims to accelerate the availability of a commercially viable and sustainable aviation biofuel industry in the United States, increase domestic energy security, establish regional supply chains, and support rural development.

The Farm to Fly initiative was launched on the back of the Biofuels Interagency Working Group's "Growing America's Fuel" report that portrayed strategies to meet the Energy Independence and Security Act (EISA) target during the Obama Administration (US USDA, 2012). The EISA target of 36 billion gallons of US biofuels per year by 2020 can be achieved through the development of first-, second-, and third-generation biofuels, inclusive of biojet fuels.

The initiative attempted to make biojet fuel, the favored choice over fossil-based aviation jet fuel from the economic and environmental point of views. As per the initiative's namesake, it will incentivize American farmers who produce energy crops for sustainable aviation biofuels. This merges rural development with clean energy innovation. It also tried to pull demand for biojet fuels from the aviation-fuel user community, instead of imposing supply side—driven biojet fuel usage. Coupled with compatible policies, the US aviation industry will be a willing buyer of competitively priced biojet fuels. The USDA and other federal agencies have a range of programs that supported the Farm to Fly initiative as tabulated in Table 1.16.

The Future of Aviation Advisory Committee (FAAC) also supported the deployment of sustainable alternative aviation fuels in December 2010. Subsequently, the USDA, the Department of Energy (DOE), and the Department of the Navy in August 2011 jointly developed and supported production facilities for both biojet fuels and marine biofuels.

The ongoing initiative was extended in April 2013 and relabeled as Farm to Fly 2.0 (ICAO, 2010a). The revised initiative targets an annual 1 billion gallons of drop-in aviation fuel by 2018. From it, CAAFI continued to push the public—private partnership efforts to develop supply chains for biojet fuels in Vermont, Maryland/Delaware, and Florida.

1.6.3 Sustainable Aviation Fuels Northwest

The Sustainable Aviation Fuels Northwest (SAFN) launched in July 2010 brought together a set of over 40 stakeholders from aviation, agriculture, forestry, biofuel producers, nonprofit advocacy, federal and state agencies, and academia. SAFN has a stated goal of mapping a flight path for a safe, sustainable, and economically viable biojet fuel industry in the Northwest region of the United States. This program is the first regional assessment in the United States on possible feedstock pathways in the four Northwest states of Washington, Oregon, Montana, and Idaho.

Table 1.16 Programs and projects supporting the Farm to Fly initiative.

Agency	Key contributions for biojet fuel	Specific programs and projects	Description
USDA Rural Development	Provides funding opportunities in the form of payments, rents, and loan guarantees for the development and commercialization of biojet fuel.	Biorefinery Assistance Program (Section 9003)	Provides loan guarantees to commercial-scale facilities for advanced biofuels from renewable biomass.
		Rural Energy for America Program	Provides grants and loan guarantees to agricultural producers to install renewable energy systems.
		Business and Industry Guaranteed Loans	Improves the economic and environmental climate in rural communities through financing and bolstering of existing private credit structure.
USDA Research and Development Programs	Focuses on improving biomass varieties and production systems of biojet fuels.	USDA Biomass Research Centers	Establishes five Biomass Research Centers to enable rural areas to participate in the biofuel supply chain and benefit economically.
		Agricultural Research Service	Invests in fundamental and applied biological science and technology. Coordinates bioenergy research programmes with focus on feedstock development, sustainable feedstock production systems and biorefining and coproducts.

Continued

Table 1.16 Programs and projects supporting the Farm to Fly initiative.—cont'd

Agency	Key contributions for biojet fuel	Specific programs and projects	Description
US Forest Service (FS)	Provides science and technology to sustainably produce high-value products such as biojet fuel.	Biobased Products and Bioenergy Research Program	Provides science and technology to sustainably produce, manage, harvest, and convert woody biomass into liquid transportation fuel, inclusive of biojet fuel.
National Institute of Food and Agriculture	Funds competitive and peer-reviewed research efforts.	Biomass Research and Development Initiatives	Makes funds available for advanced research on feedstock development and biofuels.
		Agriculture and Food Research Initiative	Offers sustainable energy grants to reduce dependence on foreign oil; have net positive social, environmental, and rural economic impacts.
		Plant Feedstock Genomics for Bioenergy	Funds research projects to utilize advanced genomics to breed energy crops on marginal lands with improved yield and quality.
Economic Research Service	Provides primary source of economic information and research in USDA.	Economic Research Service	Informs public and private stakeholders on economic and policy issues involving food, farming, natural resources, and rural development.

Table 1.16 Programs and projects supporting the Farm to Fly initiative.—cont'd

Agency	Key contributions for biojet fuel	Specific programs and projects	Description
USDA Farm Service Agency	Provides financial assistance to owners and operators of agricultural and nonindustrial private forest land for biomass feedstocks.	Biomass Crop Assistance Programs	Provides matching and "establishment and annual" payments for eligible biomass crops.
USDA Risk Management Agency	Develops new insurance products for producers of renewable, clean energy crops in the United States.	USDA Risk Management Agency Energy Crop Feasibility Study	Researches on biofuels produced from energy cane, switchgrass, and camelina.

The four feedstocks in focus were oilseed crops, forest residues, algae, and MSW, while hydroprocessing of oil and woody biomass conversion were the two primary conversion technologies analyzed (SAFN, 2011). Oilseed crops for the Pacific Northwest region include camelina, canola, rapeseed, white or yellow mustard, brown or oriental mustard, black mustard, crambe, cuphea, meadowfoam, safflower, and sunflower.

The combination of forest residues and woody biomass conversion technologies plays into the strength of the US Pacific Northwest region as compared with other regions as there is a low demand for wood chip and pellet exports to the European Union. This reduces the cost of feedstock to domestic biojet fuel producers. These set of circumstances increase the chances of SAFN to meet its ultimate aim of generating as much biojet fuel regionally. This has a twofold benefit in boosting the regional economy and also reducing the overall carbon emissions.

SAFN also recommended six priority action steps as their policy framework. One of those action steps is to ensure support for aviation fuels and promising feedstocks under the RFS2 program. SAFN recommended early support and investment from the government without advocating for permanent subsidies. In addition to the substantial public investment, SAFN identified policy support as a key element to create a biojet fuel industry in the Northwest.

The feasibility study concluded in May 2011 and resulted in the Aviation Biofuels Production bill (HB 2422) being adopted by the Washington state legislature. The bill mandates stakeholders within the jurisdiction to form an "Aviation Biofuels Work Group" for the development of biojet fuel in Washington.

1.6.4 Midwest Aviation Sustainable Biofuels Initiative

The Midwest governments and policymakers recognized the importance of the biojet fuel industry in their region. As such, the Midwest Aviation Sustainable Biofuels Initiative (MASBI) was initiated in May 2012 to link up a diverse set of stakeholders from the entire biojet fuel value chain. The initiative focused on the evaluation of the 12-state Midwest region's biojet fuel industry potential. An actionable roadmap was developed that covers feedstock, commercialization, logistic, infrastructure, and regional policies (MASBI, 2013).

MASBI categorized their 14 recommendations to advance biojet fuel development into five key areas, with four recommendations under the "Policy and Economic Development." Among the policy recommendation includes creation of long-term policies that enable investment and production, level playing field with fossil fuel, fund the Defense Production Act Title III for biojet fuels, and build regional demonstration facilities supported by municipal and state policies. MASBI opines that biofuel policy measures are often short term and does not tackle the disadvantage of biofuels over fossil fuels.

1.6.5 California Low Carbon Fuel Standard

The California Low Carbon Fuel Standard (LCFS) is a market-based measures (MBM), which uses economic incentive to reduce GHG intensity of transport fuels to meet its designated obligations. The Executive Order S-1-07 in compliance with California Assembly Bill AB 32 was enacted by the Californian governor in January 2007 to ensure that oil refineries and distributors within the Californian market meet the declining targets for GHG emissions intensity. LCFS compliance began in January 2011.

Under the policy that originated from the Global Warming Act of 2006, conventional fuels under the scope will generate LCFS deficits if its life cycle carbon intensity is above the LCFS annual target. On the other hand, alternative fuels with life cycle carbon intensity below the target will generate LCFS credits. While the LCFS covers transport, aviation fuels do not generate deficits. This could potentially change where deficits could exist if the California Air Resources Board (CARB) exerts its authority to regulate fuels for intrastate flights.

The biggest change to LCFS came in September 2018, where the 10% reduction in the carbon intensity of California's transportation fuels by 2020 is now increased to 20% by 2030. The amendment taking effect on January 1, 2019, represents the most stringent requirement in the whole country. To meet this target, the aviation sector is also included as part of the amendment. Sustainable aviation fuel producers could still opt in to generate credit if they can prove that the life cycle carbon intensity is below the 89.37 g CO_{2e}/MJ of baseline jet aviation fuel using the CA-GREET model (Davis, 2020). This was taken up by producers who obtained LCFS credits for 1.4 million gallons of biojet fuels consumed in California from registered pathways in 2019. World Energy contributed to the bulk of the SAF derived from locally sourced waste cooking oil for use in the Los Angeles International Airport.

The LCFS reporting schedule is done on a quarterly basis, and no credits can be generated for an activity from the previous quarter. Additionally, regulated entities must also submit an annual compliance report. The CARB Cap-and-Trade program has a verification program based on ISO 14064-3 and 14065 to standardize the calculation of carbon pricing mechanisms internationally.

It should be noted that currently the regulation climate will slightly favor renewable diesel as it is easier to generate higher credit due to the carbon intensity benchmark to beat being 94.17 g CO_{2e}/MJ in 2019. This gap of 4.8 g CO_{2e}/MJ makes it more economically viable to produce alternative diesel over alternative aviation fuels. In fact, the sustainable aviation fuel carbon intensity target of 89.37 g CO_{2e}/MJ in 2019 is comparable to gasoline's 89.50 g CO_{2e}/MJ in 2022 and diesel's 89.15 g CO_{2e}/MJ in 2023. However, the declining target for alternative aviation fuel is more gradual, as shown in Table 1.17 (California Air Resources Board, 2020). Parity will roughly be achieved in 2022.

Emissions reduction under "Fuel pathway crediting" for the LCFS policy can be calculated by Eq. (1.4):

$$Credits\ (tCO_2) = (CI\ of\ benchmark - CI\ of\ fuel) * Energy\ economy\ ratio * Conversion\ factor \tag{1.4}$$

Table 1.17 California's LCFS life cycle carbon intensity benchmarks for transport fuels and their substitutes.

	Carbon intensity benchmarks for transport fuels and their substitutes		
Year	**Jet fuel average (gCO_{2e}/MJ)**	**Gasoline average (gCO_{2e}/MJ)**	**Diesel average (gCO_{2e}/MJ)**
2019	89.37	93.23	94.17
2020	89.37	91.98	92.92
2021	89.37	90.74	91.66
2022	89.37	89.50	90.41
2023	89.15	88.25	89.15
2024	87.89	87.01	87.89
2025	86.64	85.77	86.64
2026	85.38	84.52	85.38
2027	84.13	83.28	84.13
2028	82.87	82.04	82.87
2029	81.62	80.80	81.62
2030 onward	80.36	79.55	80.36

The biojet fuel industry is given a boost when CARB approved a temporary fuel pathway for alternative jet fuels. This augments existing fuel–feedstock combination stated under Section 95488.9(b) (4) of the LCFS regulation. The temporary fuel pathway is only applicable to fuels produced through the hydrotreating process (California Air Resources Board, 2019). The carbon intensity values vary between feedstock with fats/oils/grease and plant oil (excluding palm oil and palm derivatives) assigned values of 50 and 70 g CO_{2e}/MJ, respectively. Other feedstock adheres to the 2010 baseline CI value for USLD. LCFS credits can be generated for fuels through this method effective from Q2 of 2019 onward.

The California LCFS as MBM is an unqualified success with credit prices soaring to near record high at the end of 2019. However, CARB has imposed a price cap on the secondary market for the LCFS credit clearance market at USD 213/t, as calculated from USD 200/t in 2016. As the renewable fuel projects remained profitable and carbon credits not hitting the cap, there are reasons to believe that the conditions are favorable economically to maintain the growth of the biojet fuel industry.

The outcome of LCFS is mixed with inroads into the formation of a biojet fuel industry being made, but also creating problems leading to lawsuits. A District Court's judge applied an injunction on California to prevent further enforcement of LCFS as it was deemed to discriminate against ethanol sources from other US states due to the counting of the transport carbon emissions of transport to California and protection of Californian farmers. However, the Ninth Circuit Court of Appeals reversed the opinion ruling that the GHG accounting system has the intention to accurately assess the carbon intensity instead of unconstitutionally breaching the "Commerce Clause" to unfairly benefit California's farmers. There is also an additional controversy arising because of the uncertainty in determining the magnitude of ILUC where feedstocks are practically classified as good and bad feedstocks. Palm oil and palm derivatives are often excluded as feedstocks that qualify for LCFS credits generation.

The LCFS credits could potentially be stacked with the relatively low-key "Tax Credit for Carbon Sequestration (Section 45Q)" tax credit. The latter that is also widely known as the 45Q tax credit enacted in February 2018 has started to lure corn ethanol producers into the industry as the fuel producers could earn up to USD 50/tCO_2 stored permanently or USD 35/tCO_2 if the CO_2 is put to use. The combination of the LCFS credits and 45Q tax credits could potentially bring the incentives to about USD 230/tCO_2, which makes for a very lucrative business. As biomass-based biojet fuel could also potentially use carbon dioxide capture and storage (CCS) in the fuel production process, biojet fuel producers might find it easier to be profitable. Biojet fuel producers that want to qualify for the 45Q scheme will need to start their CCS facilities by 2025 and will have 12 years to claim the tax credits. This might prove to be the boost that biojet fuel producers need.

1.7 Canada

The government-supported BioFuelNet Canada (BFN) is a network that connects the Canadian biofuels research community of 230 academic researchers and 152 organizations to deal with the challenges in growing the Canadian advanced biofuel industry. The network sits under the Networks of Centres of Excellence of Canada (NCE), with a funding of $25 million over a 5-year period of 2012–17. BFN wanted to harness the natural strength of Canada by focusing on using lignocellulosic feedstocks as the central piece of the puzzle in the Canadian biofuels industry.

BFN created the Aviation Task Force as one of the six task forces initiated within the second phase of BFN, namely the BFN II (BioFuelNet, 2020). The task force participant consists of Transport Canada, Environment Canada,

National Research Council, Air Canada, Airbus, International Air Transport Association, CAAFI, ASCENT (FAA Centre of Excellence), and University of Toronto.

The task force aims to resolve biojet fuel issues on deployment, production costs, feedstock costs, rigorous specification standards, and policy considerations. From a policy perspective, the BFN views that the relatively weak Canadian renewable fuel mandates and incentives when compared with other developed nations, and poor governmental coordination leading to redundancy-led wasted resources, as the main barriers to a successful biojet fuel industry. BFN aims to improve the mandate issue by developing and presenting the Canadian government with science-based evidence and economic analysis to push for stronger and more targeted mandates. Assuming success, BFN funding could be renewed past 2022 although it will be disbursed through the New Frontiers in Research Fund (NFRF).

The Canada's Biojet Supply Chain Initiative (CBSCI) is an initiative steered by Air Canada, BFN, IATA, Transport Canada, Queen's University, SkyNRG, and Waterfall. It receives funding primarily from Green Aviation Research and Development Network (GARDN), with additional project funding from BFN, IATA, and Air Canada (in-kind). CBSCI wants to enable a low carbon future for Canadian aviation through the objectives of demonstrating the operational feasibility of biojet fuels in the domestic jet fuel supply system, catalyze the development of the domestic biojet fuel sector, validate the Canadian biojet fuel supply chain elements, and generate hands-on experience with biojet handling and integration (CBSCI, 2020). The last of the objectives is practically attained through a 3-year project involving 14 stakeholders to introduce 400,000 L of blended biojet fuel into a shared fuel system at the Montréal-Trudeau Airport.

Canada has twin commitments of Carbon Neutral Growth (CNG) from 2020 onward through emissions capping and Deep Carbon Reduction arising from 50% GHG emissions reduction by 2050 as compared with 2005 levels. This two-step commitment will allow an easier quick win of CNG prior to the deeper emissions reduction required in the 2035−50 timeframe. For the Canadian aviation industry to achieve these targets, it must first produce an annual biojet fuel volume of 54 ML in 2020, increasing to 923 ML by 2035 (CBSCI, 2019).

To affirm Canada's dominant biomass for biojet fuel production, a GARDN's project led by NORAM and the University of British Columbia assessed the feasibility of forest residue as feedstock. The consortium of the ATM Project: Jet Biofuels from Forest Residuals consists of leading North

American and European groups on advanced biojet fuel. They include the Canadian Canmet ENERGY laboratories, the US Pacific National Northwest Laboratory (PNNL), (S&T)^2 Consultants, SkyNRG. Boeing, Bombardier, Air Canada, and WestJet Airlines also supported the program which is part of the broader CBSCI initiative. The use of forest residues in the context of producing sustainable biojet fuel is apt as Canada is the world leader in sustainably certified forest.

Jet fuel demand, jet fuel growth, and biojet fuel required for CNG projections from 2020 to 2035 are shown in Table 1.18 (CBSCI, 2015). The projection assumes an 81% GHG reduction from biojet fuel and biojet fuels contributing to 40% in CNG emissions reduction. It should, however, be noted that the projection was done nearly 5 years before the global COVID-19 pandemic, which is expected to reduce jet fuel demand until 2024.

While Canada lags behind major biojet fuel players such as the European Union, the United States, and Brazil in terms of biojet fuel production policies and fiscal programs, the development of a viable biojet fuel industry in Canada is still promising, as Canada has existing renewable fuels sectors

Table 1.18 Jet fuel demand, jet fuel growth, and biojet fuel required for CNG projections from 2020 to 2035.

Year	Jet fuel demand (billion liters)	Jet fuel growth (billion liters)	Biojet fuel required for Carbon Neutral Growth (billion liters)	Biojet fuel amount as a total of total demand (%)
2020	7.55	0.11	0.05	0.7
2021	7.67	0.23	0.11	1.5
2022	7.78	0.35	0.17	2.2
2023	7.90	0.46	0.23	2.9
2024	8.02	0.58	0.29	3.6
2025	8.14	0.70	0.35	4.3
2026	8.24	0.80	0.40	4.8
2027	8.35	0.91	0.45	5.4
2028	8.47	1.03	0.51	6.0
2029	8.58	1.14	0.56	6.6
2030	8.69	1.25	0.62	7.1
2031	8.80	1.37	0.68	7.7
2032	8.93	1.49	0.74	8.2
2033	9.05	1.61	0.80	8.8
2034	9.18	1.74	0.86	9.4
2035	9.30	1.87	0.92	9.4

for bioethanol, biodiesel, and advanced biofuels. Many of these measures can be adopted to set up a strong biojet fuel policy structure. The key market-based instruments and examples from road transport sector that can be adopted are summarized in Table 1.19 (CBSCI, 2015). The reworking of existing renewable fuels policies, carbon policies, and fiscal programs is less of a know-how but rather a question of political will.

1.8 Mexico

The "Flight Plan Towards Sustainable Aviation Biofuel in Mexico" initiative launched in Mexico was to identify and analyze the existing and missing supply chain elements of biojet fuel, with special emphasis being placed on the HEFA pathway. The flight plan study started in July 2010 until March 2011 with the Aeropuertos Y Servicios Auxiliares (ASA) (ICAO, 2011b), or Airports and Auxiliary Services in English leading it. Partners include the National Council for Science and Technology (CONACYT) as sponsor; strategic partners such as the Roundtable on Sustainable Biofuels (RSB), Boeing, Honeywell-UOP; and a myriad of 60 experts and over 200 stakeholders.

The Flight Plan has the ambition of leading the national efforts to develop and produce biojet fuels for the aviation industry; analyze the legal framework, feedstock availability, refining infrastructure, and economic viability of biojet fuels; and bringing together the talents of all participating sectors. Through this Flight Plan, the Mexican target is set for biojet fuels to be 1% of the national demand in 2015, raising to 15% in 2020. It was envisioned that four biojet fuel refineries will be in operation by 2020 with a throughput of 800 ML per year. It is not yet clear if the COVID-19 pandemic has caused a setback to the refineries being brought online.

The Flight Plan also recognized the shortcomings of Mexico in the form of having a feedstock quantities bottleneck, lack of biorefining infrastructure, and the need to develop appropriate legislation to support the industry. Among key successes of the flight plan includes integration of biojet fuel production in the supply chain and the pilot biojet fuel-powered flight in Mexico.

1.9 Brazil

The collection of legislations regulating the oil, natural gas, and biofuel industries in Brazil can be identified from the Brazilian National Agency of

Table 1.19 Key market-based instruments and examples from road transport sector which can be adapted for the biojet fuel industry.

Category	Instrument type	Description	Examples
Market access	Low carbon fuel standard	Policy to reduce carbon intensity of a fuel type.	- British Columbia RLCFRR - California LCFS - EU FQD
	Renewable fuel standard	Policy to set obligations for the inclusion of renewable content in a fuel sector.	- Canadian federal RFS - Canadian provincial RFS - EU RED - US RFS
	Nested mandates	Mandates for specific fuel types that exist within renewable fuel standards.	- US RFS with mandates for advanced biofuels
Fiscal measures	Capital incentives (grants, loans, guarantees, green bonds)	Incentives to reduce capital cost, reduce payback risk and enhance return profile. Can incentivise private capital investment in new sectors or novel technologies.	- Brazilian BNDES Program - EU NER 300 Program - US Title 3 DOE, DOD, ISDA
	Consumption incentive (fuel/carbon tax relief, blending incentives, credit allowance, guaranteed contract	Incentive to stimulate market use by reducing cost to end user by reducing price of low carbon fuel.	- Allowance creation under "cap and trade" system - Carbon tax exemption - Excise tax reduction - Volumetric Ethanol Excise Tax

Continued

Table 1.19 Key market-based instruments and examples from road transport sector which can be adapted for the biojet fuel industry.—cont'd

Category	Instrument type	Description	Examples
	Investment incentive (capital formation)	Incentive may include income tax relief, property tax relief, contingent taxes or royalties (after payback), investment tax credits and tax-exempt bond structures. Specific investment structures to balance project risk and reward. Can improve project cash flow in the early and payback periods.	- Accelerated depreciation - Flow-through shares - Master limited partnerships
	Innovation (technology grants, financial incentives for R&D, network of enabling institutions to participate in industry-directed research)	Public investment funds to advance a strategic sector by stimulating innovation research, reduce R&D cost, leverage private sector investment and knowledge, and continuous improvement of production assets.	- BioFuelNet Canada - Climate Change Emissions Management Corporation - Scientific Research and Experimental Development Tax Incentive Program - Sustainable Development Technology Canada

Oil, Natural Gas and Biofuels (ANP) (de Souza et al., 2018). This also means that biojet fuels fall under this regulatory agency with the fuel being defined as aviation kerosene (or querosene de aviaç ão alternativo in portugese). Biojet fuel falls under ANP Resolution No. 63 of 2014 (or RANP

Table 1.20 Regulatory acts relating to the Brazilian biojet fuel market.

Regulatory acts	Year	Description
Brazilian Law No. 9478	1997	Relates to national energy policy and activities related to the petroleum industry. Also establishes the ANP.
ANP Resolution No. 17	2006	Regulates the distribution and supply of aviation fuels.
ANP Resolution No. 37	2009	Specification for fossil jet fuel to be used pure or as blend with biojet fuel.
Brazilian Law No. 12,490	2011	Includes biofuels in the Brazilian Law No. 9478.
Brazilian Bill No. 506	2013	Creation of the National Biojet Program to encourage research and achieve sustainability of the Brazilian aviation fuels.
ANP Resolution No.63	2014	Specification for biojet and its blend with fossil jet fuel.
Brazilian Law No. 13,576	2017	Creation of the National Biofuels Policy (RenovaBio) to stimulate production and use of biofuels based on sustainability, competitiveness, and safety.

63—2014—Legislação ANP). Within it, biojet fuel is defined as a fuel produced from alternative sources such as biomass, coal, and natural gas for use in aircraft jet turbine engines with compliant production processes.

Brazil adopts the ASTM D4054 specification from ASTM International. The adoption of the internationally recognized specification has its advantage as the approved fuel must have drop-in characteristics where it could be blended or substituted directly without requiring modification to the engine and supporting infrastructures. The key regulatory acts related to the Brazilian biojet aviation fuel market are shown in Table 1.20 (de Souza et al., 2018).

The Brazilian Law No. 9478 led to the formation of the Brazilian National Agency of Petroleum, Natural Gas and Biofuels (ANP), where ANP is a federal government agency associated with the Ministry of Mines and Energy (or Ministério de Minas e Energia, MME). From there, the most prominent regulatory act for biojet fuels is the ANP Resolution No. 63 of 2014. The resolution presents the specification for the approved alternative fuels to kerosene for the aviation market. To date, alternative fuels approved by ASTM are eventually included in the resolution. The alternative fuels will have their own table of specification akin to that of the counterpart ASTM D7566.

The allowable blend levels of the approved alternative biojet fuels and petroleum-based jet fuel are also defined in the resolution. The blend is called BX Aviation Kerosene (Jet-BX) (de Souza et al., 2018) or Querosenes de Aviação Alternativos e do Querosene de Aviação B-X (QAV B-X) in Portugese. The blend follows the formula of mixing petroleum-based jet fuel with a single type of approved alternative aviation fuel, where the X refers to the volumetric amount of the alternative fuel component in the blend. Postmixing, the Jet-BX blend must still meet the Jet-A1 (QAV-1) fuel requirements as specified in the earlier ANP Resolution No. 37 of 2009. In addition to meeting the basic Jet-A1 requirements, other operational-related parameters such as aromatics, distillation for 10%, 50%, and 90% recovered volume, lubricity and viscosity also need to be met.

The tight requirement is due to the "drop-in" criteria where once the alternative and fossil-based fuels are mixed, they can no longer be separated or differentiated. Thus, it is only prudent to set up stringent criteria prior to the mixing and treat them as the conventional Jet-A1 postmixing. This very fact also makes the ANP Resolution No.17 of 2006 to be hard to police as the regulation attempts to cover the distribution of aviation fuels. In Brazil, fuel is mixed (for Jet-BX blends) prior to reaching the end customers, and the producers and distributors are allowed to import the biojet fuels. Due to the difficulty in determining the blend levels, Jet-BX cannot be exported out of Brazil. The other key Brazilian laws and bills relate to the Brazilian government attempts to promote the biojet fuel industry, increase energy security, and also reduce GHGs, in particular the RenovaBio national biofuels policy which will be discussed in detail in the next section.

1.9.1 Brazilian national biofuels policy (RenovaBio)

The then President of Brazil, Michel Temer approved the legislation creating the Brazilian National Biofuels Policy (RenovaBio), which was a new national biofuels policy. The law was officially gazetted by the Brazilian Senate in December 2017 by Law No. 13,576/2017 with the aims to increase the use of all biofuels. Additional regulations were also issued by Brazil's MME through Decree No. 9888/2019 and Ordinance No. 419 of November 20, 2019 (B3, 2019). While all biofuels inclusive of biojet fuels are part of the policy, it remained dominated by bioethanol and to a certain extent biodiesel. The clearly defined targets will bring greater degree of certainty for investment planning within the sector. The biofuels policy is also timely as Brazil is already a titan on the global stage being the world's second largest consumer and producer of biofuels (although its share of biojet fuel remain small), where bio-based energy contributes to 18% of the nation's energy mix.

As the RenovaBio is partially set up to meet the NDCs of Brazil's commitment to the Paris Agreement. For this purpose, MME, which is the ministry in charge, sets a target of 18% reduction in GHG emissions for the fuel matrix by 2030. Fuel distributors will have to meet the annual targets in accordance with their market share. To quantify the targets and account for the exact contribution of each biofuel producers, a certification system is drawn out with the formation of the Biocarbon Decarbonation Credit (CBIO). Although the target is annual, CBIO will not expire and can only be withdrawn from circulation when it is officially retired by proving that decarbonization is fulfilled.

Energy—environmental efficiency rating will also be given to biofuels producers through a certification program by inspection accredited by the ANP. This way the decarbonization credit that merges the elements of emissions reduction, and life cycle assessment for the producers can be ascertained. The credit also meant that tradable financial assets are introduced into the market for a more efficient "commodity-like" efficiency of meeting the target through financial imperatives. The main features of RenovaBio are shown in Fig. 1.5 (Denny, 2020). From the figure, it can also be seen that the replacement of fossil jet fuel with biojet fuel is part of the RenovaBio push.

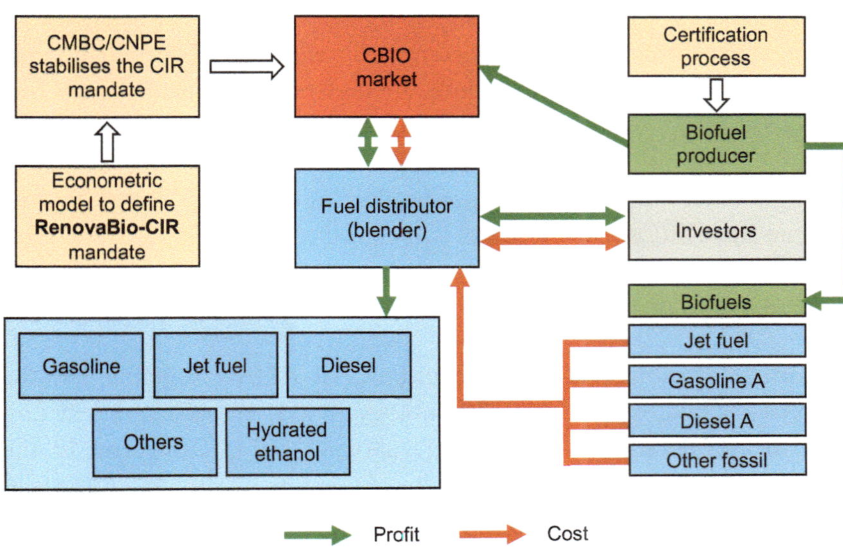

Figure 1.5 Main features of RenovaBio. *(Adapted from Denny, D.M.T., 2020. Competitive renewables as the key to energy transition—RenovaBio: the Brazilian biofuel regulation. In: The Regulation and Policy of Latin American Energy Transitions, pp. 223–242.)*

The CBIO has repercussion beyond just the renewable energy sector as it has wider implications across the wider economy. Fuel distributors can now buy CBIOs to offset the emissions produced from the sales of petroleum-based fuels. It should be noted that any parties interested to offset their carbon could purchase CBIO. In fact, CBIOs can now be freely traded in the Brazilian stock exchange, the B3—Brazil Stock Exchange, and Over-the-Counter Market (or Brasil, Bolsa, Balcão) in Sao Paulo. One CBIO is set to offset 1 ton of CO_2. It is estimated that on average, one CBIO is equivalent to 800 and 500L of bioethanol and biodiesel, respectively.

The new Brazilian carbon credit had its first trade at Real 50 per CBIO on June 12, 2020 (S&P Global Platts, 2020). In this symbolic trade, ethanol producer Adecoagro sold the first 100 units to Datagro Conference to offset carbon emissions generated during Datagro events for the year 2020. It is expected that beyond this symbolic first trade, purchasers of the credit will mainly be the mandated parties such as fuel distributors. Petrobras is mandated to purchase the largest share of the CBIO at 27.1% of total CBIO. Brazilian biojet fuel producers will remain a minor player in the CBIO market for a foreseeable time despite the standing of Brazilian biofuel industry, as none of the 189 fuel producers certified for the generation of CBIO are biojet fuel producers. The breakdown is 89.4%, 10.1%, and 0.5% for bioethanol, biodiesel, and biogas producers, respectively. The second trade on June 29, 2020, tumbled to just Real 15 per CBIO, which is below the estimated price of around Real 55 per CBIO (Argus, 2020). The CBIO trading volume remained low as players within the industry are awaiting the proposal by Brazil's MME to reduce RenovaBio's mandatory goals. The new proposal will close to halve the CBIO target from 28.7 million CBIO to 14.53 million CBIO. This is in view of the concerns by MME that existing biofuel producers might not produce and sell sufficient biofuels to generate the CBIOs.

1.10 Argentina

The Administración Nacional de Aviación Civil (ANAC) together with the National Institute of Industrial Technology (INTI) embarked on aviation biojet fuel production research in 2011 (ANAC, 2014). Concurrently, the Aerolineas Argentinas (ARSA) and the Argentine Chamber of Biofuels (CARBIO) also evaluated the feasibility of supplying its fleet with biojet fuel produced in Argentina.

The two groups formed a multidisciplinary group to combine their efforts and worked on the national production of biojet fuels, which will meet international quality standards. The standard adopted is ASTM D7566. The development of biojet fuel is supervised by the Under Secretariat of Air Transportation (or "Subsecretaria de Transporte Aerocomercial de la Secretaria de Transporte [SSTA y ST]" in Spanish). Additional members such as Yacimientos Petroliferos Fiscales (YPF) and the National Institute of Agricultural Technology (INTA) also joined the working group.

While the Argentinian biodiesel industry is well developed with supporting policies and mandates, its biojet fuel industry does not currently enjoy the same support in terms of policies and legislations. Despite that, Argentina can claim to have produced the feedstock for the first flight by an airplane powered 100% by algal biojet fuel through the Biocombustibles del Chubut S.A. plant in Puerto Madryn.

1.11 China

The Civil Aviation Administration of China (CAAC) formulated the energy conservation and emissions reduction plan as part of their mandate to promote air travel in a sustainable manner (CACC, 2008). This regulatory support from CAAC was issued through two documents in 2008, namely the "Civil Aviation Industry Energy Conservation and Emissions Reduction Plan (2005–15)" and "Circular on the Full-scale Implementation of Energy Conservation and Emissions Reductions throughout Civil Aviation Industry."

This is followed up in 2011 when CAAC released the "Guidelines to Speed Up the Promotion of Energy Conservation and Emissions Reduction Regime in Civil Aviation Industry," dovetailing with the 12th Five-Year Plan of the Civil Aviation Industry. Under the guideline, target emissions reduction is materialized as fuel consumption reduction on a 2005 RTK basis. The fuel consumption reduction targets are 11%, 15%, and 22% for Phase I (2011–12), Phase II (2013–15), and Phase III (2016–20), respectively.

The Chinese plan involves a three-phase stepping up of improving the foundation in Phase I, scaling up in Phase II, and trickling up innovation and optimization in Phase III. These targets from China are autonomous, and CAAC is given the powers to exert itself to achieve the benchmarks through joint efforts within the industry.

CAAC has several energy efficiency programs and actions. Pertinent to biojet fuel is the optimization of airspace and routes, implementation of Reduced Vertical Separation Minimum (RVSM) and gate power unit (GPU) to replace auxiliary power unit (APU). While these actions might not directly impact biojet fuel implementation, it serves to reduce the amount of aviation fuel used. In fact, an estimated 38,000 tons of jet fuel and 121,000 tons of CO_2 emissions have been saved through nonflight measures such as replacing APU by GPU. This makes it easier to increase the proportion of biojet fuel in the aviation energy mix.

China also approved the commercial use of a bio-based aviation jet fuel in February 2014. The CAAC announced that Sinopec, a leading energy and chemical company in China, was granted the first certificate of airworthiness for biojet fuel. For this, the biojet fuel needed to comply with the CTSO-2C701 standard, which applies to civilian aviation jet fuels containing synthesized hydrocarbons. Within CTSO-2C701, the synthesized component of SPK needed to conform to the ASTM D7566-11a standard.

1.11.1 Civil Aviation Development Fund

The CAAC started a Civil Aviation Development Fund (CADF) in April 2012 which is collected with the intention to develop airport facilities. The amount collected varies based on flight routes and aircraft types. The CADF is established for the development of airport facilities, typically for local airport constructions. Funds in the CADF have also been diverted for energy-saving innovative technologies and the development of biojet fuels. Companies can be granted subsidies through the CADF scheme in the 30%–60% range of total investment for emissions reduction efforts (China.org.cn, 2012).

A joint venture between the Commercial Aircraft Corp of China and Boeing has launched an aviation energy conservation and emissions reduction technology center in Beijing to convert "gutter oil" into biojet fuel. Gutter oil is the colloquial name for leftover cooking oil which is reused in the practice of restaurants. The use of gutter oil as feedstock resolves the twin issues of using recycled oil which is no longer safe for human consumption and also to reduce wastes. CAAC estimated that China may use up to 12 million tons of biojet fuel by 2020, or 30% of China's total consumption of jet aviation fuels.

The aforementioned technology center and the planned biojet fuel-powered international flight between Air China and Boeing encapsulate the increasing cooperation between China and the United States. The US Trade and Development Agency (USTDA) and China's National Energy Administration (NEA) also have announced a renewable energy agreement.

The amount collected from CADF reached an all-time high of RMB 47.672 billion in 2018, from the RMB 23.609 billion in 2012. Chinese airlines were estimated to have contributed around RMB 12 billion for the 2018 figure, although that has been reduced substantially since July 2019. Nonetheless, the amount is still substantial, and this augurs well for the development of the biojet fuel industry in China. Due to the COVID-19-induced aviation sector losses, the CAAC issued a set of relief measures in March 2020 to exempt mandatory contribution to CADF. This reduction in CADF income might have an impact on the long-term development of biojet fuels.

1.11.2 China Five-Year plans

The push for biofuels has started in 2001 through the 10th Five-Year Plan (2001–05) to mass-produce ethanol and formulate national standards for fuel ethanol and pilot projects for E10 blends (IEA Bioenergy, 2016). Further legislative pieces were passed, namely the "Renewable Energy Promotion Law" and "New Ethanol Policy" in 2005 and 2006, respectively. The former signals China's big move into renewable energy, while the latter attempts to diffuse the food versus fuel issue by mandating ethanol feedstock to not compete with grains. The ethanol industry gained a boost with ethanol subsidies of USD 0.19 per liter being put in place. The ethanol subsidy would eventually be slashed to USD 0.06 per liter in 2013, although the reduced subsidy support is still arguably better than the lack of direct biojet fuel usage subsidy.

In 2010, the national standards for B5 biodiesel fuel blend were formulated with the government also eliminating biodiesel consumption tax of 5%. Trial programs for biodiesels have also been rolled out to two counties. The 12th Five-Year Plan (2011–15) saw a new ethanol target of 4 million tons by 2015. In September 2017, several ministries issued a joint directive on the "Expansion of Ethanol Production and Promotion of Transportation Fuel" to mandate a nationwide E10 blend by 2020. In contrast, there is no biodiesel or biojet fuel mandate at national level.

China announced the 14th Five-Year Plan (2021–25) in October 2020 with aggressive goals for China to have CO_2 emissions to peak in 2030 and achieve carbon neutrality by 2060. This continues China's increased efforts in pollution prevention and control, as witnessed during the previous 13th Five-Year Plan period (2016–20). The biojet fuel industry is expected to be one of the favored industries as it supports Part 10 of the 60-point proposal to "accelerate green, low-carbon, and sustainable development" for the National Economic and Social Development and the Long-Range Objectives through the year 2035.

The danger for biojet fuel industry, as faced by the biodiesel industry with low penetration rate in China, is from the lack of available feedstock, subsidies, and clear policy support. However, China has a good track record in its Five-Year Plans in which China transformed from not producing any biofuels into the fourth largest biofuel producer in the world (USDA GAIN Report, 2018).

1.12 Malaysia

The National Biomass Strategy 2020 (AIM, 2013) was initiated in November 2011 by the National Innovation Agency (AIM). It is set up initially to assess how Malaysia could increase revenue from its palm oil industry through the broadening use of the associated biomass. This has since been expanded in 2013 to also include biomass from the forestry sector and dedicated crops on marginal land (AIM, 2013). The push for palm oil–derived biojet fuel in Malaysia will tie in well with the strategy, which in turn will benefit the industry as incentives will be provided by the government that is aligned with the strategy.

In April 2019, the Malaysian Palm Oil Council signed a memorandum of understanding in Beijing with the China Chamber of Commerce of Foodstuffs and Native Produce that would see China invests at least RM 2 billion in a Malaysian biojet fuel plant. The biojet fuel industry in Malaysia was further given a boost when budget was allocated in the Malaysian 2020 Budget to study the use of palm oil–based biojet fuel. The government allocated RM30 million for R&D matching grants for collaboration with industry and academia to develop higher value-added downstream uses of palm oil, which includes biojet fuel (Bank Negara Malaysia (BNM), Malaysia 2020 budget, 2019). The Malaysian Palm Oil Board (MPOB) will play the lead to set up a biojet production plant in Malaysia. For this, MPOB requested around RM 5–6 million for the initial biojet fuel project, which is 0.90%–1.08% of the budget allocation for R&D in Malaysia. This is reminiscence of the push in 1982 where MPOB came up with the first palm biodiesel pilot plant within 3 years, which subsequently led to Malaysia being a global player in the biodiesel industry through commercialization of plant technology, product innovation with the winter grade palm biodiesel, and production of value-added products such as palm phytonutrients.

The Malaysian National Aviation Policy was first proposed in October 2011, and the Ministry of Transport was tasked with drafting the policy. The policy was supposed to define all the key aspects of aviation, direction, objectives, and long-term strategies. The independent regulatory body Malaysian Aviation Commission (MAVCOM) proposed an Economic Master Plan for the Malaysian civil aviation sector to the Ministry of Transport in March 2016, outlining the need for a National Aviation Policy (MAVCOM, 2016). The proposed plan has the sectoral coverage of airlines, air traffic control, airport, and ground handling. It does not cover the upstream subsectors such as aircraft design, leasing companies, and "maintenance, repair, and overhaul" (MRO). It also does not cover the type of fuel and sustainability. Granted that the proposed plan only indirectly addresses environmental issues and highlights the interlinkage between environmental and economic matters, the proposal could have provided more explicit links to show how sustainable biojet fuels fit into the bigger picture and their connection to synergize with other policies. Despite the proposal, a National Aviation Policy has yet to materialize as of 2020.

1.13 Japan

The Initiatives for Next-generation Aviation Fuels (INAF) inaugurated in May 2014 is a Japanese deployment initiative to establish a supply chain for the next-generation aviation fuels in the country. INAF was set up by All Nippon Airways (ANA), Japan Airlines (JAL), Nippon Cargo Airlines (NCA), Boeing Japan, Narita Airport, Japan Petroleum Exploration, and University of Tokyo. In all, INAF comprised 46 organizations from the government, industry, and academia collaborating to roll out biojet fuel by 2020.

For the Japanese situation, INAF suggested the adoption of six raw materials such as municipal waste, microalgae, natural oil, waste cooking oil, nonedible biomass, and woody biomass. It is upon the development of supply chain involving these feedstocks will Japan be ready to start producing biojet fuel by financial year 2020. A summary of the feedstock-centric roadmap is shown in Table 1.21.

Table 1.21 Feedstocks and stages of development for the production of biojet fuels as identified by INAF.

	Fiscal year						
Stages	**2015**	**2016**	**2017**	**2018**	**2019**	**2020**	**2021**
Formulate business plan	MW MA NO WCO NEB WB	MA WCO WB					
Design, construction, and trial operation of demonstration plant for production of biojet fuel		MW NO NEB	MW MA NO WCO NEB WB	MW MA NO WCO NEB WB	MA WCO WB		
Produce next-generation aviation fuel (biojet fuel) through demonstration projects				MW NEB	MW NO	MW MA WCO WB	
Expand scale of production (commercial project)			MW NEB	MW NEB	MW NO NEB	MW MA NO NEB	MW MA NEB WB

MA, microalgae; *MW*, municipal waste; *NEB*, nonedible biomass; *NO*, natural oil; *WB*, woody biomass; *WCO*, waste cooking oil.

INAF acknowledges that it will be extremely difficult to produce economically viable alternative aviation fuel below 2015's crude oil price of around USD 65 per barrel. Given that the crude oil price postpandemic in 2020 traded around the USD 35–50 per barrel band, it is likely that the business cost for Japanese biojet fuel producers will exceed the conventional aviation fuel prices. Supportive policies will have to be drafted to provide subsidies, specially recognized depreciation of capital investment, reduction of aircraft fuel tax, and "petroleum and coal" taxes when using biojet fuel. INAF also conceded that fuel prices are cyclical and fluctuates, which makes it difficult for the nascent biojet fuel supply chain to be viable.

However, INAF is of the opinion that the long lead time is required for stable supply to be achieved, so demonstration projects are the key to provide confidence to businesses for the long-haul regardless of the prevailing aviation fuel supply–demand levels. Continuing technological development and securing a margin of investment are seen as crucial moves to ensure the long-term success of the biojet fuel industry in Japan.

The INAF committee met on a biannual basis from July 2015 until April 2018. However, meeting was suspended until members have good prospect of a concrete plan to introduce biojet fuel during the 2020 Tokyo Olympic and Paralympic Games.

1.14 Indonesia

The Indonesian government established the Indonesian Aviation Biofuels and Renewable Energy Task Force (ABRETF) to reduce aviation GHG emissions and increase the use of biojet fuel in the blend mix. The task force was set up in August 2014 under DG Decree No. 517 K/73/DJE/2014 and KP. 429 Year 2014. The ABREFT jointly sits under the Ministry of Transportation's Directorate General of Civil Aviation (DGCA) and the Ministry of Energy and Mineral Resources's Directorate General of Renewable Energy and Energy Conservation (EBTKE). The initiative covers the full value-chain focusing on HEFA initially, with stakeholders covering the government, the massive feedstock producers, oil companies, airlines, airports, aviation associations, and universities. It should be noted that the inclusion of feedstock producers is crucial as Indonesia is the largest palm oil producer in the world and the feedstock is earmarked as one of the possible feedstocks through the HEFA pathway.

This ties in well with the mandate from the Indonesian Ministry of Energy and Mineral Resources under Decree No. 25 Year 2013 to use 2% of biojet fuel in aviation fuel blends nationally in 2016. The mandate also specified an increase to 3% and 5% by 2020 and 2025 (Wei et al., 2019), respectively. Although the 5% amount looks low at first glance, the ambitions of the Indonesian government are comparable with other major initiatives from the European Union and China to have drop-in biofuel quantity of 4% and 3% in 2020, respectively. Also, to provide additional context, air transport consumed 4% of fuel consumption for transport, or 1.9% of total national fuel in 2015. The sustainable aviation fuel from vegetable oil, primarily palm oil used in Indonesia, is labeled as Bioavtur.

ABREFT has conceded that the 2016 goal of 2% biojet fuel in aviation fuel blends will not be achieved due to national circumstances. However, oil producers in Indonesia have since reiterated their commitment to have a production capacity of 257,000 kL/year (ICAO, 2014a).

1.15 Australia

The Australian Initiative for Sustainable Aviation Fuels (AISAF) was formed in 2012. Key public–private partners include the Australian government; airliners such as Qantas Airways, Virgin Australia, Air New Zealand; aircraft supply chain such as Airbus, Boeing, GE; academia and research entities such as CSIRO, the US Studies Centre at the University of Sydney; aviation groups such as Queensland Sustainable Aviation Fuels Initiative and CAAFI; and others such as Baker McKenzie. AISAF sets a long-term target of 50% sustainable biojet fuel by 2050. This target is not legally binding and does not have any obligated parties.

A year later, AISAF joined forces with Aviation/Aerospace Australia (A/AA) to contribute to the long-term sustainability of the country's aviation and aerospace sector. There are four working groups within AISAF, namely "Research, Development and Demonstration—feedstocks and conversion technologies," "Fuel Certification and Qualification," "Environmental Impacts," and "Commercialization." The groups were tasked to identify research gap for the production and commercialization of SAF, improve integration of activities leading to the development of SAF, and implement the Australia–US Memorandum of Understanding on Sustainable Aviation Fuels. The program in this format ended in 2015 when it was absorbed into the University of Sydney-based US Studies Centre's Alternative Transport Fuel Initiative.

Simultaneously, there were also a few other projects running in tandem with the AISAF. One of those is the "Qantas and Shell Aviation Biofuel Feasibility Study" in 2013 to identify environmental and economic challenges of the development of a commercially viable SAF industry in Australia. The study stated that a commercially viable SAF industry can be formed if access to substantial and ratable volume of feedstock can be obtained at submarket prices, ramping up emerging and nonfood domestic feedstock production, and policy that incentivizes the production of any renewable transport fuel (including biojet fuel). The modeling work predicted that between 5% and 35% of Qantas' domestic flights could be flown using a 50% biojet fuel blends from FT or HEFA pathway.

In 2014, the Sustainable Mallee Jet Fuel project commissioned by Airbus evaluated the possibility of using mallee, which is a dominant vegetation in the semiarid areas of Australia. The use of mallee will avoid the food versus fuel debate altogether as arable lands and food feedstock need not be used. In this study, a value chain was proposed according to the Roundtable for Sustainable Biomass (RSB). It includes the production of biojet fuels and sustainability certification. The results are promising, with the pyrolysis of the lignocellulosic-based mallee reducing emissions by 40% as compared with conventional aviation fuel. However, it narrowly missed the 50% emissions reduction threshold of the RSB standard. Additionally, mallee-derived biojet fuels were not price-competitive against that of its fossil counterpart; it was projected that mallee biomass could be competitive by 2021. The plunging demand for air transport since the COVID-19 pandemic might affect this estimation.

The development of biojet fuel is important to Australia as it currently imports about 93% of its commercial aviation fuel. The high import level, closure of refineries, and low stockpile of jet fuel at 23 days represent an energy security concern. Notably, the share of aviation fuel in the transport fuel mix increased the most in recent decades as compared with petrol and diesel fuel. With this in mind, the Australian government should set mandatory biojet fuel blends and reignite the fervor of the early 2010s when multiple biojet fuel initiatives were introduced.

1.16 Summary

As commercial biojet fuels assume greater significance in the global aviation industry, it is crucial that policies and legislations are passed to support biojet fuel initiatives to meet the roadmap goals. Presently, governments around the world are increasingly implementing decarbonization targets concerning the aviation sector and introducing mandates for mandatory blending. Also, more countries are voluntarily signing up to international carbon reduction schemes such as CORSIA. The need for both intranational and international measures is apparent as the major polluting flights often cross national borders, starting the flight in a country and ending in another. The biojet fuel industry is currently developing technological maturity and attempting to swing the pendulum toward profitability. The industry can follow the footsteps of the successful biodiesel and bioethanol industries. Among steps include supportive public policies to protect and nurture the industry, encourage academia to collaborate with industry to develop the technology, and have long-term targets and transparent mandates to improve supply chain for better risk management and financial projections. The public will also need convincing on the safety aspects. This can be conveyed explicitly by requiring biojet fuels to meet the standards prior to usage. It is encouraging that biojet fuel-related policies are steering actions, mandates are ensuring biojet fuel usage, legislations set the boundaries of law, and roadmaps are showing the end goals. Early government-steered initiatives in collaboration with industrial stakeholders are also gaining momentum. The present public policy measures are top-down heavy, relying on vertical integration with industrial players and academia. Nonetheless, for biojet fuel to be the de facto aviation fuel and supplanting conventional jet fuel in the long term, citizen involvement in public policymaking will increase the motivation to adopt biojet fuel for the common good and foster accountability.

References

AIM, 2013. National Biomass Strategy 2020: New Wealth Creation for Malaysia's Biomass Industry.

ANAC, 2014. Administración Nacional de Aviación Civil, Argentina's Action Plan on CO_2 emissions reduction. https://www.icao.int/environmental-protection/Lists/Action-Plan/Attachments/53/Argentina_Action%20Plan%202014.pdf.

Argus, 2020. Brazil Carbon Credits Miss Price Expectations. https://www.argusmedia.com/en/news/2118789-brazil-carbon-credits-miss-price-expectations.

B3, 2019. Decarbonization Credit. http://www.b3.com.br/en_us/products-and-services/additional-services/informational-service/decarbonization-credit-cbio/.

Bank Negara Malaysia (BNM), Malaysia 2020 budget, 2019. https://bnmold.bnm.gov.my/documents/budget/bs2020.pdf.

BioFuelNet, 2020. Task Forces. http://www.biofuelnet.ca/nce/phase-ii/phase-ii-task-forces/.

CACC, 2008. China's Action Plan to Limit and Reduce CO_2 Emissions from International Aviation. https://www.icao.int/environmental-protection/Lists/ActionPlan/Attachments/5/China_en.pdf.

California Air Resources Board, 2019. Low Carbon Fuel Standard Proposed New Temporary Fuel Pathway. https://ww2.arb.ca.gov/sites/default/files/classic//fuels/lcfs/fuelpathways/comments/ajf_temp.pdf?utm_medium=email&utm_source=govdelivery.

California Air Resources Board, 2020. Low Carbon Fuel Standard. https://ww2.arb.ca.gov/sites/default/files/2020-09/basics-notes.pdf.

CBSCI, 2015. Feasibility Study of Canadian Biojet Fuel Supply Chain Phase 2: Creating a Biojet Market. https://cbsci.ca/wp-content/uploads/Aviation_CTI_Biojet_Phase_2.pdf.

CBSCI, 2019. Feasibility Study of Canadian Biojet Fuel Supply Chain Executive Summary. https://cbsci.ca/wp-content/uploads/CTI_Biojet_Combined_Executive_Summary.pdf.

CBSCI, 2020. Canada's Biojet Supply Chain Initiative. https://cbsci.ca/about/.

Celignis Analytical, 2020. Renewable Identification Numbers (RINs) Credit. https://www.celignis.com/RINs-credits.php.

Chiaramonti, D., Goumas, T., 2019. Impacts on industrial-scale market deployment of advanced biofuels and recycled carbon fuels from the EU renewable energy directive II. Appl. Energy 251, 113351.
China.org.cn, 2012. Subsidies for Going Green in Aviation. http://www.china.org.cn/business/2012-08/22/content_26302422.htm.
Davis, U.C., 2020. Sustainable Aviation Fuels under a Low Carbon Fuel Standard. https://policyinstitute.ucdavis.edu/wp-content/uploads/Murphy-ITF-LCFS-and-SAF-Today.pdf.
de Souza, L.M., Mendes, P.A.S., Aranda, D.A.G., 2018. Assessing the current scenario of the Brazilian biojet market. Renew. Sustain. Energy Rev. 98, 426–438.
Deane, J.P., Pye, S., 2018. Europe's ambition for biofuels in aviation - a strategic review of challenges and opportunities. Energy Strategy Rev. 20, 1–5.
Deane, P., Gallachóir, B.Ó., Shea, R.O., R. O., 2017. Biofuels for aviation. In: Europe's Energy Transition - Insights for Policy Making, pp. 79–88.
Denny, D.M.T., 2020. Competitive renewables as the key to energy transition—RenovaBio: the Brazilian biofuel regulation. In: The Regulation and Policy of Latin American Energy Transitions, pp. 223–242.
DfT, U., 2017. The Renewable Transport Fuel Obligations Order. https://assets.publishing.service.gov.uk/government/uploads/system/uploads/attachment_data/file/644843/renewable-transport-fuel-obligations-order-government-response-to-consultations-on-amendments.pdf.
DfT, U., 2018. Renewable Transport Fuel Obligation Statistics. https://assets.publishing.service.gov.uk/government/uploads/system/uploads/attachment_data/file/752990/notes-and-definitions.pdf.
European Commission, 2009. Summary of the Member State Forecast Documents. https://ec.europa.eu/energy/sites/ener/files/dir_2009_0028_article_4_3_forecast_by_ms_summary.pdf.
European Commission, 2011a. Flightpath 2050 - Europe's Vision for Aviation (Report of the High-Level Group on Aviation Research) Luxembourg. Publications Office of the European Union 2011, p. 25. https://doi.org/10.2777/50266.
European Commission, 2011b. White Paper on Transport. https://ec.europa.eu/transport/sites/transport/files/themes/strategies/doc/2011_white_paper/white-paper-illustrated-brochure_en.pdf.
European Commission, 2013. 2 Million Tons Per Year: A Performing Biofuels Supply Chain for EU Aviation. https://ec.europa.eu/energy/sites/ener/files/20130911_a_performing_biofuels_supply_chain.pdf.
European Commission, 2017. Evaluation of Directive 98/70/EC of the European Parliament and of the Council Relating to the Quality of Petrol and Diesel Fuels ('Fuel Quality Directive'). https://ec.europa.eu/transport/sites/transport/files/swd20170178-fuelqualitydirective.pdf.
European Commission, 2018. Renewable Energy Directive. https://ec.europa.eu/energy/topics/renewable-energy/renewable-energy-directive/overview_en.
European Commission, 2020. Reducing Emissions from Aviation. https://ec.europa.eu/clima/policies/transport/aviation_en.
IATA, 2019. Economic Per Formance of the Airline Industry. https://www.iata.org/en/iata-repository/publications/economic-reports/airline-industry-economic-performance—december-2019—report/.
ICAO, 2010a. Farm-to-Fly. https://www.icao.int/environmental-protection/GFAAF/Pages/InitiativesAndProjectsDetails.aspx?ProjectID=32.
ICAO, 2011a. European Advanced Biofuels Flight Path. https://www.icao.int/environmental-protection/GFAAF/Pages/Project.aspx?ProjectID=9.

ICAO, 2011b. Flight Plan Towards Sustainable Aviation Biofuel in Mexico. https://www.icao.int/environmental-protection/GFAAF/Pages/InitiativesAndProjectsDetails.aspx?ProjectID=23.
ICAO, 2012. Action Plan of Iceland. https://www.icao.int/environmental-protection/Documents/ActionPlan/Iceland_en.pdf.
ICAO, 2013. Assembly Resolutions in Force. https://www.icao.int/meetings/glads-2015/documents/a38-18.pdf.
ICAO, 2014a. Indonesian Aviation Biofuels and Renewable Energy Task Force (ABRETF). https://www.icao.int/environmental-protection/GFAAF/Pages/Project.aspx?ProjectID=40.
ICAO, 2014b. Nordic Initiative for Sustainable Aviation (NISA). https://www.icao.int/environmental-protection/GFAAF/Pages/Project.aspx?ProjectID=25.
ICAO, 2016. Initiative Towards Sustainable Kerosene for Aviation (ITAKA). https://www.icao.int/environmental-protection/GFAAF/Pages/Project.aspx?ProjectID=19.
ICAO, 2018. ICAO Vision. https://www.icao.int/environmental-protection/GFAAF/pages/ICAO-Vision.aspx.
ICAO, 2019a. Committee on Aviation Environmental Protection (CAEP). https://www.icao.int/environmental-protection/CORSIA/Documents/CAEP_Analysis%20on%20-the%20estimation%20of%20CO2%20emissions%20reductions%20and%20costs%20-from%20CORSIA.pdf.
ICAO, 2019b. CORSIA at a Glance Series. https://www.icao.int/environmental-protection/CORSIA/Documents/CORSIA%20Leaflets/CorsiaLeaflet-EN-9-WEB.pdf.
ICAO, 2019c. CORSIA Default Life Cycle Emissions Values for CORSIA Eligible Fuels. https://www.icao.int/environmental-protection/CORSIA/Documents/ICAO%20document%2006%20-%20Default%20Life%20Cycle%20Emissions.pdf.
ICAO, 2019d. CORSIA Methodology for Calculating Actual Life Cycle Emissions Values. https://www.icao.int/environmental-protection/CORSIA/Documents/ICAO%20document%2007%20-%20Methodology%20for%20Actual%20Life%20Cycle%20Emissions.pdf.
ICAO, 2019e. ICAO CORSIA CO2 Estimation and Reporting Tool (CERT). www.icao.int/environmental-protection/CORSIA/CERTTool/ICAO%20CORSIA%20CERT%20-version%202019%20-%20Design,%20Development%20and%20Validation.pdf.
ICAO, 2019f. ICAO Strategy to Deal with Aviation Emissions. https://www.icao.int/EURNAT/Other%20Meetings%20Seminars%20and%20Workshops/ICAO%20EUR%20Environment%20Task%20Force%20Meetings/Second%20ICAO%20EUR%20Environment%20Task%20Force%20Meeting/1.1.presentationICAO.2019.pdf.
ICAO, 2020a. Climate Change: Adaptation. https://www.icao.int/environmental-protection/pages/adaptation.aspx.
ICAO, 2020b. Sustainable Aviation Fuels (SAF). https://www.icao.int/environmental-protection/pages/SAF.aspx.
ICAO Secretariat, 2019. Environemental Report, an Overview of CORSIA Eligible Fuels (CEF). https://www.icao.int/environmental-protection/Documents/EnvironmentalReports/2019/ENVReport2019_pg228-231.pdf.
IEA Bioenergy, 2016. The Potential of Biofuels in China. https://www.ieabioenergy.com/wp-content/uploads/2017/01/The-Potential-of-biofuels-in-China-IEA-Bioenergy-Task-39-September-2016.pdf.
JRC, E.U., 2020. RES Target for EU Member State. https://visualise.jrc.ec.europa.eu/t/NREAPs/views/RESsharetrajectory/RES_shares_trajectories?:isGuestRedirectFromVizportal=y&:embed=y. (Accessed 14 November 2020).
MASBI, 2013. Fueling a Sustainable Future for Aviation. http://www.masbi.org/content/assets/MASBI_Report.pdf.
MAVCOM, 2016. Malaysian Aviation Commission. https://www.mavcom.my/wp-content/uploads/2020/02/Executive-Summary.pdf.

Meijerink, O., 2016. Report, the Voluntary RED Opt-In for Aviation Biofuels. https://skynrg.com/wp-content/uploads/2019/03/Publications-The-voluntary-RED-opt-in-for-aviation-biofuels.pdf.
NER, 2020. Nordic Energy Research, Sustainable Jet Fuel for Aviation. https://www.nordicenergy.org/wp-content/uploads/2020/01/Sustainable-Jet-Fuel-Update-Final-NER.pdf.
Openairlines, 2018. CORSIA: Who Needs to Be Participating in the Scheme? https://blog.openairlines.com/corsia-who-needs-to-be-participating.
RICARDO, 2017. Future Fuels for Flight and Freight Competition (F4C). https://ee.ricardo.com/transport/case-studies/f4c.
S&P Global Platts, 2020. Brazilian Carbon Credit First Trade at Near $10/CBIO. https://www.spglobal.com/platts/en/market-insights/latest-news/oil/061220-brazilian-carbon-credit-first-trade-at-near-10cbio.
SAFN, 2011. Sustainable Aviation Fuels Northwest: Powering the Next Generation of Flight. https://www.climatesolutions.org/sites/default/files/uploads/safn_2011report.pdf.
Scarlat, N., Dallemand, J.-F., 2019. Future role of bioenergy. In: The Role of Bioenergy in the Bioeconomy, pp. 435–547.
US USDA, 2012. U.S. Department of Agriculture, Agriculture and Aviation: Partners in Prosperity. https://www.usda.gov/sites/default/files/documents/usda-farm-to-fly-report-jan-2012.pdf.
US EPA, 2020a. Environmental Protection Agency, Approved Pathways for Renewable Fuel. https://www.epa.gov/renewable-fuel-standard-program/approved-pathways-renewable-fuel.
US EPA, 2020b. Environmental Protection Agency, what Is a Fuel Pathway? https://www.epa.gov/renewable-fuel-standard-program/what-fuel-pathway.
US EPA, 2020c. RIN Trades and Price Information. https://www.epa.gov/fuels-registration-reporting-and-compliance-help/rin-trades-and-price-information.
US EPA, 2020d. Total Available RINs to Date Report. https://www.epa.gov/fuels-registration-reporting-and-compliance-help/available-rins.
USDA GAIN Report, 2018. China Biofuels Annual 2018. https://apps.fas.usda.gov/newgainapi/api/report/downloadreportbyfilename?filename=Biofuels%20Annual_Beijing_China%20-%20Peoples%20Republic%20of_7-25-2018.pdf.
USDA GAIN Report, 2019. Global Agricultural Information Network, Biofuel Mandates in the EU by Member State in 2019. https://apps.fas.usda.gov/newgainapi/api/report/downloadreportbyfilename?filename=Biofuel%20Mandates%20in%20the%20EU%20by%20Member%20State%20in%202019_Berlin_EU-28_6-27-2019.pdf.
Wei, H., Liu, W., Chen, X., Yang, Q., Li, J., Chen, H., 2019. Renewable bio-jet fuel production for aviation: a review. Fuel 254, 115599.

CHAPTER 2
Biojet fuel production pathways

2.1 Introduction

Biojet fuel production is the process of producing renewable liquid fuel for aviation applications. The renewable fuel can be produced from various biomass from plants, animals, wastes, and residues. Presently, the dominant method of production is via the hydroprocessed ester and fatty acid (HEFA) pathway using first-generation feedstocks, primarily from edible oils. Unlike the more mature biofuel like biodiesel where the transesterification process is the de facto production method, the best pathway for biojet fuel production is still up for debates. Alongside HEFA, there are eight other pathways, which might still prove to be viable when the technologies mature.

Biojet fuel production pathways can be categorized into four main types: (1) oil-to-jet (OTJ) that uses lipids and biooils, (2) gas-to-jet (GTJ) that utilizes syngases, (3) alcohol-to-jet (ATJ) that consumes small carbon-chain alcohols, and (4) sugar-to-jet (STJ) that uses sugars and starches. The categories are denoted by the raw materials used in the key processes of the pathways. Fig. 2.1 illustrates the broad categories of the various biojet fuel production pathways.

The different production pathways have different feedstock requirements, convert to different intermediates, yield different product compositions, and operate on different sets of conditions and time scales.

2.2 Oil-to-jet

The common oil-to-jet (OTJ) conversion pathways are HEFA, catalytic hydrothermolysis (CH), and hydroprocessed depolymerized cellolusic jet (HDJC). The typical feedstocks used for HEFA, CH, and hydrogenated pyrolysis are fatty acid and esters, triglycerides-rich oil, and wastes/ligno-celluloses, respectively (Casas-Godoy et al., 2020). For a pathway to be

Biojet Fuel in Aviation Applications
ISBN 978-0-12-822854-8
https://doi.org/10.1016/B978-0-12-822854-8.00003-2

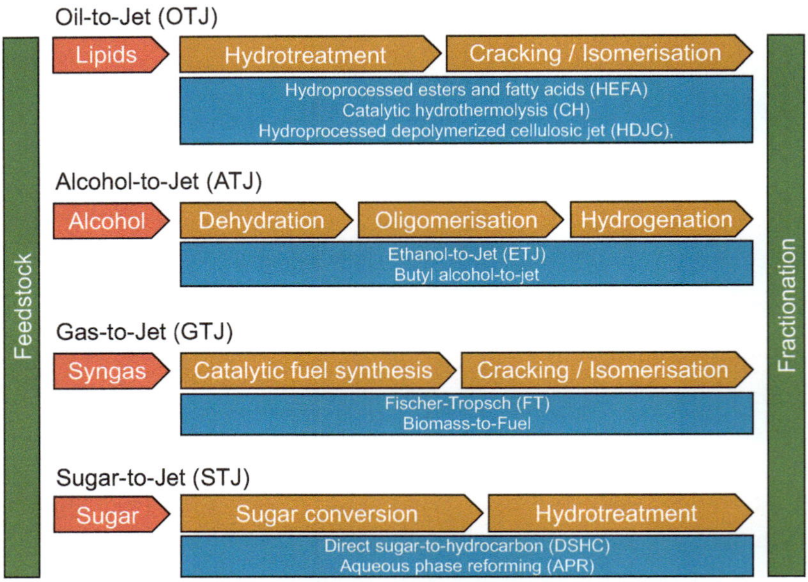

Figure 2.1 Biojet fuel conversion pathways by category.

considered as OTJ, the starting feedstock need not necessarily be lipid-based, instead the main conversion process will convert oil into biojet fuel.

2.2.1 Hydroprocessed esters and fatty acids

HEFA is also commonly known by its other names such as hydrotreated vegetable oil (HVO) or hydrotreated renewable jet (HRJ) (Dayton and Foust, 2020). HEFA biojet fuels are straight-chain paraffinic hydrocarbons produced from the hydroprocessing of lipids such as oils and fats. As such, HEFA fuels have high cetane numbers and do not generally contain oxygen, sulfur, and aromatics. They are also microbial growth-resistant and are stable for storage. HEFA biojet fuels are unique in the sense that they can be used for aviation in jet engines, road in diesel cars, and marine in ships.

The hydrotreating process for the HEFA conversion pathway as developed by UOP Honeywell in 2009 is illustrated in Fig. 2.2. The basic HEFA pathway involves three main processes of deoxygenation, cracking/isomerization, and distillation. Even in the early days, the quantum of biojet fuel produced was in thousands of gallons, from feedstocks covering multiple generations such as palm, soybean, camelina, jatropha, and algae oils (Gutiérrez-Antonio et al., 2015). This led to demonstration flights being

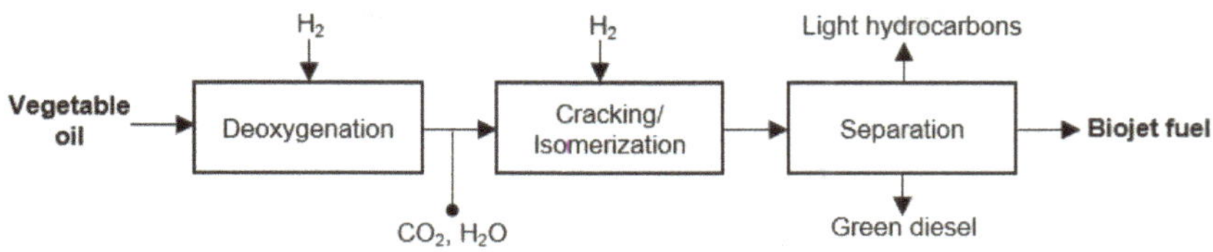

Figure 2.2 Hydrotreating process for production of HEFA biojet fuel. *HEFA*, hydroprocessed ester and fatty acid.

conducted to prove that the biojet fuel achieved parity in performance with conventional jet fuel, culminating in the approval of HEFA biojet fuel for commercial flights by ASTM International in July 2011. HEFA fuels meeting the ASTM D7566 specifications can be blended with conventional jet aviation fuel up to a volumetric blend ratio of 50% (Yilmaz and Atmanli, 2017). This followed the approval of Fischer–Tropsch (FT)–synthesized paraffinic kerosene (FT-SPK) as a conventional aviation fuel in 2009. Despite FT-SPK gaining approval ahead of HEFA, HEFA has the highest technology readiness level (TRL) among the ASTM-certified pathways at TRL 9. This level of TRL denotes advanced commercialization level of technology maturity. Fig. 2.3 shows the technology maturity versus performance curve of various biojet fuel conversion pathways (Vásquez et al., 2017).

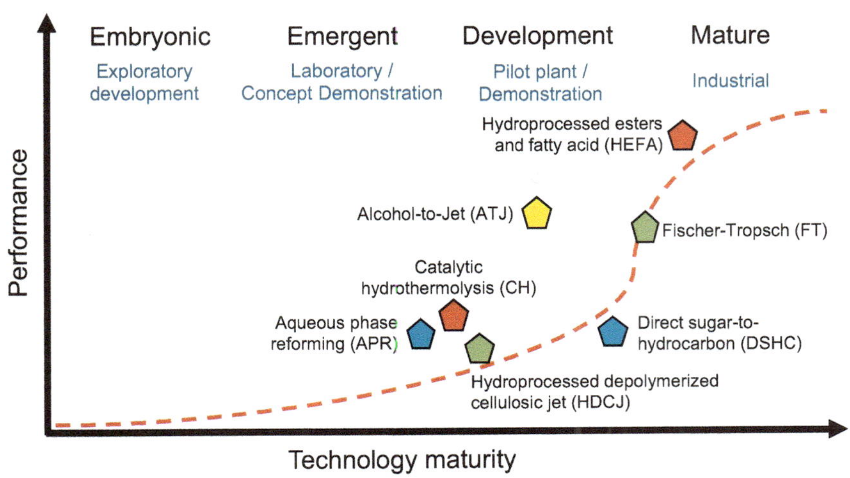

Figure 2.3 Technology maturity versus performance curve of various biojet fuel conversion pathways. *(Adapted from Vásquez, M.C., Silva, E.E., Castillo, E.F., 2017. Hydrotreatment of vegetable oils: a review of the technologies and its developments for jet biofuel production. Biomass Bioenergy 105, 197–206.)*

Prior to the first chemical reaction step of the HEFA process chain, pretreatment procedures such as oil pressing, oil extraction, and prerefining are first conducted. The prerefining step is the most crucial for waste cooking oil or discarded animal fats as contaminants from these feedstocks will bring detrimental effects to the catalyst-controlled HEFA pathway.

Upon completing the physical pretreatment processes, the conditioned feedstocks can undergo the first chemical step, the hydrogenation of lipids. The operating conditions vary greatly with temperature ranging from 250 to 450°C, while hydrogen pressure is in the 10–300 bar range. The act of introducing hydrogen gas in the hydrogenation process will saturate existing double bonds and form water molecules with the feedstock-bound oxygen contents. This converts the triglycerides into hydrocarbons such as alkanes. The sequence of reactions can be summed as (1) hydrogenation of the C=C bonds in the unsaturated fatty acids units of triglycerides, (2) hydrogenolysis of the triglycerides into fatty acids, and (3) deoxygenation of the fatty acids into straight-chained paraffins (Lee et al., 2019). The reactions in the hydrogenation reactor are controlled by catalysts. Common catalysts are supported metal-type catalysts (from Pt, Pd, or Ni) or MoS_2-type catalysts. It is likely that undesirable molecules such as CO and CO_2 be produced as a by-product (Neuling and Kaltschmitt, 2018). The inorganic products of CO, CO_2, and H_2O will be removed, primarily to avoid CO from poisoning the catalysts.

While the straight-chained products at this point have a higher energy density due to the removal of oxygen, the freezing point is still out of the range required for aviation applications. For this, a cracking and isomerization process is needed to produce branched isoalkanes. Typically, metal–acid bifunctional catalysts are used for the hydrocracking process in this stage. The balance between cracking and isomerization can be controlled based on the operating temperature and pressure. Cracking reactions are dominant at higher temperature and lower pressure, while the isomerization reactions are predominant for the inversed conditions of lower temperature and higher pressure. Upon the completion of the process, the products are physically separated via distillation columns. By-products of water and gaseous components are removed. Typically, the liquid distillates produced through the HEFA pathway are diesel, biojet fuel, and naphtha. The C8–C16 paraffins are conventionally considered as the biojet fuel range. Commercial producers such as Neste Oil and UOP Honeywell can achieve biojet fuel product yields of 82.3%–83.96% (w/w) and 82% (w/w), respectively (Pujan et al., 2017).

Table 2.1 compares the physicochemical properties of HEFA biojet fuels, FT biojet fuels, and conventional JP-8 fuels. The biojet fuels are all from commercial entities such as Syntroleum Corporation, UoP, Shell, Sasol, and Rentech. The Syntroleum Corporation's R-8 uses waste fats and greases as feedstocks with the compositions of poultry fat (46%), yellow grease (18%), brown grease (18%), floatation grease (9%), and prepared food (9%). In general, both the HEFA and FT biojet fuels meet all criteria of the standards with the notable exception of specific gravity. It will not be an issue if blended, but if used in neat form the relatively low specific gravity might impact aircraft range. The impacts are dependent on the limitations, be it weight or volume limited. HEFA biojet fuels showed better performance than JP-8 for total acid number, freezing point, smoke point, and heat of combustion. This shows that the HEFA production pathway independent of feedstock is able to provide paraffinic fuels for use as drop-in fuels in compliance with ASTM 7566.

Table 2.2 tabulates the aromatic species analysis, hydrocarbon type analysis, and proportion of *n*-paraffins of HEFA biojet fuels, FT biojet fuels, and conventional JP-8 fuel. The low concentration of longer-chain *n*-paraffins for HEFA from tallow and camelina led to improved low temperature properties as compared with that of JP-8. This is proven where the R-8 HEFA biojet fuel has identical freeze point of −49°C with JP-8 and both have higher concentration of C14−C19 *n*-paraffins.

HEFA fuels also have low to negligible amount of aromatics as compared with JP-8. This will affect the swelling of O-rings used in aircraft fuel systems, as O-ring seals will shrink, harden, and fail to function without aromatics exposure. It was found that the volume swell of extracted nitrile rubber is in the range of 7.0%−9.1% for HEFA biojet fuels as compared with 16.6% for JP-8 aviation fuel. Lighter fuels swell more than heavier fuels, and linear molecules are more mobile than branched molecules. From this, it can be concluded that HEFA biojet fuels will not shrink O-ring seals, although it does not have the same ability as JP-8 fuel to impart volume swell.

Direct conversion of oil into biojet fuel is difficult due to the production of catalyst-poisoning CO during the deoxygenation reaction of the hydrogenation process. This will rapidly deactivate the metal−acid bifunctional catalyst and cause undesired overcracking. Lee et al. (2019) resolved this by using a novel CO-tolerant catalyst by supporting the bimetallic PtRe on ultrastable Y (USY) zeolite as acidic support. The PtRe/USY showed tolerance to CO as it has weakened interactions with

Table 2.1 Physicochemical properties of HEFA biojet fuels, FT biojet fuels, and conventional JP-8 fuel (Corporan et al., 2011).

ASTM tests	Standards requirement	JP-8	SC R-8 HEFA	UOP tallow HEFA	UOP camelina HEFA	Shell FT	Sasol FT	Rentech FT
Total acid number, mg KOH/g (D3242)	Max 0.015	0.005	0.002	0.002	0.002	0.002	0.002	0.004
Aromatics, % vol (D1319)	Max 25.0	17.2	0.0	0.4	0.0	0.0	0.4	1.7
Total sulfur, % wt (D4294 or D2622)	Max 0.30	0.064	<0.001	<0.0003	0.0018	<0.001	<0.001	<0.001
Distillation, initial boiling point (IBP), °C (D86)	Report	152	158	165	151	146	149	152
10% recovered, °C (D86)	Max 205	173	175	179	161	162	166	168
20% recovered, °C (D86)	Report	179	185	185	166	162	170	179
50% recovered, °C (D86)	Report	198	215	210	182	169	180	216
90% recovered, °C (D86)	Report	239	260	243	237	184	208	263
Final boiling point, °C (D86)	Max 300	260	274	255	259	198	228	275
Distillation residue, % vol (D86)	Max 1.5	1.1	0.8	1.2	1.1	1.0	1.4	1.0
Loss, % vol (D86)	Max 1.5	0.2	0.2	0.8	0.9	0.4	0.5	0.8
Freeze point, °C (D5972)	Max −47	−49	−49	−62	<−77	−55	<−77	−50
Existent gum, mg/100 mL (D381)	Max 7.0	0.4	<1	<1	<1	1.6	1.4	<1
Viscosity @ −20°C, cSt (D445)	Max 8.0	4.1	5.5	5.3	3.3	2.6	3.8	5.1

Lubricity test (BOCLE) (D5001) wear scar mm	Report	0.54	0.92	0.76	0.76	0.75	0.87	0.82
Specific gravity (D4052)	0.775–0.840	0.799	0.762	0.758	0.751	0.737	0.762	0.763
Smoke point, mm (D1322)	Min 19.0	25	>40	>40	50	40.0	>40	>40
Flash point °C (D93)	Min 38	48	48	55	43	44	44	44
Heat of combustion, MJ/kg (D3338)	Min 42.8	43	44.1	44.1	44.1	44.1	44.2	44.2
Hydrogen content, % mass (D3343)	13.4	13.9	15.3	15.3	15.4	15.6	15.1	15.2

Table 2.2 Aromatic species analysis, hydrocarbon type analysis, and proportion of n-paraffins of HEFA biojet fuels, FT biojet fuels, and conventional JP-8 fuel (Corporan et al., 2011).

Category	Subcategory	JP-8	SC R-8 HEFA	UOP tallow HEFA	UOP camelina HEFA	Shell FT	Sasol FT	Rentech FT
Aromatics (vol %)	Monoaromatics	15.1	0.3	<0.2	<0.2	<0.2	0.4	1.5
	Diaromatics	1.6	<0.1	<0.1	<0.1	<0.1	<0.1	<0.1
	Total aromatics	16.7	0.3	<0.2	<0.2	<0.2	0.4	1.5
	Total saturates	83.3	99.7	>99.8	>99.8	>99.8	99.6	98.5
Hydrocarbon type (vol %)	Paraffins (normal + iso)	50	92	98	90	>99	88	92
	Cyclo-paraffins	34	8[a]	2	10	<1	12[a]	7[a]
	Alkylbenzenes	12	<0.3	<0.3	<0.3	<0.3	0.4	1.3
	Indans and tetralins	3	<0.3	<0.3	<0.3	<0.3	<0.3	<0.3
	Indenes and C_nH_{2n-10}	0.4	<0.3	<0.3	<0.3	<0.3	<0.3	<0.3
	Naphthalene	<0.3	<0.3	<0.3	<0.3	<0.3	<0.3	<0.3
	Naphthalenes	1.4	<0.3	<0.3	<0.3	<0.3	<0.3	<0.3
	Acenaphthenes	<0.3	<0.3	<0.3	<0.3	<0.3	<0.3	<0.3
	Acenaphthylenes	<0.3	<0.3	<0.3	<0.3	<0.3	<0.3	<0.3
	Tricyclic aromatics	<0.3	<0.3	<0.3	<0.3	<0.3	<0.3	<0.3

n-Paraffins (wt %)	*n*-Heptane	0.10	0.13	<0.001	0.017	0.012	<0.001	0.007
	n-Octane	0.34	0.80	0.12	0.71	1.63	<0.01	1.88
	n-Nonane	1.21	2.28	2.01	3.20	22.4	<0.05	2.75
	n-Decane	3.48	2.47	1.88	2.80	25.1	<0.03	2.22
	n-Undecane	4.24	2.10	1.52	1.20	3.78	<0.02	1.81
	n-Dodecane	3.71	1.64	1.25	0.87	0.29	<0.01	1.52
	n-Tridecane	2.84	1.23	0.82	0.60	0.003	<0.01	1.40
	n-Tetradecane	1.79	0.92	0.86	0.41	0.001	<0.01	1.05
	n-Pentadecane	0.87	0.80	0.35	0.37	<0.001	<0.005	0.90
	n-Hexadecane	0.27	0.60	0.004	0.061	<0.001	<0.003	0.64
	n-Heptadecane	0.089	0.052	<0.001	0.015	<0.001	<0.001	0.071
	n-Octadecane	0.024	0.026	<0.001	0.006	<0.001	<0.001	0.002
	n-Nonadecane	0.008	<0.001	<0.001	0.001	<0.001	<0.001	<0.001
	Total *n*-paraffins	19.0	13.1	8.8	10.2	53.3	<0.2	14.3
	C_7–C_9	1.7	3.2	4.0	6.7	24.1	<0.06	4.6
	C_{10}–C_{13}	14.3	7.4	2.9	4.9	29.2	<0.1	7.0
	C_{14}–C_{16}	2.9	2.3	0.35	0.45	<0.003	<0.02	2.6
	C_{17}–C_{19}	0.12	0.079	<0.001	0.022	<0.003	<0.003	0.07

[a]The ASTM D2425 test method may overpredict cycloparaffins for highly branched SPKs.

CO. Rather than removing CO, the chemisorbed CO could be converted into CH_4 and H_2O through the methanation process. Using the novel catalyst, 41 wt% of biojet fuel could be obtained from palm oil through direct hydroconversion. The properties of the biojet fuel met all requirements of the standard.

Similarly, Verma et al. (2011) successfully produced biojet fuels via a single-step route using hierarchical mesoporous zeolites from algae and jatropha oils. Key operating conditions are reaction temperature of 330–430°C, pressure from 10 to 80 bar, and liquid hourly space velocity of 1.0–1.1/h. Using jatropha oil as feedstock and Ni–W catalyst supported on an acidic zeolitic ZSM-5 support, biojet fuel yield in the range of 40%–45% with high isomerization selectivity of isomer/*n*-alkane around 2–6 was achieved. Yield was further increased to 40%–50% with an even greater isomerization selectivity ratio of ~3–13 for the Ni–Mo catalyst. The catalysts have intracrystalline mesoporosity with hierarchical structure. The greatest yield of 77% was achieved when algal oil and sulfided Ni–Mo catalyst supported on semicrystalline ZSM-5 were used, although isomerization selectivity is reduced to 2.5. The biojet fuel products from this single-step route met all of the basic requirements including freezing point, density, flash point, heat of combustion, and viscosity. The used catalysts could also be regenerated and reused without noticeable degradation in performance after resulfidation.

Eller et al. (2016) improved the technical grade coconut oil HEFA conversion pathway with special hydrocracking, utilising lower chemical and energy costs by maintaining the sulfide state of the catalyst. The choice of coconut oil as feedstock reduced H_2 consumed due to the high proportion of saturated fatty acids with carbon number in the jet fuel range. In this study, presulfided $NiMo/Al_2O_3$ was used as the catalyst at operating pressure of 30 bar, temperature range of 280–380°C, H_2/feedstock volume ratio of 600 Nm^3/m^3, and liquid hourly space velocity of 1.0–3.0/h. The two methods employed to maintain sulfide state of catalyst are through the use of sulfidation agent such as dimethyl-disulfide (DMDS) and H_2S containing H_2 gas. Table 2.3 shows the main properties and yields of the biojet fuel products from special hydrocracking and after isomerization using the two sulfidation agents.

The product yields for the H_2S in H_2 gas sulfidation method are higher by 0.5%–2.0% than those of DMDS. This can be attributed to the lower sulfur content, which poisons the active sites of the catalyst to a smaller

Table 2.3 Main properties and yields of the biojet fuel products from special hydrocracking and after isomerization using the two sulfidation agents (Eller et al., 2016).

		Sulfidation agent	
Process	**Properties and yields**	**Dimethyl-disulfide (DMDS) in liquid feedstock**	**H_2S containing H_2 gas**
From special hydrocracking	Target product yield (%)	57.85	58.75
	Approaching of the theoretical yield (%)	85.49	86.82
	Density (kg/m^3)	0.7646	0.7601
	Sulfur content (mg/kg)	8	2
	Aromatic content (%)	<0.1	<0.1
	Freezing point (°C)	−8	−11
	Smoke point (mm)	34	36
After isomerization	Liquid product yield (%)	91.6	91.5
	Isoparaffin content of product (%)	72.3	77.6
	Sulfur content of feedstock (mg/kg)	8	1.5
	Aromatic content (%)	<0.1	<0.1
	Freezing point (°C)	−41	−45
	Freezing point after cold flow improver additive (°C)	−48	−49
	Cold flow improver additive (mg/kg)	20	15

extent. While both sulfidation agents obtained yields above 85% of the theoretical yields and have products with excellent oxidation stability, the biojet fuel fractions after the isomerization process failed to meet the required freezing point. This was remedied through the use of cold flow improvers, where the freezing points for the DMDS and H_2S in H_2 gas cases improved to −48 and −49°C, respectively. The results proved that H_2S in H_2 gas can be used as a sulfidation agent to maintain the sulfide state of catalysts. Such a solution is also attractive as H_2S will no longer be required to be removed from the gas stream.

The typical biojet fuel fractions from the HEFA conversion pathway do not contain aromatics. Aromatics are crucial in aviation fuels to maintain the fuel system elastomers. Rabaev et al. (2015) resolved this by converting

Table 2.4 Organic liquid yields, aromatic contents of the organic liquid product, and fatty acid contents of the different vegetable oils (Rabaev et al., 2015).

Vegetable oils	Organic liquid product weight yield (%)	Aromatics (wt%)	Polyunsaturated fatty acids		
			Di- (linoleic acid)	Tri- (linolenic acid)	Total polyunsaturated
Sunflower	84.6	15.5	63.8	0.1	63.9
Soybean	83.8	15.0	51.7	5.9	57.6
Camelina	83.2	14.8	19.5	37.0	56.5
Castor	78.2	2.0	5.0	<0.1	5.0
Palm	81.9	1.7	11.7	0.5	12.2
Tallow	85.0	1.0	3.0	1.0	4.0

six different vegetable oils using a novel $Pt/Al_2O_3/SAPO\text{-}11$ catalyst to diesel and jet fuels, which contain aromatics. Table 2.4 tabulates the organic liquid yields, aromatic contents of the organic liquid product, and fatty acid contents of the different vegetable oils. It was found that the formation of aromatic during the hydrotreating process is dependent on total polyunsaturated proportion of the fatty acid. Ratios between di-(linoleic acid) and tri-(linolenic acid) were determined to be a nonfactor. From this study, it was found that biojet fuels produced from sunflower, soybean, and camelina oil have the potential to be drop-in fuels as they meet the minimum 8 wt% aromatic contents.

Gutiérrez-Antonio et al. (2015) intensified the HEFA conversion pathway by using thermally coupled distillation for the purification stage. Table 2.5 summarizes the optimum heat duties of conventional and thermally coupled distillation column for the process. The optimized heat duty values are modeled using a multiobjective genetic algorithm method based on the criteria of lowest heat duty and optimum number of stages for the distillation columns. As compared with conventional distillation methods, the thermally coupled methods have lower energy requirements by an average of 21%. The direct thermally coupled (DTC) distillation column only requires 18,358.08 W with a total 129 stages, while the indirect thermally coupled (ITC) method requires 10.5% more energy with 130 stages. The use of Petlyuk sequence (PS) and dividing wall column (DWC) distillation columns is not recommended for the separation stage due to the substantially greater energy requirement. The use of DTC will improve the sustainability of the HEFA production process while also reducing cost due to lower energy usage.

Table 2.5 Heat duties of conventional and thermally coupled distillation columns for HEFA conversion pathway (Gutiérrez-Antonio et al., 2015).

Heat duty	Conventional		Intensified—Thermally coupled			
	Direct conventional	Indirect conventional	Direct thermally coupled	Indirect thermally coupled	Petlyuk sequence	Dividing wall column
Minor heat duty (W)	21,972	23,208	18,355	20,295	29,154	269,300
Similar number of total stages (W)	23,494	23,213	18,358	20,295	31,176	270,265
Minor number of stages (W)	29,747	26,015	19,295	22,322	53,304	293,601

The HEFA conversion pathway should be pursued as the main conversion pathway at present, due to the possibility of leveraging the lower technology complexity of the process, level of consolidation of this technology around the world, and availability of potential feedstock to immediately scale-up production. Thus, more investment should be made in the direction to develop this production route (de Souza et al., 2018).

2.2.2 Catalytic hydrothermolysis

Catalytic hydrothermolysis (CH) is also commonly known as hydrothermal liquefaction (HTL). The process utilizes subcritical water to convert wet biomass to carbon-rich biocrude (Capaz and Seabra, 2016). The biocrude is also often referred as biooil. Unlike fast pyrolysis that relies on dry biomass, CH could use wet biomass like algae or oil seeds. This allows the CH process to avoid the energy-intensive feedstock drying process. With respect to biojet fuel, the CH pathway is mainly developed for algae as feedstock, as the method is robust to allow the use of species with lower lipid content as lipid content can increase 5%–30% (Cremonez et al., 2015). Fig. 2.4 illustrates the CH biojet fuel conversion pathway.

In general, the CH biojet fuel conversion pathway consists of the pretreatment, CH conversion, and postrefining steps (Wei et al., 2019). At first, wet biomass is fed to the process as slurry (Gírio, 2019). The pretreatment step consists of the cleaning of the slurry, followed by conjugation, cyclization, and cross-linking. The molecular structures of the feedstock are improved through the pretreatment steps. After pretreatment, the CH step comprising initially of the cracking and hydrolysis reactions occurs at a mild temperature range of 250–380°C and operating pressure of 5–30 MPa in the presence of water assisted by catalysts. The aforemention steps involving CH can also take place without catalyst [71]. In general, CH can use organic materials such as lipids, proteins, and carbohydrates. Lignin and cellulose cannot be used as feedstock for the CH pathway. Assuming algal biomass is used as feedstock, then the free sugars, amino acids, and residual polysaccharides will be hydrolyzed (Yoo et al., 2015). The free fatty acids, derivative of steroids and pigments mixed with asphaltenes, will form

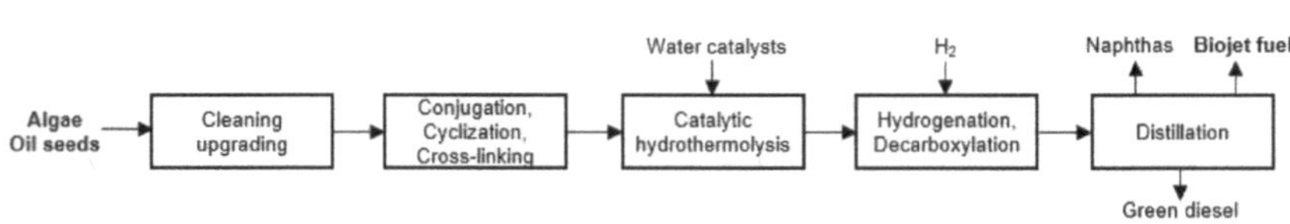

Figure 2.4 Catalytic hydrothermolysis (CH) biojet fuel conversion pathway.

the hydrophobic biooil. The two reactions are followed by catalytic decarboxylation and dehydration reactions. Finally, postrefining hydrotreating and fractionation reactions will produce alkanes through the conversion of straight-chain, branched, and cycloolefins.

Tzanetis et al. (2017) simulated biojet fuel production through the CH conversion pathway by varying catalysts, catalyst-to-biomass ratio, and operating temperature. Catalysts choice includes water (or no catalyst), Na_2Co_3(aq), and Fe(aq), with catalyst loading as high as 33%. Operating temperature ranges from 280 to 340°C. The best case obtained was for the 10% loading of Fe(aq) catalyst at 340°C. Table 2.6 compares the best CH biojet fuel cases, with and without catalyst. While the Fe(aq) catalyst case

Table 2.6 Comparison of catalytic hydrothermolysis and biooil upgrading with Fe(aq) as catalyst and water as catalyst (Tzanetis et al., 2017).

CH pathway stage	Categories	Properties	FE340	W300
Catalytic hydrothermolysis (to obtain biooil)	Operating conditions	Catalyst	Fe(aq)	Water
		Catalytic loading ($kg_{catalyst}/kg_{biomass}$ in %)	10	—
		Thermochemical conversion temperature (°C)	340	300
	Yields	Biocrude oil (%)	36	27
		Solid residue (%)	10	20
		Dissolved organics (%)	15	16
		Water (%)	21	26
		Gas (%)	18	11
		Overall biooil yield from biomass (%)	37.5	28.6
	Properties	Biooil HHV (MJ/kg)	35.2	35.2
		Biooil LHV (MJ/kg)	33.1	33.1
		Biooil oxygen content (wt%)	16.6	16.7
		Biooil water content (wt%)	1.2	1.2

Continued

Table 2.6 Comparison of catalytic hydrothermolysis and biooil upgrading with Fe(aq) as catalyst and water as catalyst (Tzanetis et al., 2017).—cont'd

CH pathway stage	Categories	Properties	FE340	W300
Biooil upgrading (to obtain biojet fuel and other coproducts in the product mix)	Operating conditions	Biooil upgrading temperature (°C)	400	400
		Biooil upgrading pressure (MPa)	0.85	0.85
	Yields	Renewable jet fuel yield from biomass (%)	10.4	8.1
		Upgraded oil yield from biomass (%)	26.6	20.3
		Renewable jet fuel yield from biooil (%)	28.4	28.3
		Upgraded oil yield from biooil (%)	70.9	70.9
	Properties	Product mix HHV (MJ/kg)	49.3	49.3
		Product mix LHV (MJ/kg)	46.2	46.2
		Product mix oxygen content (wt%)	0.1	0.1
		Product mix water content (wt%)	0.2	0.2
		Product mix density at 15°C (kg/L)	0.77	0.77

showed greater yield of 10.4% biojet fuel from biomass, as compared with 8.1% for the water case, the water case without catalytic loading is compelling as it resolves two big issues of a typical chemical process, which are high energy usage and specialized catalyst requirement.

While it is apparent that the CH pathway is best used for algae, the pathway remains to be more suitable for microalgae than macroalgae as the biooil yields are generally higher as tabulated in Table 2.7. Catalytic hydrothermolysis of *Dunaliella tertiolecta* with 78.4% moisture content as feedstock under the operating temperature of 300°C and pressure of

Table 2.7 Effects of catalytic hydrothermolysis on microalgae and macroalgae for biooil production (Kumar et al., 2016).

Algae	Species	Water type	Bio-oil yield (% w/w)	Dry biomass-to-water ratio (w/v)	Higher heating value (MJ/kg)	Temperature (°C)	Catalysts
Microalgae	*Spirulina*	Freshwater	31	Moisture 78.4% (w/w)	35–37	300	–
	Scenedesmus	Freshwater	45.4	Moisture 78.4% (w/w)	35.5	300	–
	LEA of *Scenedesmus*	Freshwater	36	Moisture 78.4% (w/w)	35.3	300	–
	Dunaliella tertiolecta	Marine	37	Moisture 78.4% (w/w)	36	340	With and without
	Desmodesmus	Freshwater	49	~1:13	22–36	375	With and without
	Nannochloropsis sp.	Marine	57	1:18	38	350	Pd/C, Pt/C, Ru/C, Ni/SiO_2–Al_2O_3, CoMo/γ-Al_2O_3 (sulfided), and zeolite
	Botryococcus braunii	Freshwater	64	3:2	~50	300	Na_2CO_3

Continued

Table 2.7 Effects of catalytic hydrothermolysis on microalgae and macroalgae for biooil production (Kumar et al., 2016).—cont'd

Algae	Species	Water type	Bio-oil yield (% w/w)	Dry biomass-to-water ratio (w/v)	Higher heating value (MJ/kg)	Temperature (°C)	Catalysts
Macroalgae	*Laminaria saccharina*	Marine	79	1:10	35.97	350	–
	L. saccharina	Marine	19.3	1:10	36.5	350	KOH
	Enteromorpha prolifera	Marine	23	2:15	28–30	220–320	Na_2CO_3
	Oedogonium	Freshwater	26.2	1:14	33.7	330–340	–
	Cladophora	Freshwater	19.7	1:14	33.5	330–340	–
	Cladophora	Marine	13.5	1:14	33.3	330–340	–
	Derbesia	Marine	19.7	1:14	33.2	330–340	–
	Ulva	Marine	18.7	1:14	33.8	330–340	–

10 MPa resulted in 37% biooil yield (Kumar et al., 2016). Attempts were made to increase biooil yield through catalysts, but the presence or type of catalysts did not make significant differences.

Presently, actual data from large-scale pilot or commercial endeavors are not available as the technology maturity is just after "emergent" and at the early stage of pilot studies.

2.2.3 Hydroprocessed depolymerized cellulosic jet

Hydroprocessed depolymerized cellulosic jet (HDJC) is an oil upgrading process to convert biooils produced from the pyrolysis of lignocellulosic feedstocks into biojet fuels. It is also called hydrogenated pyrolysis oil (HPO kerosene) or pyrolysis-to-jet (PTJ). Fig. 2.5 shows the HDJC process. Through this process, biooils from the pyrolysis process will shed their undesirable physicochemical properties such as high oxygen content, low energy density, high corrosivity, and poor thermal instability (Wei et al., 2019).

Factoring in only the postpyrolysis process, or the actual biooil upgrading process, the HDJC pathway is a two-step hydrotreating process. The first step is a catalyst-assisted hydrotreating process under mild conditions for hydrodeoxygenation of biooil. It is followed by a more conventional high-temperature hydrogenation catalyst-controlled process to obtain hydrocarbon fuels. For HDJC, the ZSM-5 and Raney nickel catalysts are typically used for the fast pyrolysis and hydrotreating processes of the pathway, respectively.

A cooperative research and development work among UOP, the National Renewable Energy Laboratory (NREL) and the Pacific Northwest National Laboratory (PNNL), resulted in a pilot-scale biorefinery to upgrade biooil (Gutiérrez-Antonio et al., 2017). The biorefinery has a capacity of one ton of dry biomass per day. The facility combined fast pyrolysis and hydrotreating to produce biojet fuel alongside other coproducts of diesel and green gasoline. The fast pyrolysis section of the biorefinery operates at a high temperature of 450–600°C for a yield range of 65–75 wt% of biooil from dried woody biomass as feedstock. The biooil is then upgraded using

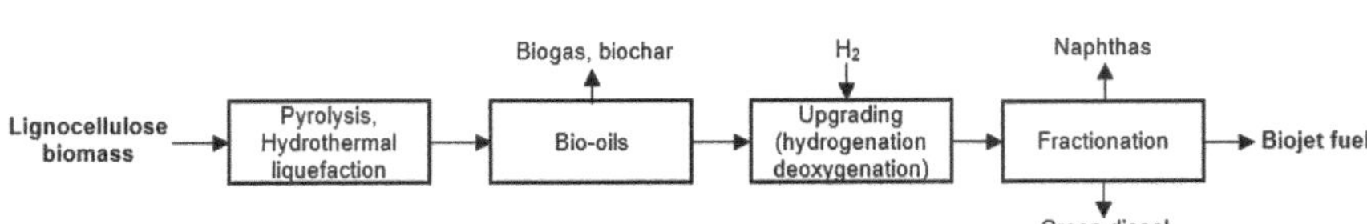

Figure 2.5 Hydroprocessed depolymerized cellulosic jet (HDJC) biojet fuel conversion pathway.

the UOP Ecofining technology to produce biojet fuel. Although classified under oil-to-jet pathway, this method allows the use of low-cost lignocellulose feedstocks instead of the more expensive edible oil.

Table 2.8 tabulates studies related to the HDJC conversion pathways. It was found that high product yield is obtained when there is a positive synergy for the aromatics in the catalytic microwave copyrolysis and performance of the Raney nickel catalyst. It should be noted that biojet fuels produced through the HDJC conversion pathway will contain some aromatic compounds, which are required to avoid sealing problems in jet engines. This is in contrast with the more common HEFA or FT method where aromatics are produced in negligible quantities (Kousoulidou and Lonza, 2016). As such, HDJC biojet fuels has the potential to be used in neat form or drop-in fuel as long as other key physicochemical properties meet the requirement from the standards.

2.2.4 Commercial flights from oil-based feedstocks

The intergovernmental coordinated push to decarbonize global economy has in recent years included the aviation industry. As compared with road transport, the aviation industry is severely lagging in displacing conventional jet fuel. The lag is not from the lack of interests; instead it is waiting for the biojet fuel production technologies to mature to the point of profitability. The interests from airline companies are triggered by legislation, economics, and "environmental, social, and corporate governance" (ESG) concerns. Governments in meeting legally binding international targets like COP 21 have set national mandates to blend biojet fuels with conventional jet fuels. Airlines anxious of impending carbon tax are looking for low-carbon alternative power sources. Simultaneously, ESG issues are taking center stage in the aviation sector, which sparks the "flight shaming" movement, pressured by institutional funds to only invest in airliners with good ESG scores. All these led to airliners with foresight to set up trial flights using biojet fuels as early as 2007. Table 2.9 shows key demonstration flights powered with biojet fuel from oil-based feedstocks.

These demonstration flights have shown progressive ambitions and increase in scale. Since the first flight involving biojet fuel in 2007 by GreenFlight International, over a quarter million flights have involved the use of biojet fuel, and more than 40 airlines have experience dealing with biojet fuels. Companies such as Neste Oil Company, SkyNRG, UOP Honeywell, ENI, Galp Energia, Hawai'i Bio Energy, Alt Air, and Byogy

Table 2.8 Studies involving the hydroprocessed depolymerized cellulosic jet (HDJC) conversion pathway to produce biojet fuels (Gutiérrez-Antonio et al., 2017).

Feedstock (at pyrolysis stage)	Operating temperature (°C)	Catalyst for fast pyrolysis	Catalyst for hydrotreating	Yield (wt%)	Remarks
Douglas fir pellets	375	ZSM-5	5–10 wt% Raney nickel	12.63	Process under mild conditions operation
Lignocellulosic biomass	500 (fast pyrolysis) 200 (hydrotreating)	ZSM-5	10–20 wt% Raney nickel	24.68	Up to 84.59% selectivity of jet fuel range cycloalkanes from intact biomass under very mild conditions
Biomass-to-plastic ratio of 0.75	375 (fast pyrolysis)	ZSM-5	Home-made Raney nickel	34.20	Process under mild conditions operation

Table 2.9 Key demonstration flights powered with biojet fuel from oil-based feedstocks (Zhao et al., 2019).

Date	Airline company	Aircraft	Feedstock	Remark
October 2007	GreenFlight International	Aero, L-29 Delfin	Waste vegetable oil	The very first flight of an aircraft powered entirely by neat biojet fuel.
February 2008	Virgin Atlantic	Boeing 747	Coconut and babassu	Biofuel test flight between London and Amsterdam using a 20% volumetric blend of biojet fuels in one of its engines.
January 2009	Continental Airlines	Boeing 737	Algae and jatropha	First flight of an algae-fueled jet. The pilots conducted a series of tests at 12,000 m (or 38,000 feet), inclusive of a midflight engine shutdown.
April 2010	US Navy	F/A-18 (the "Green Hornet")	Camelina	Results indicated that the aircraft performed as expected through its full flight envelope with no degradation of capability.
June 2011	Boeing	Boeing 747-8F	Camelina	The company flew the new 747-8F model to the Paris Air Show with all four engines burning a 15% mix of camelina biojet fuel.
August 2011	US Navy	T-45	Camelina	Training aircraft successfully flew using a camelina biojet fuel blend with petroleum-based JP-5 at 50:50 volumetric blend levels.
September 2011	US Navy	AV-88	Camelina	First biojet fuel flight test in an AV-8B Harrier from Air Test and Evaluation Squadron 31.

Table 2.9 Key demonstration flights powered with biojet fuel from oil-based feedstocks (Zhao et al., 2019).—cont'd

Date	Airline company	Aircraft	Feedstock	Remark
October 2011	Air China	Boeing 747-400	Jatropha	China's first flight using biojet fuel. Biojet fuel was produced from Chinese grown jatropha from PetroChina. The 2-h flight above Beijing used 50% biojet fuel in one engine.
April 2012	Jetstar Airways	Airbus A320	Refined cooking oil	SkyNRG supplied the biojet fuel for the 75-min flight JQ 705 from Melbourne to Hobart.
March 2013	KLM	Boeing 777-206 ER	Waste vegetable oil	Weekly flight between John F. Kennedy Airport in New York and Schiphol Airport in Amsterdam using biojet fuel supplied by SkyNRG.
August 2014	Gol Transportes Aéreos	Boeing 737-700	Inedible corn oil and waste vegetable oil	A 4% blend of biojet fuel flight, Gol Flight 2152 from Rio Santos Dumont Airport toward Brasilia.
March 2015	Hainan Airlines	Boeing 737-800	Waste vegetable oil	China's first commercial flight carrying 156 passengers from Shanghai to Beijing. The Sinopec-supplied biojet fuel was blended at 50% level with conventional petroleum jet fuel.

Continued

Table 2.9 Key demonstration flights powered with biojet fuel from oil-based feedstocks (Zhao et al., 2019).—cont'd

Date	Airline company	Aircraft	Feedstock	Remark
September 2016	KLM	Boeing 737-400	Waste vegetable oil	SkyNRG supplied the biojet fuel made from used cooking oil by AltAir Fuels in Los Angeles.
May 2017	Singapore Airlines	Airbus A350-900	Used cooking oil	The airline begun a series of biojet fuel-powered flights using an A350-900 aircraft on nonstop trans-Pacific flights between Singapore and San Francisco.

have either established or have plans to construct hydrotreating plants for biojet fuel worldwide. These show not only the commitment by the aviation sector to implement biojet fuels as an alternative to conventional jet fuels but also a testament to the performance and safety of flight achieved by biojet fuel-powered flights around the world.

2.3 Alcohol-to-jet

Synthesis of jet fuel from alcohols can be performed by several conversion pathways, including sugar fermentation with yeast or microbes, starch hydrolyzation—fermentation, hydrolyzation—fermentation of lignocellulosic feedstock or thermochemical conversion, and fermentation via catalytic hydrogenation. Conversion of lignocellulosic biomass into alcohol typically requires hydrolysis, followed by fermentation or thermochemical conversion process. Another possible production route is via gasification, followed by fermentation to produce alcohol (Wei et al., 2019). Different types of alcohol such as methanol, ethanol, or higher alcohols can be used to produce biofuels through a series of reaction: dehydration, oligomerization, hydrogenation, and fractionation (Yang et al., 2019), as illustrated in Fig. 2.6. The sugars are fermented to produce isobutanol or ethanol, which

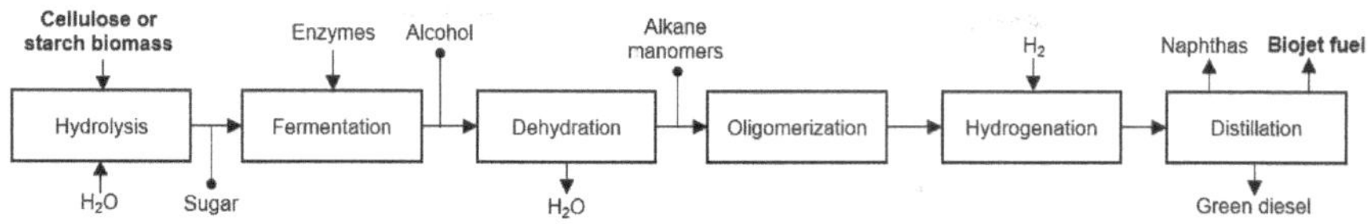

Figure 2.6 Process of converting cellulose and starch biomass into biojet fuel via the alcohol-to-jet (ATJ) pathway.

will be catalytically dehydrated into isobutylene or ethylene. The oligomerization process creates a carbon chain length suitable for fractionation into fuel components after hydrogenation (Geleynse et al., 2020). Although various alcohols or different intermediate pathways are feasible for production of jet fuels, the most commonly utilized ATJ production pathways are via ethanol or butanol. The following sections discuss the progress in the ATJ conversion technology.

2.3.1 Ethanol-to-jet

The conversion of alcohol to SPK jet fuel can be performed by using a variety of alcohols and oxygenated intermediates. Ethanol emerges as a readily available feedstock due to the established technology of bioethanol produced from biomass, which is primarily used as transportation fuel to replace gasoline. Ethanol is first dehydrated to produce ethylene, which is a versatile component used prevalently in industrial and consumer products, e.g., plastic manufacturing, polyethylene production and surfactant fabrication (Cheng et al., 2020). Dehydration of ethanol can occur in two ways: direct dehydration into ethylene or formation of diethyl ether followed by cracking into ethylene. At low temperatures (<300°C), diethyl ether is formed and cracked into ethylene and water when facilitated by strong acidity (Bokade and Yadav, 2011). The catalytic dehydration of ethanol can be achieved by using various catalysts such as transition metals (TiO_2, Fe_2O_3, Mn_2O_3, Cr_2O_3), heteropoly acids ($H_3PW_{12}O_{40}$, montmorillonite K10 (mont K10) impregnated with dodecatungstophosphoric acid [DTP]), gamma alumina (γ-Al_2O_3), and zeolites (microporous HZSM-5), with the latter two commonly being used in industrial ethanol dehydration (Cheng et al., 2020; Zhang et al., 2010b; de Reviere et al., 2020). The dehydration of ethanol is strongly dependent on the acid sites in the chosen catalyst. Brønsted's strong acidity can facilitate both ethanol dehydration and diethyl ether (DEE) cracking. Table 2.10 shows the various catalysts utilized for ethanol dehydration. Phung et al. (2015) compared the ethanol

Table 2.10 Various catalysts utilized for ethanol dehydration.

Catalyst	Reaction conditions	Ethylene selectivity (%)	References
Fabricated gamma alumina sheet (γ-Al_2O_3)	0.06 g catalyst with ethanol pumping at the rate of 2 μL/min for 30 min at 350°C	99.4	Chen et al. (2018)
Mesoporous SBA-15	0.3 g catalyst with initial ethanol concentration of 50 wt% and LHSV (liquid hourly space velocity) of 16 mL/g·h for 5 h at 400°C	84.7	Cheng et al. (2020)
Alumina-silica composite 60Al-SSP (60 mol% of Al mixed with 1TEOS: 0.3CTAB: $11NH_3$: 58Ethanol: $144H_2O$)	0.05 g of catalyst fed with vaporized pure ethanol (99.98%) from 1 h up to 10 h at 400°C	99	Krutpijit et al. (2020)

dehydration efficiency using zeolites, alumina, and silica alumina as catalyst. H zeolites were found to be more active than silica alumina and alumina on catalyst weight base to convert ethanol into DEE and ethylene. This is due to the presence of Brønsted acidic bridging hydroxyl groups only in the zeolite cavities. The highest ethylene yield of 99.9% was obtained by using H-FER and faujasite at 573K.

The ethylene dehydrated from ethanol can be converted into jet fuel via oligomerization process. The use of nickel-exchanged silica—alumina catalysts showed that 41% of C10+ products can be obtained from oligomerization of ethylene at relatively low temperature of 100—120°C and pressure of 35 bar (Heveling et al., 1998). Production of higher oligomers is possible, but the downsides are long duration needed and low conversion yield. Thus, a two-step approach can be adopted, first by converting the ethylene into intermediate olefins, e.g., butene or hexene, followed by the subsequent oligomerization process into jet fuel-length olefins. For short-chain olefins production, homogenous catalyst system based on titanium tetrabutoxide $Ti(OC_4H_9)_4$/triethylaluminum (TEA)/tetrahydrofuran (THF)/EDC has been shown to produce high selectivity of butene (80%) and hexene (~15%—20%) with conversion efficiency of >94%

(Mahdaviani et al., 2010). Much interests have been focused on transition metal such as nickel- and zeolite-based materials for catalytic oligomerization of ethylene into high stability and selectivity to liquid oligomers. The use of heterogenous nickel-exchanged mesostructured materials with MCM-41 catalysts was shown to have high selectivity toward C4, C6, C8, and C10 olefins at 150°C and 3.5 MPa, with n-heptane as solvent (Lallemand et al., 2007). Ethylene oligomerization using nano-sized HZSM-5 zeolites with different Si/Al ratio was performed in a fixed-bed reactor at 275–300°C and 3 MPa. Result showed that Si/Al ratio of 80 produced 64.3% C4+ olefins and 13.3% α-olefins. Brønsted acid sites were reported to promote secondary reactions such as cracking, isomerization, hydrogen transfer, and condensation reactions, but the excess of Brønsted acid will lead to deactivation of the catalyst (Zhang et al., 2020).

Ni loaded on nanocrystalline zeolite H-Beta was reported to effectively convert ethylene into C10+ oligomers up to 40 wt% at 87.2% ethylene conversion (Martínez et al., 2013). The $SiO_2-Al_2O_3$-supported nickel phosphide (Ni_2P) catalyst was reported to be capable of transforming ethylene into higher olefins (Shin et al., 2020). Ni_2P/SiO_2 had a higher ratio of Brønsted to Lewis acid sites (B/L ratio), which leads to higher catalytic activity due to the POeH groups generated on the catalyst surface, where the acidity induced the isomerization of terminal olefins and cationic oligomerization. The effect of the catalyst particle size is also pronounced, where smaller particles tend to increase the interaction between Ni and P with the support, leading to higher oligomerization efficiency (Shin et al., 2020). Lee et al. (2018) examined the performance of nickel supported by amorphous silica–alumina (SIRAL-30) with high Brønsted acid site density. A 4 wt% loading on the SIRAL-30 was found to achieve optimal ethylene conversion with C10+ selectivity of 18% at 200°C and 10 bar. The heterogenous catalyst can be recycled by reheating at 550°C and was able to produce 16% of C10+ olefins.

The butene produced can subsequently be oligomerized into jet fuel range oligomers. N-butene can be trimerized into C12 hydrocarbon via Ni-doped HZSM-5 catalysts at 148 h, 420°C, weight hour space velocity (WHSV) = 2/h, and 1.0 MPa. About 77.5 wt.% conversion of n-butene and 50.5 wt.% selectivity of C12 trimers were obtained using the 1NiHZSM-5(320) catalyst (Zhang et al., 2009). Zirconium-based catalyst was shown to effectively convert 1-butene into jet fuel range oligomers dominated by C8 and C12 oligomers, followed by hydrotreatment process to produce hydrocarbons with good cold flow property and heating value similar to jet fuel (Wright et al., 2008). Kim et al. (2015) found that the C8 olefin selectivity was maximized at temperature below 473K when using H-ferrierite as catalyst to oligomerize 1-butene. Cofeeding hexane as cosolvent to 1-butene can shift the product selectivity from heavier to lighter species. The obtained heavy hydrocarbons (>C12) are mostly highly branched. Díaz et al. (2020) also utilized HZSM-5 zeolite to oligomerize 1-butene. Result showed that C5-11 and C12-20 produced at the optimum condition of 250–275 °C and 30 bar were >40% and >20%, respectively. High conversion level of >60% with 50% C8-11 and >30% C12+ can be achieved at 20–30 bar. The heavy oligomers can be converted into branched alkanes via hydrotreating and isomerization process, followed by distillation process to obtain jet fuel.

2.3.2 Butyl alcohols-to-jet

Another pathway of ATJ production apart from ethanol is through butyl alcohol. The chemical synthesis technology of isobutanol (carbonylation and aldol condensation) has been flourishing since 1950s and is currently the main production method of isobutyl alcohol in the world (Guo et al., 2020). Companies such as UOP, Gevo, and Cobalt/US Navy have developed methods to produce alternative jet fuel based on butanol. In general, dehydration of butyl alcohol can be divided into n-butanol and isobutanol. Dehydration of iso-butanol produces olefins, which include 1-butene, cis-2-butene, trans-2-butene, and iso-butene (Wei et al., 2019). Isobutanol is mostly dehydrated over mildly acidic α-Al_2O_3 catalysts, but other catalysts such as inorganic acids, metal oxides, zeolites, and acidic resins, among others, have reported to be feasible. For n-butanol, dehydration can occur at lower temperatures over acid catalyst to produce butene, but higher temperatures are required for the skeletal isomerization of n-butanol to occur (Gunst et al., 2017). Fig. 2.7 illustrates the differences in intermediate products produced by using different alcohols but ultimately can lead to the production of jet fuel.

Recent progress in the ATJ production pathway has focused on the development of new catalyst for converting isobutanol into jet fuel. Zhang et al. (2010a) compared the performance of three zeolite catalysts (Theta-1, ferrierite, and ZSM-23) for a one-step dehydration and isomerization of n-butanol to isobutene and found that ferrierite can lead to the highest yield of isobutene (33.8 wt%) but showed poor catalytic performance over time as the presence of water lowers the acidity of the catalyst and causes

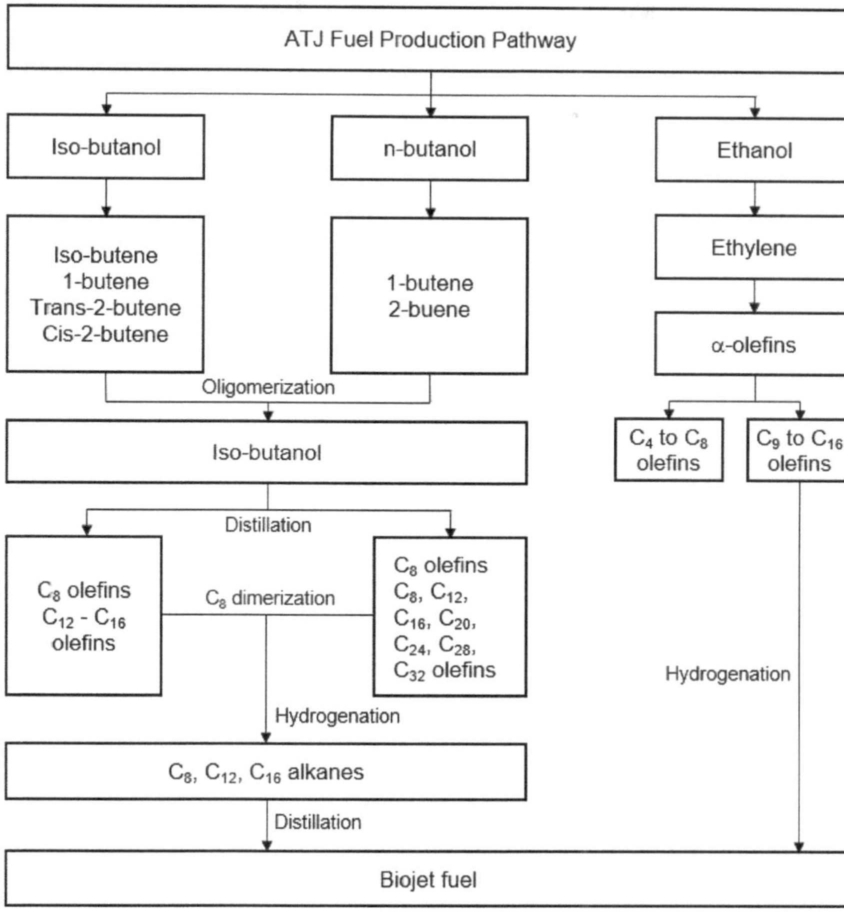

Figure 2.7 Difference of ATJ pathways from isobutanol, n-butanol, and ethanol (Vela-García et al., 2020). *ATJ*, alcohol-to-jet.

dealumination. On the other hand, ZSM-23 was found to produce low yields of isobutene (28.2 wt%) due to its two-dimensional channel network, which provides intersections for the formation of bulky intermediates, but high isomerization activity due to the zeolite's high acidity. One study investigated the oligomerization of isobutyl alcohol to jet fuels using various dealuminated methods of zeolite beta and found that zeolite beta (treated by HCl twice) exhibited the highest conversion of 98% and highest C8-16 selectivity of 59% (Xu et al., 2020). Another study investigated the production of tri-isobutane as an ATJ fuel from isobutanol. The dehydration of isobutanol (325°C, 0.62 MPa, and WHSV of 5/h) achieved a 99.1% isobutene yield, whereas the subsequent oligomerization (100°C, 0.20 MPa, space−time τ' of 5.5 g h/L) produced 90% tri-isobutene, and further subjected to hydrogenation to finally produce the tri-isobutane (Vela-García et al., 2020). The production of tri-isobutane also exhibited an estimated 28% lesser greenhouse gas emission than Jet-A1 production.

2.3.3 Challenges and prospects

Currently, ATJ fuels must adhere to the standard specifications of ASTM D7566 Annex 5 as the minimum property requirements for aviation turbine fuel that contain synthesized hydrocarbons. Conversion of jet fuel by companies such UOP, Gevo, and Cobalt/US Navy is based on butanol, which involves the basic process of dehydration, oligomerization, and hydrogenation. Isobutene is another producer material that has been approved for the production of jet fuel (ASTM D7566-19b, 2019), such as those used by Gevo (Chemicals Technology, 2010). LanzaTech and Byogy are companies that focus on the production of alternative jet fuel using the ethanol-to-jet (ETJ) approach. LanzaTech and PNNL have developed an ATJ pathway that converts ethanol to synthetic paraffinic jet fuel by using the waste gas from steel mill as feedstock (PNNL, 2018). The waste gas is first fermented into ethanol via a biological conversion process to produce alcohol. The advantage of the process is its ability to use different ethanol feedstock for jet fuel production, including those produced from municipal solid waste and waste gases. Fig. 2.8 shows the LanzaTech ATJ process, which selectively builds up jet fuel hydrocarbons from smaller compounds. The physicochemical properties of the LanzaTech's jet fuels are shown in Table 2.11. To date, LanzaTech's first commercial plant in China has produced over 10 M gallons of ethanol from the waste gas of recycled steel mills (Lanzatech, 2019). The ATJ fuel blends have been tested by both the US Airforce and US Navy (Zhang et al., 2016). LanzaTech has partnered with Japan New Energy and Industrial Technology Development Organization (NEDO) to conduct a feasibility study on scaling the LanzaTech ATJ platform in Japan (Lanzatech, 2019). Lanzajet is a company that will start producing sustainable aviation fuel from 2022 at the integrated biorefinery at LanzaTech's Freedom Pines site in Soperton, Georgia (LanzaTech, 2020).

The high production cost is a major issue that makes ethanol-derived ATJ fuel market-uncompetitive compared with fossil jet fuel (Silva Braz and Pinto Mariano, 2018). The advantage of the ATJ method is that

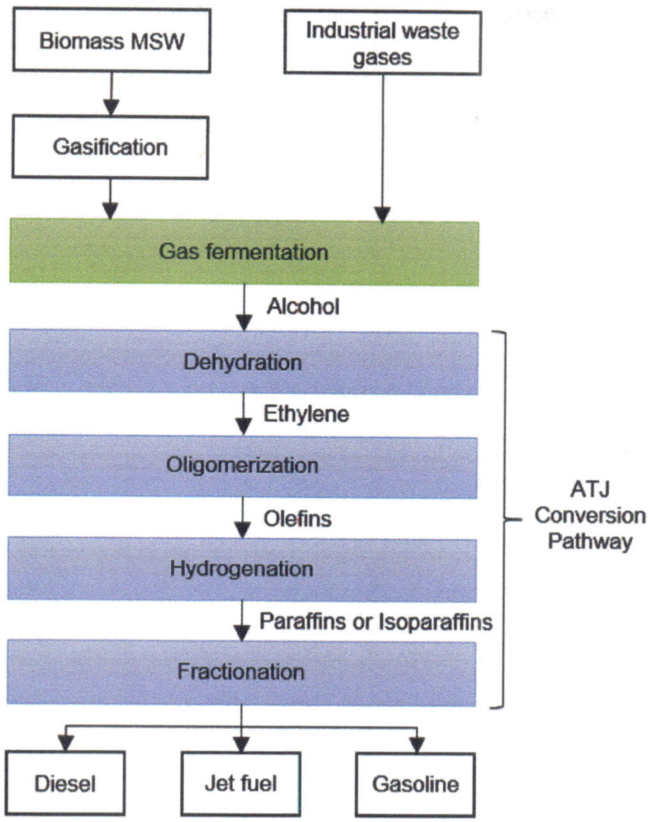

Figure 2.8 Production of synthetic jet fuel via the ATJ pathway by LanzaTech (Green Car Congress, 2018). *ATJ*, alcohol-to-jet.

available infrastructure for ethanol production can be utilized to drive down investment cost. Vertimass LLC received $1.4 million from the Bioenergy Technologies Office within the US Department of Energy's Office of Energy Efficiency and Renewable Energy to optimize jet fuel production from ethanol with emphasis on single step conversion of ethanol into hydrocarbon blend. The aim is to conduct the process without hydrogen addition at a relatively low temperature and atmospheric pressure, while producing minimal amounts of light gases. The technology is expected to expand upon the current liquid biofuels market beyond the constraints (Biomass Magazine, 2019). Presently, the US ethanol production plants have the capacity to produce approximately 16 billion gallons per year, which is at its limits if 10% of blends is used with gasoline. The

Table 2.11 Physico-chemical properties of the ATJ jet fuel produced by LanzaTech.

Property	ASTM test method	ASTM D7566	ATJ-SPK (July 2015)
Hydrogen content, mass %	D7171	n/a	15.4
Freeze point, °C	D5972	≤−40	<−75
Flash point, °C	D93	≥38	44
Density at 15°C, kg/L	D4052	0.751 −0.770	0.763
Viscosity at −20°C, cSt		<8	4.42
at −40°C, cSt		<12	9.30
Heat of combustion, MJ/kg	D4809	≥42.8	43.89
Thermal stability (325°C)	D3241	2/25	1/0 (pass)
Distillation	D86		
10%		≤205	170
Final boiling point		≤300	263
T90-T10, °C		>22	68
Hydrocarbon type analysis			
Aromatics, vol %	D6379	≤0.5	<0.01
Paraffins, mass%	D2425	Report	97.45

ATJ, alcohol-to-jet.

incorporation of existing ethanol dehydration studies with potential hydroprocessing catalysts to produce ATJ fuel range hydrocarbons is expected to further lower the production cost as the technology becomes matured.

2.4 Gas-to-jet

2.4.1 Fischer–Tropsch

FT process is the process of converting a mixture of carbon monoxide and hydrogen (synthesis gas) into transportation fuels and other liquid products of higher molecular weight hydrocarbons. The process was developed by German researchers Franz Fischer and Hans Tropsch in 1922 as a method for making liquid fuels from coal with alkalized iron chips at 400°C and pressures above 100 bar (Liu et al., 2013). The industrial use of natural gas as FT feedstock is also economically attractive especially in the context of stranded natural gas and shale gas (Eschemann and de Jong, 2015), with minimal contaminants being produced (Luque et al., 2012). Biomass has received considerable attention as potential feedstock for gasification to produce synthesis gas via the FT process to produce hydrocarbon fuels, as shown in the process flow in Fig. 2.9. The FT process can be thought of as a catalytic polymerization of carbon monoxide accompanied by reaction with hydrogen to make the methylene (CH_2) units of paraffins, which comprises two general reactions (Santos and Alencar, 2020),

$$\text{Alkane formation: } n\text{CO} + (2n+1)\,\text{H}_2 \leftrightarrow \text{C}_n\,\text{H}_{2n+2} + n\text{H}_2\text{O} \quad (3.1)$$

$$\text{Alkene formation: } n\text{CO} + 2n\,\text{H}_2 \leftrightarrow \text{C}_n\,\text{H}_{2n} + n\text{H}_2\text{O} \quad (3.2)$$

The product composition from the FT process varies depending on the hydrocarbon to carbon monoxide ratio, catalyst, and process conditions. The FT synthesis generally requires H_2 and CO at a ratio near 2.1:1, depending on the selectivity, and operates at pressure ranging from 20 to 40 bar and 180–250°C. The selectivity of the FT product is also influenced by the catalyst, types of catalyst support, and reactor used (Tijmensen et al., 2002). Additional processing of the raw product of FT synthesis is usually needed to further process into acceptable fuel. Such processes include the cracking the long chains into smaller units before rearranging some of the atoms via isomerization to obtain the desired fuel properties. The upgrading process typically produces liquid hydrocarbon product with a wide boiling range that consists of naphtha, kerosene, and diesel, which are subsequently distilled to obtain the final products. The European aviation industry predicts that approximately 140 kilotons of FT fuel could be produced annually based on the pilot plants planned (Kousoulidou and Lonza, 2016).

The products derived from FT process are usually free from sulfur or nitrogen compounds, which is advantageous from the combustion perspective as no contaminants such as sulfur dioxide or sulfuric acids are produced. FT fuels have been shown to emit 2.4% less CO_2, 50%–90% less PM, and no sulfur compared with fossil jet fuels (Zhang et al., 2016). Besides, the lack of aromatic results in cleaner burning with low level of soot produced. FT fuels were reported to display decreased contrail

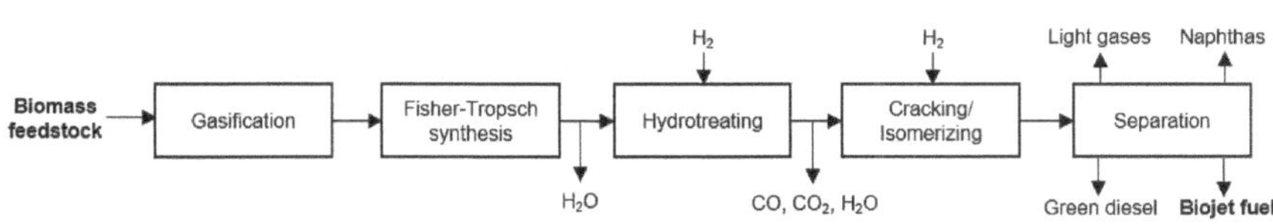

Figure 2.9 Process flow of converting lignocellulosic biomass into biojet fuel through gasification and Fischer–Tropsch synthesis.

formation and lesser soot emissions, which decreases the potential of the fuel to act as a cloud condensation nuclei (Jürgens et al., 2019). However, FT fuels with low aromatic content have caused fuel leakage problems in the engine due to shrinkage of elastomer and have lower energy efficiency (Kandaramath Hari et al., 2015; Wei et al., 2019). These issues can be solved by blending the FT fuel with conventional jet fuel to maintain a certain level of aromatics in the jet fuel. The ASTM D7566 standard has specified that a minimum of 8% aromatics have to be maintained in aviation turbine fuel, regardless of any type of synthetic jet fuel used as blend (ASTM D7566-19b, 2019). Other significant drawbacks of the FT process include the high gasification costs and relative higher CO_2 emissions compared with crude oil refining (Marsh, 2008). The chain growth mechanism also produces approximately 25—45 wt% of FT wax, which has a boiling point above 360°C and reduces the cost-effectiveness of the process (Tomasek et al., 2020).

The advantage of the FT process is the versatility of the feedstock that can be used, including coal, natural gas, and biomass. Commercial-scale application of FT products has been shown feasible by Sasol and Shell (Tijmensen et al., 2002). Sasol has produced three types of SPK fuels, namely the Sasol IPK derived from coal, while Sasol GTL-1 and GTL 2 are derived from natural gas. The Sasol GTL-1 is a distillate cut from the GTL fuel produced at the Oryx plant in Qatar, while the Sasol GTL-2 is an upgrade from GTL-1 with reduced paraffinic fraction with wider boiling range. Shell GTL is produced from natural gas in Bintulu, Malaysia, and is used by the US Air Force. The S-8 FT fuel produced by Syntroleum is derived from natural gas and is used by the US Air Force for test flights by blending with JP-8. Comparison of the compositional analysis of the synthetic paraffinic fuels is shown in Table 2.12. The FT SPKs consists primarily of isoparaffins and normal paraffins with a small fraction of cycloparaffins. The S-8 has slightly higher mass fraction of cycloparaffins compared with other SPKs. There is virtually no aromatics in the FT fuels. In spite of the similar carbon-to-hydrogen ratio for all the FT fuels, the distribution of hydrocarbon by carbon number can be quite different, as shown in Fig. 2.10.

Variation in the hydrocarbon chain length and the ratio of normal paraffins to isoparaffins can be attributed to the processing differences. Sasol IPK has a narrow range of hydrocarbon chain between C8 and C15 consisting of isoparaffins with practically no normal paraffins. Although the Sasol GTL-1 contains similar carbon numbers as Sasol IPK, the majority of

Table 2.12 Comparison of the compositional analysis of different FT-SPK (Moses, 2008).

Property	Test method	Limit (ASTM D7566)	Sasol IPK	S-8	Shell GTL	Sasol GTL-1	Sasol GTL-2
Hydrocarbon composition, mass %							
Aromatics	D2425	≤0.5	0	0	0	0	0.3
Cycloparaffins	D2425	≤15	2.6	9.0	4.0	2.6	7.7
Iso + n-paraffins	D2425	Report	97.4	91.0	96.0	97.4	92.0
Carbon and hydrogen content, mass %							
Hydrogen	D5291	–	84.33	83.99	85.00	84.45	84.69
Carbon	D5291	–	15.38	15.58	15.71	15.40	15.50

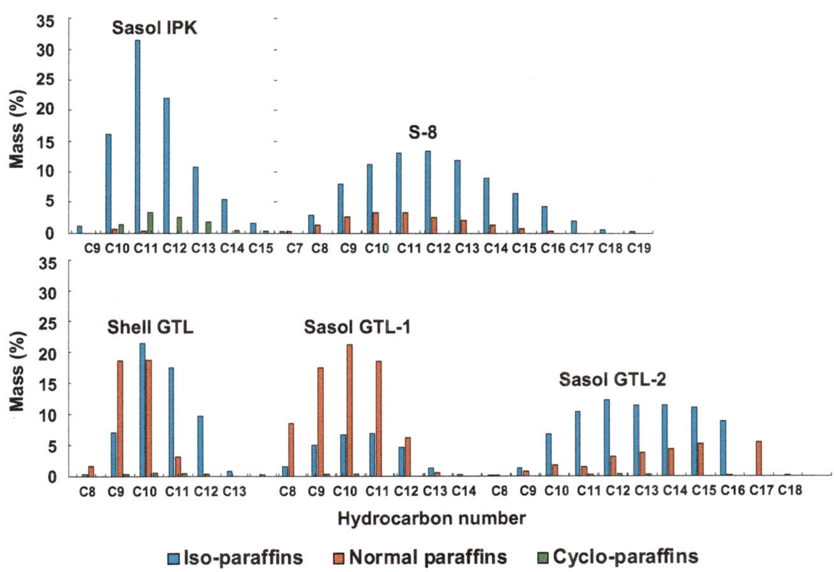

Figure 2.10 Distribution of hydrocarbons for different SPK fuels (Moses, 2008). *SPK*, synthesized paraffinic kerosene.

the composition is normal paraffins and a small fraction of isoparaffins. The Sasol IPK has about 7–13% of cycloparaffins, but other FT fuels have less than 1% cycloparaffins. The S-8 and Sasol GTL-2 contain a wider range of 8–9 carbon numbers dominated by isoparaffins, but the Sasol GTL-2 contains normal paraffins with longer hydrocarbon chain. The difference in the hydrocarbon groups and chain length leads to the slight variation in

Table 2.13 Properties of different FT fuels (Moses, 2008).

Property	Limits ASTM D7566	Sasol IPK	S-8	Shell GTL	Sasol GTL-1	Sasol GTL-2
Total acid number (mg KOH/g)	≤0.015	0.004	0.004	0.003	0.002	0.003
Initial boiling point (°C)	Report	174	144	154.1	144	179
Final boiling point (°C)	≤300	232	275	195.2	208	266
Freezing point (°C)	≤−40	<−65	−51	−53.8	−52.5	−62
Existent gum (mg/100 mL)	≤7	0.6	0.6	4.2	0.9	0.6
Viscosity at −20°C (cSt)	≤8	3.23	4.9	2.49	2.43	6.09
Density @ 15°C, kg/m^3	730−770	765	756	736	735	762
Smoke point (mm)	≥25	42	42	>50	29	28
Flash point (°C)	≥38	53	45	43	48	70
Heat of combustion (MJ/kg)	≥42.8	44.0	43.9	44.2	44.3	44.2
Water, mg/kg	≤75	25	22	28	40	32
Sulfur, mg/kg	≤15	0.7	0.6	0.6	0.6	0.6

FT, Fischer−Tropsch.

physicochemical properties as shown in Table 2.13. The boiling point distribution curve is noticeably different among the fuels. Shell GTL has narrower range of boiling point. The S-8 and Sasol GTL-2 fuels are distilled to have boiling point slopes that are typical of conventional jet fuel, while the rest are relatively flat. The relatively high final boiling point and viscosity for Sasol-2 imply its less volatile characteristic; thus the flash point can be seen to be higher. All the FT fuels have excellent freeze point characteristics and conform to the batch requirement of ASTM D7566. The heat of combustion for all FT fuels are almost similar, but the sooting tendencies are considerably higher for Sasol 1 and Sasol 2. Corporan et al. (2011) reported that the FT-SPK fuels produced by Sasol, Shell, and Rentech possess superior thermal stability and produced less pollutant emissions when fueled in a T63 engine.

2.4.2 Biomass-to-fuel

Crop-based biomass to liquid fuel has gained popularity as CO_2 emitted during the combustion of fuel is offset by the CO_2 absorbed during the crop growing process (Marsh, 2008). The biomass FT process can be split into six different steps, which include the pretreatment, gasification, gas conditioning, acid gas removal, FT processing, and syncrude refining (Wei et al., 2019). Fig. 2.11 shows the general FT pathways using biomass,

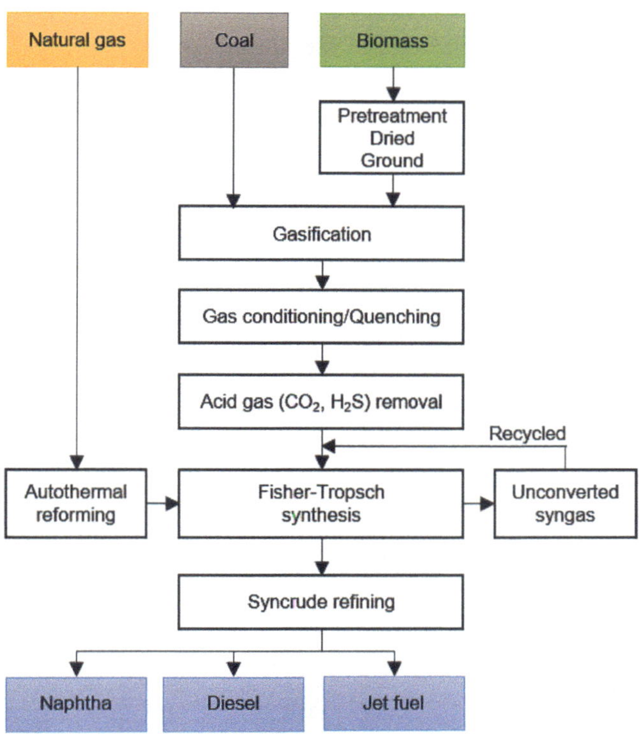

Figure 2.11 Production of hydrocarbon fuels from biomass, coal, and natural gas via the Fischer–Tropsch synthesis.

coal, and natural gas as feedstocks. Initially, the biomass is pretreated, dried, and ground before being converted into synthesis gas in a gasifier with the addition of oxygen (Hillestad et al., 2018). The syngas produced by the gasification process contains different kinds of contaminants such as particulates, condensable tars, alkali compounds, H_2S, HCl, NH_3, and HCN, which may reduce FT synthesis efficiency (Tijmensen et al., 2002). The gas is subjected to quenching, while components such as COS, CS_2, and HCN are hydrolyzed to form H_2S, NH_3, and CO_2. Tar and ash are moved followed by acid gas treatment to remove the CO_2 (can be used to improve the kinetics and economics of the downstream process), H_2S (avoids catalyst poisoning), and sulfide (Wei et al., 2019). To remove the contaminants, the syngas is subjected to a cleaning process, typically via the "wet low temperature cleaning" as shown in Fig. 2.12. Ash particulates are removed by the cyclone separator and bag filters, while several stages of scrubbing are used to remove different contaminants. Scrubber with H_2SO_4 or Sulfinol D solvent is used to HCN and NH_3. The sulfuric compounds such as H_2S or carbonyl sulfide (COS) are removed through scrubber with Sulfinol D, COS hydrolysation unit, and ZnO guard bed. The Cl compound such as hydrogen chloride (HCl) can either be absorbed by dolomite in the tar cracker, reacts with particulates in the bag filter, or removed by scrubber with NaOH solvent. Alkalis and tars will condense on particulates or vessel when syngas is cooled below 500°C (Tijmensen et al., 2002).

Another option to clean the syngas is via the advanced "dry hot gas cleaning" method (Mitchell, 1998). The process consists of several filters and separation units without cooling the syngas, which could lead to increased efficiency and lower operational costs as the hot syngas fed to the shift reactor or a reformer requires high inlet temperature. However, this method is still not commercially viable due to the challenge of removing contaminants at high temperature (Tijmensen et al., 2002). Finally, the remaining syngas is fed into the FT synthesis reactor to be converted into hydrocarbons. Conventional refinery processes, such as hydrocracking,

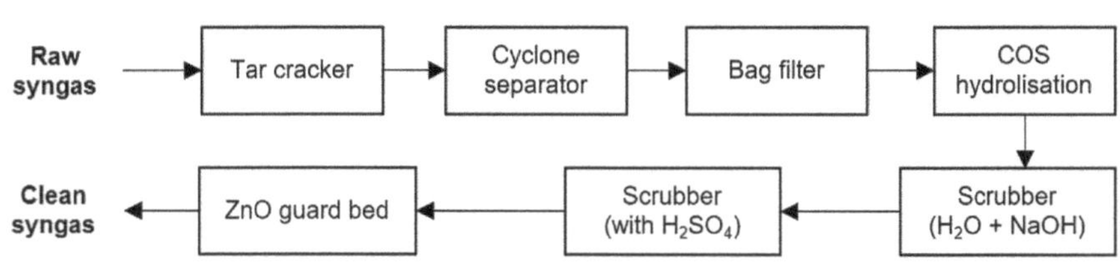

Figure 2.12 Schematic view of "wet" low temperature cleaning (Tijmensen et al., 2002).

isomerization, hydrogenation, and fractionation, are then applied to upgrade the FT synthesis product to high-quality biojet fuel range hydrocarbons (Wang and Tao, 2016).

Biomass gasification has been reported to be rather different than coal and natural gas gasification. Syngas produced from biomass gasification often has low H_2/CO ratios (less than one) (Ostadi et al., 2019). The inconsistent moisture level, density, energy content, complex lignocellulosic structure, and size of biomass make it difficult for uniform feed rates, and the relatively high oxygen and moisture content of the feedstock result in a fuel with high methane content, low heating value fuel, and low hydrogen content (Luque et al., 2012). Conversion of syngas from biomass is often only 50%–60% effective, with tail gas consisting of lighter hydrocarbons and unconverted syngas (Hillestad et al., 2018). Besides, the feedstock composition influences the production composition, and a large amount of carbon is often lost as CO_2 (Ostadi et al., 2019). Nevertheless, the FT process is still an attractive option for biomass utilization as the gasification product (syngas) and solid by-product (biochar) are valuable for various applications. Table 2.14 shows the installation and production capacity of different biomass FT plants.

2.4.3 Advances in Fischer–Tropsch technology

2.4.3.1 Biomass gasification technology

Different gasification technologies have been developed to accommodate different feedstocks. Samiran et al. (2016) presented a comprehensive review on the types of gasifier used to gasify biomass, such as the fixed bed, fluidized bed, entrained flow, and transport reactor gasifiers. Each type of gasifier has its own technoeconomic merits and drawbacks, depending on the operational need and application. The fixed bed–type gasifier uses either updraft (Gunarathne et al., 2014) or downdraft (Olgun et al., 2011) gasification air introduced from below or above a constant depth of feedstock bed, as shown in Fig. 2.13. The solid fuel is supplied from the top of the gasifier, while the reaction zone is supported by a grate. The updraft method is known to produce higher level of tar compared with the downdraft configuration (Samiran et al., 2016). The fixed-bed gasifier is generally cost-effective, but the syngas produced needs to be cleaned separately owing to the high tar content.

The fluidized-bed gasifier employs blowing air through a bed of solid particles to maintain the particles in suspension state and to mix and react with the feedstock at elevated temperature. The bubbling-type fluidized

Table 2.14 Installations of different of biomass FT plants (Ail and Dasappa, 2016; Green Car Congress, 2009; Shahabuddin et al., 2020).

Organisation	Feedstock	Year	Gasifier	Reactor details
Solena Fuels, Green Sky (Essex, United Kingdom)	Municipal and commercial waste	2015	Solena plasma gasification	• 1157 barrel per day (bpd) jet fuel • Cocatalyst
Red Rock Biofuels (Oregon, United States)	Forest and saw mill waste (460 t/d biomass feed)	2017	TRI steam reformer	• 1100 bpd jet fuel • Cocatalyst
Sierra Biofuels, Fulkrum Bio-energy (Nevada, United States)	Municipal solid waste	2016	TRI steam reformer	• 657 bpd jet fuel • Cocatalyst
Centerpoint Biofuels (Indiana, United States)	Municipal solid waste	2018	–	• 700,000 tons of waste from Chicago • 33 million gallons fuel/year
SYNDIESE, CEA (Nevada, USA)	Forest and agriculture waste	2015	Entrained flow, O_2 blown, high pressure gasifier	• 205 t/d biomass feed • 530 bpd liquid fuel
Velocys (Gussing, Austria)	150 t/d dry biomass (pilot scale)	2010	Dual fluidized bed gasifier	• 5000 bpd jet fuel • Cocatalyst
French CEA (Nancy, France)	Forest and agriculture waste	2009	Gasification provided by CHOREN	• 75,000 t/year of feedstock • 23,000 t/year of biofuel

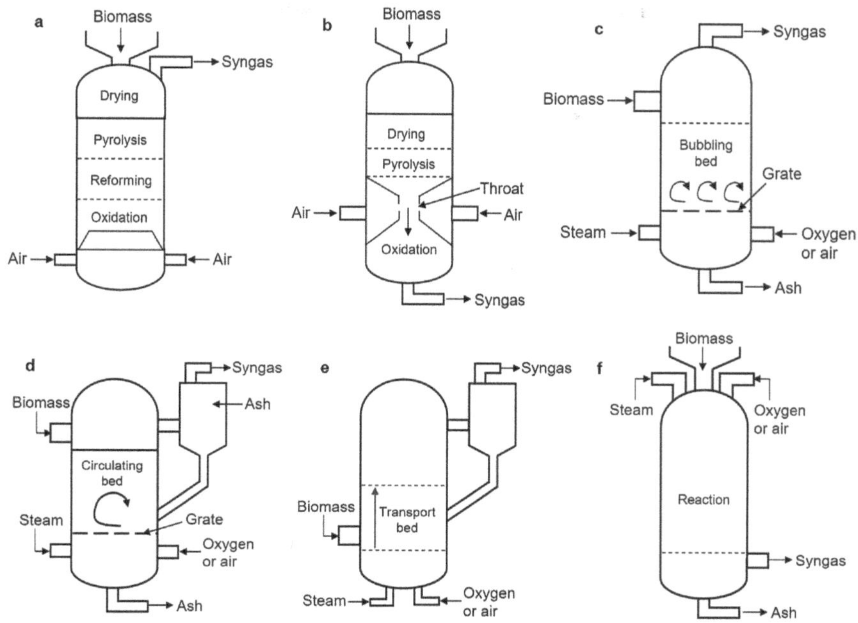

Figure 2.13 Schematics of the (A) fixed-bed updraft and (B) downdraft, (C) fluidized bubbling (D) recirculating, (E) entrained, and (F) transport-type gasifier (Richards and Casleton, 2010; Samiran et al., 2016; Breault, 2010).

bed introduces air of low velocity (<5 m/s) through the gate at the bottom to the bed material. The bed is heated externally to provide energy for the endothermic steam reforming reaction process. A strong vortex of gas—solid flow is introduced to intensify the fluid motion in the reactor to promote homogenous temperature for biomass reaction (Udomsirichakorn et al., 2013). Another variation is the circulating fluidized-bed gasifier, where a circulation process of the bed materials with products takes place between the reaction vessel and a cyclone reactor as shown in Fig. 2.13D. The bed material and char are recirculated back to the combustion zone, while the ash is removed through cyclone reactor. Fluidized bed is more commonly opted in the industry due to medium operating cost and lower tar content. Finland's Valmet has developed a circulating fluidized-bed gasification unit to gasify a variety of biomass. The combustion temperature (850—900°C) is below the melting point of ash, thus minimizing fouling and slagging of heat surface. The solids recirculation provides a long residence time and enables high combustion efficiency (Valmet, 2017). Another merit of the fluidized-bed gasifier is the reduced formation of tar in the syngas (Matsuoka et al., 2013). The entrained gasifier operates at high temperature of 700−1500°C to oxidize fine biomass particles in a short residence time of 1−5 s. For the transport reactor gasifier, feedstock enters with the oxidizer stream into an upward flow to react and fluidize the bed of feedstock. The gasifier reactor needs to operate at high velocity (15 m/s) to transport all bed materials up the reactor. The feedstock is first devolatilized in the fluidized-bed mixer, followed by char combustion in the combustor riser (Breault, 2010; Samiran et al., 2016). The transport and entrained gasifiers have the potential to produce higher-quality syngas but demand stricter requirement on the feedstock size and operating conditions.

Other emerging gasification technologies such as plasma gasification, microwave gasification, and supercritical water gasification can be utilized to gasify biomass to produce syngas (Shahabuddin et al., 2020). Plasma gasifier operates at a much higher temperature regime compared with the conventional fixed bed or fluidized bed. Plasma torches are used to melt the biomass at ~1600°C, which is high enough to melt any inorganic material in waste. Furthermore, the high temperature syngas (950°C) can gasify the tar and convert them into smaller molecules such as CO_2, CH_4, CO, and H_2. Zhang et al. (2012) gasified municipal solid waste under plasma gasification condition in an updraft moving gasifier at a temperature up to 6000°C. An increase in the plasma power resulted in the increase of H_2/CO ratio from 1.5 to 2.0 owing to the increased cracking of tar. The advantage of plasma gasification is its flexibility in handling hazardous waste and feedstocks with high moisture content up to 40% (Mountouris et al., 2006), but the energy intensive process is cost-prohibitive for commercial production. Fig. 2.14A shows the schematic of a plasma gasifier.

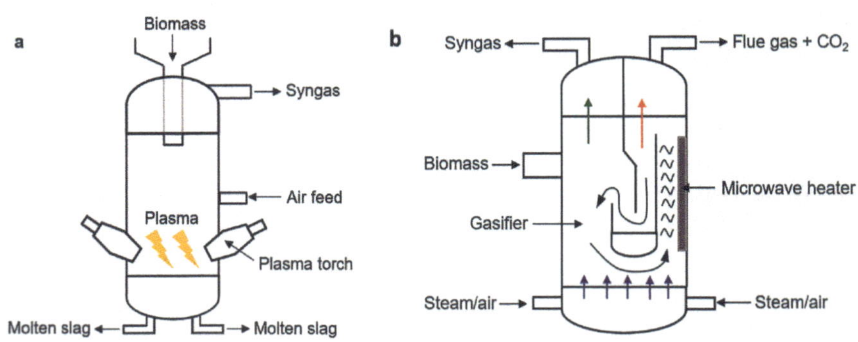

Figure 2.14 (A) Plasma and (B) microwave-assisted dual-fluidized-bed gasifier (Shahabuddin et al., 2020; Xie et al., 2014).

The use of microwave as a heating source was shown to be effective in gasifying corn stover for syngas production with Ni/Al_2O_3 as catalyst (Xie et al., 2014). Extreme high temperature (>1200°C) can be obtained by transforming the electromagnetic energy in microwave into thermal energy at molecular level with the aid of microwave absorbents, thus enabling rapid, efficient heating for syngas production which consumes lesser energy compared with conventional fluidized-bed gasifier. The study showed the potential of incorporating microwave heating with dual-fluidized-bed gasifier for industrial-scale biomass gasification, as shown in Fig. 2.14B. Improvement on the syngas quality can be achieved by adding steam to the microwave gasification unit. Xie et al. (2014) reported an increase in syngas yield up to 83.91% and a reduction of tar content to 5.11% with steam addition. Warsita et al. (2017) examined the cracking of tar with steam in a microwave heating system using naphthalene and toluene as model tar. Naphthalene and toluene are the main constituents in tar, although other compounds such as phenol and pyrene are also found in biomass tar. The effect of steam on tar removal was found to be pronounced, with >95% tar removal efficiency achieved at temperature of >1000°C at water to tar ratio of 0.3. The tar steam reforming reactions are

$$C_7H_8 + 7H_2O \leftrightarrow 7CO + 11H_2 \quad \Delta H_{393K} = +881.7\ kJ/mol \tag{3.3}$$

$$C_{10}H_8 + 10H_2O \leftrightarrow 10CO + 14H_2 \quad \Delta H_{393K} = +1177.8\ kJ/mol \tag{3.4}$$

The steam reforming reaction of tar led to the production of CO and H_2, which explains the increase in yield of syngas when steam was added (Xie et al., 2014). Zhou et al. (2020) demonstrated the production of syngas produced from biomass using a microwave-assisted pyrolysis system. About 67% volume of high-quality syngas (18 MJ/Nm^3) was produced at the temperature of 800°C. Lesser tar (2.7 wt%) was produced compared with conventional pyrolysis process operating at 900–1000°C (3.0–6.9 wt% tar).

Supercritical water gasification is achieved by heating water to a temperature above its critical temperature (647K) and pressure (22.1 MPa) (Matsumura and Minowa, 2003). It is a form of hydrothermal gasification that has good raw material adaptability including biomass with high moisture content. The use of high volume of water for the production of combustible gases at supercritical condition avoids the energy-intensive drying process (Chen et al., 2020). The main reactions that occur for biomass in supercritical water are (Osada et al., 2006)

$$C + H_2O \rightarrow CO + H_2 \quad \text{water gas reaction} \tag{3.5}$$

$$CO + H_2O \rightarrow CO_2 + H_2 \quad \text{water–gas shift reaction} \tag{3.6}$$

$$CO + 3H_2 \rightarrow CH_4 + H_2O \quad \text{methanation reaction} \tag{3.7}$$

The reactions in supercritical water gasifier lead to the production of hydrogen and carbon dioxide as major products, while methane and carbon monoxide are present as minor products (Osada et al., 2006; Guo and Jin, 2013). Varying the operating parameters including the temperature, biomass/water ratio, and catalyst can change the output. Osada et al. (2006) showed that gasification at high temperature region (773–973K) can effectively inhibit the formation of char, while the use of alkali catalyst can promote water–gas shift reaction, thereby increasing the yield of H_2 and lowering the CO yield (Kruse, 2008). Kruse and Dahmen (2015) reported that higher temperature favors the production of hydrogen, while lower temperature results in more CH_4 production. Application of this gasification method for production of synthetic jet fuel requires reforming of the CH_4 into synthesis gas for FT reactions.

The use of renewable energy such as solar power to produce synthesis gas is a promising route for zero-carbon fuel. An EU-funded project known as SOLAR-JET has synthesized the first "solar" kerosene from water and CO_2 via a high-temperature solar reactor to produce H_2 and CO, leaving O_2 as the purge as at the outlet (SOLAR-JET, 2015). The schematic of the solar reactor is shown in Fig. 2.15. The two-step solar thermochemical cycle based on ceria redox reactions to produce synthesis gas, which are the inputs needed for FT synthesis to produce jet fuel. The project was led by ETH-Zurich in cooperation with DLR, Shell, and Bauhaus Luftfahrt. The project was later succeeded by a project known as SUN-to-LIQUID in 2016 aiming to scale up the production of solar-derived jet fuel. High-flux solar concentrating system consisting of heliostat array was used to focus the sunlight onto a reactor core made of cerium oxide to produce syngas (SUN-to-LIQUID, 2016). The advantage of the solar-derived jet fuel is its carbon neutrality, and no changes are needed for the current infrastructure including storing, transporting, and distribution systems. However, the high cost of production (~\$9/gallon) meant that it is still not market-competitive for commercial usage (Swain, 2020).

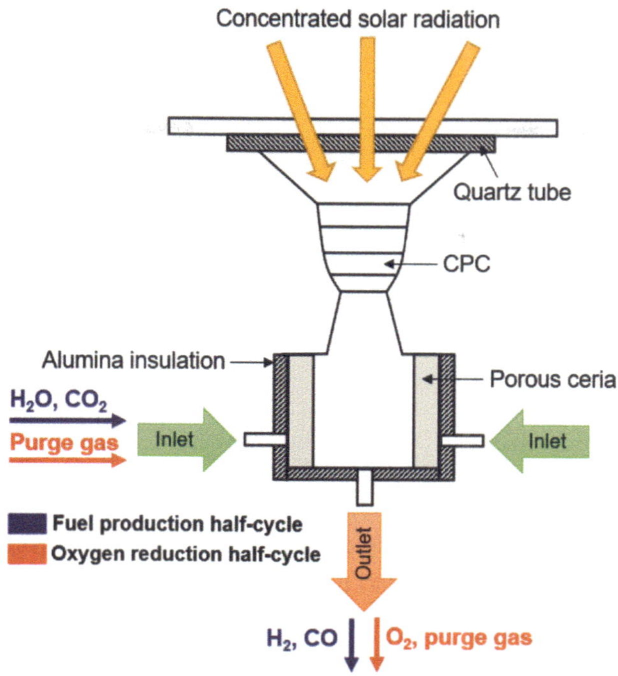

Figure 2.15 Schematic of the solar reactor configuration for the two-step solar-driven thermochemical production of fuels (SUN-to-LIQUID, 2016).

2.4.3.2 Fischer–Tropsch reactor

The types of reactors commonly used for FT process include fluidized-bed reactor, fixed-bed reactor, and the slurry-phase reactor. Fixed-bed reactor typically requires periodical catalyst replacement, thus incurring maintenance cost and long downtime. Conversion efficiency up to 80% with >90% selectivity of C5+ is possible, although pressure drop can be as high as 3–7 bar. Nonetheless, this technology is proven, and production can be scaled up easily. The slurry-type reactor requires lesser catalyst and can achieve high conversion efficiency with <1 bar of pressure drop. The main disadvantage of the slurry reactor is the need for catalyst and wax separation (Tijmensen et al., 2002). Fig. 2.16A and B show the schematics of the tubular fixed-bed and slurry-bed FT reactor. According to Sasol, the wax selectivity from a fixed-bed reactor using an iron catalyst is approximately 50%–55%, while that for a slurry reactor using a similar catalyst is ~55%–60% in low temperature condition. The slurry reactor was shown

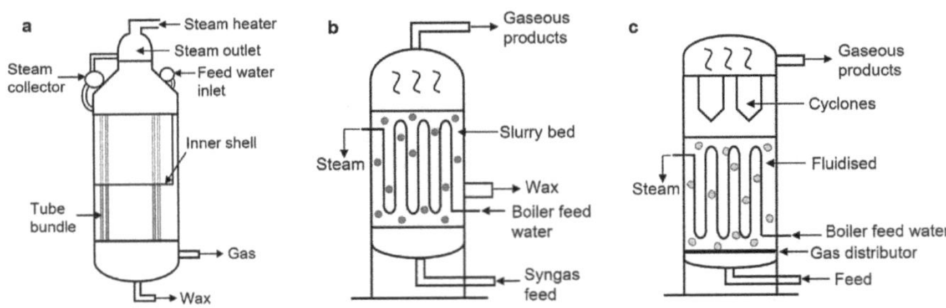

Figure 2.16 (A) Tubular fixed bed, (B) slurry bed, and (C) SAS reactor (Geerlings and Wilson, 1999; Steynberg et al., 1999).

to produce 50% higher amount of olefins with selectivity in the range of C5−C18 compared with the fixed-bed reactor (Espinoza et al., 1999). They later deployed a fluidized bed−type reactor or Sasol Advanced Synthol (SAS) reactor that operates at pressure of 20−40 bar and at high temperature of 340°C with iron as catalyst. Compared with the Synthol CFB reactor, the SAS reactor has improved production rate due to higher catalyst/gas ratio, which increases throughput, consumes lesser catalysts, and is more energy efficient (Steynberg et al., 1999). The schematic of the SAS reactor is shown in Fig. 2.16C. The increase of the partial pressure of H_2 and CO tends to lead to higher selectivity of C5+, whereas the presence of inert gas will lower the selectivity of C5+ (Tijmensen et al., 2002). To obtain jet fuel-quality fuel, the produced FT-liquid requires catalytic hydrocracking. Hydrogen is added to remove the double bond, and the desired final product can be obtained by altering the hydrocracking conditions.

2.4.4 Scientific advances

As a technique to improve the H_2/CO ratios of the syngas, external hydrogen or steam is often added into the process. The introduction of steam improves thermal reforming of the hydrocarbons and water−gas shift of carbon monoxide, thus increasing the hydrogen content in the gaseous product (Hillestad et al., 2018). According to Hillestad et al. (2018), carbon efficiency of the process can be improved from 38% to more than 90% with the use of hydrogen sourced from renewable energy. Adding hydrogen results in higher carbon efficiency and hydrocarbon production rate (Ostadi et al., 2019). Chiodini et al. (2017) investigated the gasification of two different biomasses (forest residue and Triticale crop) for direct FT application. It was found that using a bed material constituted of active material such as magnesium oxide can produce syngas with H_2/CO ratio of 2 at 1020K. The produced syngas does not require any water–gas shift section within the process and hence would save up to 10% of production cost.

Reaction parameters that affect the range of hydrocarbons produced include pressure, temperature, and catalyst used. Given the highly exothermic nature of the FT process, the reaction temperature should be monitored to avoid the formation of hot spots and the runaway phenomenon generation, which would be detrimental to both the catalyst and reactor (Méndez and Ancheyta, 2020). Furthermore, low temperatures during the FT process also result in increased production of methane. High pressures also increase the conversion rate and favor the formation of long-chain alkanes (Luque et al., 2012). A good catalyst should have strong catalytic activity, selectivity, high space–time yield, good carbon chain growth ability, cheap, and long catalytic life (Wei et al., 2019). Common catalysts used for the FT process include Ni, Ru, Co, and Fe and have been extensively reviewed by various researchers (Wei et al., 2019; Xu et al., 2020; Abbaslou et al., 2009). Ni is often avoided because it is prone to methane formation and cracking of higher hydrocarbon, while Ru is expensive and rare. Both Co and Fe have high productivity and stability, but the latter is often used for its cheap price and ability to be deployed in environment with temperature above 300°C (Benedetti et al., 2020). Sasol produced the FT liquid fuels by utilizing iron as catalyst at the operating temperature of 340°C (Steynberg et al., 1999). Kumabe et al. (2010) utilized Fe-based catalyst to synthesis FT fuels using a fixed-bed reactor under the temperature of 533–573K and a pressure of 3.0 MPa. The selectivity of CO to the C11–C14 hydrocarbons equivalent to jet fuel kerosene was found to be the second highest. The highest yield of kerosene was obtained at the feeding gas H_2:CO:N_2 with a ratio of 2:1:3 and at the reaction temperature of 553K with neat Fe as catalyst.

Folkedahl et al. (2011) demonstrated a pilot-scale FT reactor system that is capable of processing up to 9 kg/h of coal and biomass to produce synthesis gas using a fluidized-bed gasifier. Fe-based catalyst was used to produce wax and liquid products, which were collected and subsequently processed into synthetic isoparaffinic kerosene. The FT products were catalytically hydrodeoxygenated to remove the alcohols and oxygenated compounds, followed by isomerization and distillation processes. The SPK produced is close to the specification of military-grade jet fuel. Hanaoka et al. (2015) investigated the hydrocracking behaviors of the FT products derived from biomass (n-$C_{28}H_{58}$ and n-$C_{36}H_{74}$), using Pt-loaded β-type zeolite catalysts with constant Pt content, acid amount, and pore parameters. The experiments were conducted at temperature 250°C, initial H_2 pressure 1 MPa, and reaction time 1 h. Pt-loaded β-type zeolite catalysts with Pt particle sizes of 2.3–13.1 nm and higher acid amounts led to high jet fuel yields. The maximum jet fuel yield of 29.1 C-mol% was obtained with Pt particle size 7.6 nm.

2.5 Sugar-to-jet

The common sugar-to-jet (STJ) conversion pathways are biological-type direct sugar-to-hydrocarbon (DSHC) and catalytic conversion aqueous phase reforming (APR). Feedstocks containing sugar and starches can all be used for the STJ conversion pathways. For a pathway to be considered as STJ, the starting feedstock should be sugar and typically does not require a dedicated stage to convert alcohol into the final biojet fuel product.

2.5.1 Direct sugar-to-hydrocarbon

The DSHC method shares a few similarities with the ATJ pathway, where typically biochemical fermentation or catalytic conversion of sugar to hydrocarbon fuels occurs (Hari et al., 2015). The feedstock for DSCH can be directly from sugar sources such as sugarcane, beets and maize, or lignocellulosic biomass (Wei et al., 2019). The key difference is that DSHC process does not require an alcohol intermediate. The technology is developed due to advances in genetic engineering and screening technologies, enhancing how microbes metabolize sugar. A more specific route of DSHC involves the production of farnesane from sugar, called hydroprocessed fermented sugars to synthetic iosparaffins (HFS–SIP), and has been approved by ASTM in 2015. However, the terms HFS–SIP and DSHC are typically used in an interchangeable way in literature.

Fig. 2.17 illustrates the DSHC biojet fuel conversion pathway. In the following description, emphasis is placed on the HFS–SIP specific pathway as it is the ASTM approved pathway. In general, the DSHC pathway

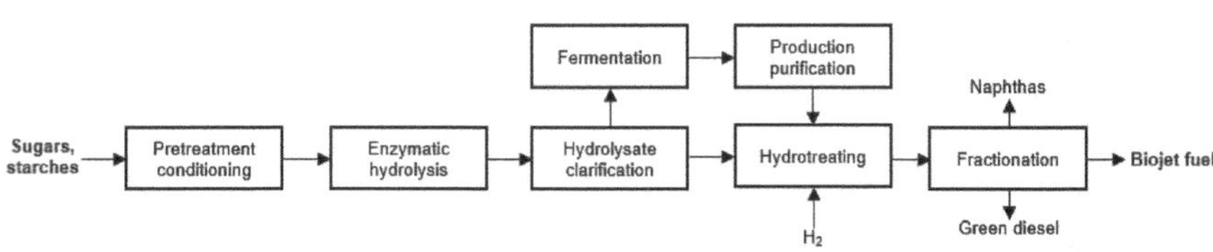

Figure 2.17 Direct sugar-to-hydrocarbon (DSHC) biojet fuel conversion pathway.

involves six main steps and one postprocessing fractionation step. It starts with the conditioning of the raw material to improve feedstock quality prior to the enzymatic hydrolysis step. Enzyme mixtures or catalytic proteins work together to break down cellulose fibers into cellobiose and soluble glucooligomers, then into glucose monomers (Davis et al., 2013). The solubilized C5 and C6 sugars are separated and concentrated (CORSIA, 2019). This process usually lasts 3.5 days, at slightly elevated temperature of 48°C and a cellulase loading of 10 mg protein/g cellulose. Conversion efficiency values of glucan to glucose oligomer, cellobiose, and glucose are 4.0%, 1.2%, and 90.0%, respectively. The conversion of cellobiose to glucose is 100%. The enzymatic hydrolysis reactions are

$$(\text{Glucan})_n \rightarrow n\text{Glucose oligomer} \tag{3.8}$$

$$(\text{Glucan})_n \rightarrow 1/2\ n\text{Cellobiose} \tag{3.9}$$

$$(\text{Glucan})_n\ nH_2O \rightarrow n\text{Glucose} \tag{3.10}$$

$$\text{Cellobiose} + H_2O \rightarrow 2\ \text{Glucose} \tag{3.11}$$

The hydrolysate or hydrolysis product undergoes clarification for unwanted particle removal and removing turbidity. This step differentiates the pathway from ETJ pathways where sugars are converted into ethanol by temperature reduction and fermentation initiation, without any intermediate conditioning. Using vacuum belt filters for the hydrolysate separation, 99% of soluble sugar can be recovered. Wash ratio for such a system is 2.5:1 (L water: L liquor in filter cake).

The clarified hydrolysate then goes through biological conversion of its sugar and through fermentation by microorganisms to produce farnesene. The fermentation process takes typically up to 69 h, with a vessel turnaround time of 79 h. The purified fermentation products are reacted with hydrogen through hydrotreating in a catalytic bed. This serves to convert the farnesene into a saturated alkane, namely farnesane (2,6,10-trimethyldodecane), which is distilled to produce aviation grade biojet fuel (Zschocke et al., 2012).

Ideally, only farnesane is produced as the final product, but the main contaminant hexahydrofarsenol (3,7,11-trimethyl-do-decan-1-ol), or simply HHF, is also produced (Buffon and Stradiotto, 2019). Degradation of the fuel system occurs in the presence of HHF as the contaminant will form polymeric chains and sediment within the system components. According to the ASTM D7566 standards for HFS–SIP, HHF amount must not exceed 1.5% (m:m). In practical terms, there will also be trace amounts of farnesene and olefins (partially hydrogenated farnesene). The produced long-chained farnesane (C15) biojet fuel has relatively high viscosity and poorer combustion performance in aviation turbine engines, as compared with biojet fuels from other methods (Yang et al., 2019).

For the bioconversion or fermentation of sugar, farnesene is not the best product pathway class. In fact, it is the worst among all possible product pathway classes as shown in Table 2.15. This shows the difficulty faced by DSHC to achieve parity as compared with other product pathway classes. It is important to understand that the DSHC pathway is still in nascent stages of development relative to ethanol and other product pathway classes; hence, improvements can still be made for the fermentation process to allow DSHC be a viable process.

Key companies working on DSHC biojet fuel pathway include Amyris, Total, Solazyme, and LS9. The American company, Amyris, engineered *Saccharomyces cerevisiae*, a species of yeast, for industrial production of isoprenoid artemisinic acid for antimalarial treatment. It was then reengineered for large amount production of isoprenoid beta-farnesene, which has the potential to be used for biojet fuel through the DSHC/HFS—SIP pathway (Gírio, 2019). Amyris also developed a DSHC fermentation pathway where sugars are aerobically fermented into a farnesene intermediate using the mevalonate pathway in yeasts. The Amyris process obtained a maximum farnesene yield of 16.8 g farnesene/100 g sugar at a productivity level of 16.9 g/L/d. The process has a carbon efficiency of 60%. The 300 L capacity

Table 2.15 Theoretical metabolic yields for various product pathway classes through bioconversion of sugar (Davis et al., 2013).

Product pathway classes	Mass yield (%)	Carbon yield (%)	Energy yield—HHV basis (%)
Farnesene (mevalonic acid pathway)	25	56	74
Farnesene (1-deoxy-D-xylulose 5-phosphate pathway)	29	64	85
Pentadecane	29	62	88
Fatty alcohol (hexadecanol)	34	67	93
FAEE (ethyl palmitate)	35	67	90
Fatty acid (palmitic acid)	36	67	89
Ethanol	51	67	98

demonstration project could convert both C5 and C6 sugars from corn stover and have a farnesane recovery of 95% and purity of 97% (Wang et al., 2016).

LS9 also made huge progress to commercialize the DSHC pathway using biological conversion (Wang et al., 2016). The process uses *Escherichia coli* for fatty acids production through the anaerobic-based fatty acid biosynthesis pathway. In addition to producing fatty acids, other coproducts such as fatty alcohols, fatty esters, and alkanes can be produced. The alkanes can then be further processed to produce biojet fuel. LS9 has also been researching on the direct conversion of sugar to alkanes without the additional step of hydrogenation. This will make it similar to the farnesane production pathway. The purified fatty acids at the end point of the fermentation will be hydrotreated and hydroisomerized for the production of biojet fuel.

While DSHC is a technically viable method to produce biojet fuel, the complexity, long residence time, and cost of production make it more suitable to be used for the production of high-value chemicals rather than sustainable aviation fuel. Another strike against DSCH biojet fuels is that the official HFS—SIP fuel can only be blended with fossil jet fuel up to 10% by volume as compared with 50% volumetrically for the HEFA and FT pathways.

2.5.2 Aqueous phase reforming

The aqueous phase reforming (APR) pathway uses a technology to convert soluble plant sugars to biojet fuel range hydrocarbons (Wei et al., 2019). Before obtaining the final biojet fuel products, plant sugars will be converted first into chemical intermediates such as acids, alcohols, aldehydes, furans, ketones, and other oxygenated hydrocarbons. Fig. 2.18 shows the APR biojet fuel conversion pathway.

The APR conversion pathway consists of five main steps. Pretreatment of feedstock is required to disrupt the matrix of polymeric compounds that are bonded within the lignocellulosic wall structures. They include hemicellulose, lignin, and cellulose microfibrils (Davis et al., 2015). The enzymatic hydrolysis process will break down cellulose fibers into cellobiose and

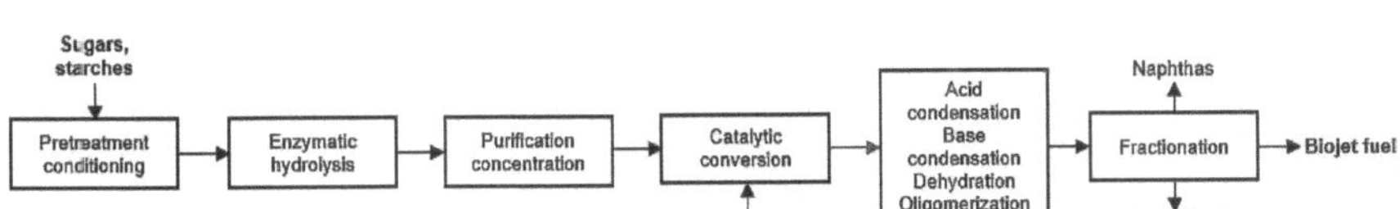

Figure 2.18 Aqueous phase reforming (APR) biojet fuel conversion pathway.

soluble glucooligomers and finally into glucose monomers. The resulting glucose and other sugars hydrolyzed during pretreatment are purified through microfiltration and ion exchange.

At this point, the process is still identical to that of DSHC. The catalytic conversion stage is where APR diverges with DSHC. This step removes oxygen to remove the functions of carbohydrates to convert into diesel range hydrocarbons. The catalytic process has a set of two reforming reactors for hydrogenation and APR steps, followed by condensation and oligomerization, and ending with hydrotreating. The reactors typically have operating temperature of 350°C and pressure of 7.24 MPa. In the condensation stage, intermediates from the previous stage go through the C—C bond forming reactions to form longer continuous carbon chains. In the condensation reactor, the reaction temperature is 262 and 300°C at the inlet and outlet, respectively, while retaining pressure at 6.21 MPa. Here, average carbon chain length increases from $<C_6$ to the C_8-C_{24} range. The range matches those of biojet fuel and diesel fuel.

Table 2.16 shows the overall product carbon yields from the APR pathway. Unlike the DSHC pathway that favors ethanol production, the APR pathway favors the biojet fuel range or the C_8-C_{14} range. The APR pathway is equally adept at producing diesel range fuels.

Key companies developing the APR pathway for biojet fuel production include Virent, Shell, and Virdia. Nonetheless, the APR pathway is still a new and novel approach to biojet fuel production. Understanding of the technology, in particular the catalytic conversion process is still rudimentary with minimal literature available in the research and public domain as compared with the other more mature pathways such as HEFA and FT. More research have to be conducted to bring it past the emergent technology stage and into pilot plant stage.

Table 2.16 Overall product carbon yields from the APR pathway (Davis et al., 2015).

	Proportion of feed C to hydrogenation/APR (%)	
Products by carbon number	**Patent from Virent Inc**	**Model results from NREL**
C_1-C_7	23	22
C_8-C_{14}	50	50
$C_{15}-C_{24}$	23	23
C_{24+}	1	1

APR, aqueous phase reforming.

Table 2.17 Summary of the various biojet fuel conversion pathways.

Conversion category	Pathways	Main feedstocks	Intermediates	Economic costs	Major companies	Technology readiness level
Oil-to-jet	Hydroprocessed esters and fatty acids	Vegetable oil, animal fats, waste cooking oil, algal oil	Biooil	Low production cost but feedstock sensitive	Agrisoma Biosciences, AltAir Fuels, ASA, Neste Oil, PetroChina, Sapphire Energy, SG Biofuels, Syntroleum, Tyson Food, UOP	8–9
	Catalytic hydrothermolysis	Algae, oil seeds	Biooil	Potentially low production cost	Aemetis, Applied Research Assoc., Chevron Lummus Global	5–6
	Hydroprocessed depolymerized cellulosic jet	Lignocellulosic biomass	Biooil or pyrolysis oil	Moderate production cost	Dynamotive, Envergent, GTI, Hunt Refining, Kior, Petrotech	6
Alcohol-to-jet (ATJ)	Ethanol-to-jet	Sugar and starch crops, municipal waste	Ethanol	High production cost	Coskata, LanzaTech, MixAlco, Swedish Biofuels, Terrabon	6–7
	Butyl ATJ	Butyl alcohol	Isobutanol or *n*-butanol	High production cost	Albemarie, Byogy, Cobalt, Gevo, Solazyme	~6

Continued

Table 2.17 Summary of the various biojet fuel conversion pathways.—cont'd

Conversion category	Pathways	Main feedstocks	Intermediates	Economic costs	Major companies	Technology readiness level
Gas-to-jet	Fischer–Tropsch	Lignocellulosic biomass, municipal waste	Syngas	High production cost	Rentech, Shell, Solena, SynFuels, Syntroleum	7–8
	Biomass-to-fuel	Lignocellulosic biomass, municipal waste	Syngas	Moderate production cost	Coskata, INEOS Bio, LanzaTech, Swedish Biofuels	~6
Sugar-to-jet	Direct sugar-to-hydrocarbon	Sugar and starch crops	Farnesene and fatty acids	Not suitable for biojet fuel as higher value chemicals can be produced	Amyris, LS9, Solazyme, Total	6
	Aqueous phase reforming	Sugar and starch crops	Hydrocarbons	Moderate production cost	Shell, Virdia, Virent	4–5

2.6 Summary

The production methods of biojet fuels are varied with many competing pathways as shown in Table 2.17. This is in contrast with biodiesel production where the transesterification method is the most dominant pathway by far. Among all, the HEFA pathway has the most mature technology and lowest average cost at present. The HEFA pathway has one distinct disadvantage in being the most cost-sensitive to feedstocks. The dominance of HEFA might not last long as other conversion pathways are developing fast and going up the technology readiness level. FT pathway has the potential to join HEFA as a commercially mature production method.

References

Abbaslou, R.M.M., Tavassoli, A., Soltan, J., Dalai, A.K., 2009. Iron catalysts supported on carbon nanotubes for Fischer–Tropsch synthesis: effect of catalytic site position. Appl. Catal. A Gen. 367 1, 47–52.

Ail, S.S., Dasappa, S., 2016. Biomass to liquid transportation fuel via Fischer Tropsch synthesis – technology review and current scenario. Renew. Sustain. Energy Rev. 58, 267–286.

ASTM D7566-19b, 2019. ASTM International, Standard Specification for Aviation Turbine Fuel Containing Synthesized Hydrocarbons.

Benedetti, V., Ail, S.S., Patuzzi, F., Cristofori, D., Rauch, R., Baratieri, M., 2020. Investigating the feasibility of valorizing residual char from biomass gasification as catalyst support in Fischer-Tropsch synthesis. Renew. Energy 147, 884–894.

Biomass Magazine, 2019. Vertimass Wins $1.4 Million for Ethanol to Jet Fuel Process. In: http://www.biomassmagazine.com/articles/16582/vertimass-wins-1-4-million-for-ethanol-to-jet-fuel-process.

Bokade, V.V., Yadav, G.D., 2011. Heteropolyacid supported on montmorillonite catalyst for dehydration of dilute bio-ethanol. Appl. Clay Sci. 53 2, 263–271.

Breault, R.W., 2010. Gasification processes old and new: a basic review of the major technologies. Energies 3, 216–240.

Buffon, E., Stradiotto, N.R., 2019. Electrochemical sensor based on molecularly imprinted poly(ortho-phenylenediamine) for determination of hexahydrofarnesol in aviation biokerosene. Sensor. Actuator. B Chem. 287, 371–379.

Capaz, R.S., Seabra, J.E.A., 2016. Life cycle assessment of biojet fuels. In: Biofuels for Aviation, pp. 279–294.

Casas-Godoy, L., Barrera-Martínez, I., Ayala-Mendivil, N., Aguilar-Juárez, O., Arellano-García, L., Reyes, A.L., et al., 2020. Chapter 4 - Biofuels in Biobased Products and Industries, pp. 125–170.

Chemicals Technology, 2010. Lanxess Invests in Isobutene Production. https://www.chemicals-technology.com/uncategorised/news86770-html/.

Chen, S., Chen, J., Qu, T., Xiang, K., Zhang, Y., Hao, P., et al., 2018. Crown ether induced assembly to γ-Al_2O_3 nanosheets with rich pentacoordinate Al^{3+} sites and high ethanol dehydration activity. Appl. Surf. Sci. 457, 626–632.

Chen, J., Liang, J., Xu, Z., E, J., 2020. Assessment of supercritical water gasification process for combustible gas production from thermodynamic, environmental and techno-economic perspectives: a review. Energy Convers. Manag. 226, 113497.

Cheng, Y.W., Chong, C.C., Cheng, C.K., Ng, K.H., Witoon, T., Juan, J.C., 2020. Ethylene production from ethanol dehydration over mesoporous SBA-15 catalyst derived from palm oil clinker waste. J. Clean. Prod. 249, 119323.

Chiodini, A., Bua, L., Carnelli, L., Zwart, R., Vreugdenhil, B., Vocciante, M., 2017. Enhancements in Biomass-to-Liquid processes: gasification aiming at high hydrogen/ carbon monoxide ratios for direct Fischer-Tropsch synthesis applications. Biomass Bioenergy 106, 104–114.

Corporan, E., Edwards, T., Shafer, L., DeWitt, M.J., Kingshirn, C., Zanarnick, S., et al., 2011. Chemical, thermal stability, seal swell, and emissions studies of alternative jet fuels. Energy Fuels 25, 955–966.

CORSIA, 2019. CORSIA Supporting Document - Life Cycle Assessment Methodology. https://www.icao.int/environmental-protection/CORSIA/Pages/default.aspx.

Cremonez, P.A., Feroldi, M., Araújo, A.V., Borges, M.N., Meier, T.W., Feiden, A., Teleken, J.G., 2015. Biofuels in Brazilian aviation: current scenario and prospects. Renew. Sustain. Energy Rev. 43, 1063–1072.

Davis, R., Tao, L., Tan, E.C.D., Biddy, M.J., Beckham, G.T., Scarlata, C., 2013. Process Design and Economics for the Conversion of Lignocellulosic Biomass to Hydrocarbons: Dilute-Acid and Enzymatic Deconstruction of Biomass to Sugars and Biological Conversion of Sugars to Hydrocarbons. NREL Technical Report NREL/TP-5100-60223. https://www.nrel.gov/docs/fy14osti/60223.pdf.

Davis, R., Tao, L., Scarlata, C., Tan, E.C.D., Ross, J., Lukas, J., Sexton, D., 2015. Process Design and Economics for the Conversion of Lignocellulosic Biomass to Hydrocarbons: Dilute-Acid and Enzymatic Deconstruction of Biomass to Sugars and Catalytic Conversion of Sugars to Hydrocarbons. Technical Report NREL/TP-5100-62498. https://www.nrel.gov/docs/fy15osti/62498.pdf.

Dayton, D.C., Foust, T.D., 2020. Alternative jet fuels. In: Analytical Methods for Biomass Characterization and Conversion, pp. 147–165.

de Reviere, A., Gunst, D., Sabbe, M., Verberckmoes, A., 2020. Sustainable short-chain olefin production through simultaneous dehydration of mixtures of 1-butanol and ethanol over HZSM-5 and γ-Al2O3. J. Ind. Eng. Chem. 89, 257–272. https://doi.org/10.1016/j.jiec.2020.05.022.

de Souza, L.M., Mendes, P.A.S., Aranda, D.A.G., 2018. Assessing the current scenario of the Brazilian biojet market. Renew. Sustain. Energy Rev. 98, 426–438.

Díaz, M., Epelde, E., Tabernilla, Z., Ateka, A., Aguayo, A.T., Bilbao, J., 2020. Operating Conditions to Maximize Clean Liquid Fuels Yield by Oligomerization of 1-butene on HZSM-5 Zeolite Catalysts. Energy, p. 118317.

Eller, Z., Varga, Z., Hancsók, J., 2016. Advanced production process of jet fuel components from technical grade coconut oil with special hydrocracking. Fuel 182, 713–720.

Eschemann, T.O., de Jong, K.P., 2015. Deactivation behavior of Co/TiO_2 catalysts during Fischer–Tropsch synthesis. ACS Catal. 5 6, 3181–3188.

Espinoza, R.L., Steynberg, A.P., Jager, B., Vosloo, A.C., 1999. Low temperature Fischer–Tropsch synthesis from a Sasol perspective. Appl. Catal. A Gen. 186, 13–26.

Folkedahl, B.C., Snyder, A.C., Strege, J.R., Bjorgaard, S.J., 2011. Process development and demonstration of coal and biomass indirect liquefaction to synthetic iso-paraffinic kerosene. Fuel Process. Technol. 92 (10), 1939–1945.

Geerlings, J.J.C., Wilson, J.H., 1999. Fischer-Tropsch technology-from active site to commercial process. Appl. Catal. A Gen. 186, 27–40.

Geleynse, S., Jiang, Z., Brandt, K., Garcia-Perez, M., Wolcott, M., Zhang, X., 2020. Pulp mill integration with alcohol-to-jet conversion technology. Fuel Process. Technol. 201, 106338.

Gírio, F., 2019. Innovation on Bioenergy. In: The Role of Bioenergy in the Bioeconomy, pp. 405–433.

Green Car Congress, 2009. French CEA Launches Biomass-to-Liquids Project; Hydrogen Used to Optimize Mass Yield. https://www.greencarcongress.com/2009/12/bure-saudron-20091225.html.

Green Car Congress, 2018. ASTM Greenlights Ethanol as Feedstock for Alcohol-to-Jet Synthetic Fuel; Blend Level up to 50%. https://www.greencarcongress.com/2018/04/20180416-atj.html.

Gunarathne, D.S., Mueller, A., Fleck, S., Kolb, T., Chmielewski, J.K., Yang, W., Blasiak, W., 2014. Gasification characteristics of steam exploded biomass in an updraft pilot scale gasifier. Energy 71, 496–506.

Gunst, D., Alexopoulos, K., Van Der Borght, K., John, M., Galvita, V., Reyniers, M.-F., Verberckmoes, A., 2017. Study of butanol conversion to butenes over H-ZSM-5: effect of chemical structure on activity, selectivity and reaction pathways. Appl. Catal. A Gen. 539, 1–12.

Guo, L., Jin, H., 2013. Boiling coal in water: hydrogen production and power generation system with zero net CO_2 emission based on coal and supercritical water gasification. Int. J. Hydrogen Energy 38, 12953–12967.

Guo, X., Guo, L., Zeng, Y., Kosol, R., Gao, X., Yoneyama, Y., et al., 2020. Catalytic oligomerization of isobutyl alcohol to jet fuels over dealuminated zeolite Beta. Catal. Today. https://doi.org/10.1016/j.cattod.2020.04.047.

Gutiérrez-Antonio, C., Gómez-Castro, F.I., Hernández, S., Briones-Ramírez, A., 2015. Intensification of a hydrotreating process to produce biojet fuel using thermally coupled distillation. Chem. Eng. Process Process Intensif. 88, 29–36.

Gutiérrez-Antonio, C., Gómez-Castro, F.I., de Lira-Flores, J.A., Hernández, S., 2017. A review on the production processes of renewable jet fuel. Renew. Sustain. Energy Rev. 79, 709–729.

Hanaoka, T., Miyazawa, T., Shimura, K., Hirata, S., 2015. Jet fuel synthesis in hydrocracking of Fischer–Tropsch product over Pt-loaded zeolite catalysts prepared using microemulsions. Fuel Process. Technol. 129, 139–146.

Hari, K., Yaakob, T.Z., Binitha, N.N., 2015. Aviation biofuel from renewable resources: routes, opportunities and challenges. Renew. Sustain. Energy Rev. 42, 1234–1244.

Heveling, J., Nicolaides, C.P., Scurrell, M.S., 1998. Catalysts and conditions for the highly efficient, selective and stable heterogeneous oligomerisation of ethylene. Appl. Catal. A Gen. 173, 1–9.

Hillestad, M., Ostadi, M., Alamo Serrano, G.d., Rytter, E., Austbø, B., Pharoah, J.G., Burheim, O.S., 2018. Improving carbon efficiency and profitability of the biomass to liquid process with hydrogen from renewable power. Fuel 234, 1431–1451.

Jürgens, S., Oßwald, P., Selinsek, M., Piermartini, P., Schwab, J., Pfeifer, P., et al., 2019. Assessment of combustion properties of non-hydroprocessed Fischer-Tropsch fuels for aviation. Fuel Process. Technol. 193, 232–243.

Kandaramath Hari, T., Yaakob, Z., Binitha, N.N., 2015. Aviation biofuel from renewable resources: routes, opportunities and challenges. Renew. Sustain. Energy Rev. 42, 1234–1244.

Kim, Y.T., Chada, J.P., Xu, Z., Pagan-Torres, Y.J., Rosenfeld, D.C., Winniford, W.L., et al., 2015. Low-temperature oligomerization of 1-butene with H-ferrierite. J. Catal. 323, 33–44.

Kousoulidou, M., Lonza, L., 2016. Biofuels in aviation: fuel demand and CO_2 emissions evolution in Europe toward 2030. Transport. Res. D Transport Environ. 46, 166–181.

Kruse, A., 2008. Supercritical water gasification. Biofuels Bioprod. Biorefining 2, 415–437.

Kruse, A., Dahmen, N., 2015. Water-A magic solvent for biomass conversion. J. Supercrit. Fluids 96, 36–45.

Krutpijit, C., Tochaeng, P., Jongsomjit, B., 2020. Temperature and ethanol concentration effects on catalytic ethanol dehydration behaviors over alumina-spherical silica particle composite catalysts. Catal. Commun. 145, 106102.

Kumabe, K., Sato, T., Matsumoto, K., Ishida, Y., Hasegawa, T., 2010. Production of hydrocarbons in Fischer–Tropsch synthesis with Fe-based catalyst: investigations of primary kerosene yield and carbon mass balance. Fuel 89, 2088–2095.

Kumar, K., Ghosh, S., Angelidaki, I., Holdt, S.L., Karakashev, D.B., Alvarado Morales, M.A., Das, D., 2016. Recent developments on biofuels production from microalgae and macroalgae. Renew. Sustain. Energy Rev. 65, 235–249.

Lallemand, M., Finiels, A., Fajula, F., Hulea, V., 2007. Ethylene oligomerization over Ni-containing mesostructured catalysts with MCM-41, MCM-48 and SBA-15 topologies. Stud. Surf. Sci. Catal. 170, 1863–1869.

Lanzatech, 2019. LanzaTech Moves Forward on Sustainable Aviation Scale up in the USA and Japan. https://www.lanzatech.com/2019/11/22/lanzatech-moves-forward-on-sustainable-aviation-scale-up-in-the-usa-and-japan/.

LanzaTech, 2020. LanzaJet Takes off!. https://www.lanzatech.com/2020/06/02/lanzajet-takes-off/.

Lee, M., Yoon, J.W., Kim, Y., Yoon, J.S., Chae, H.-J., Han, Y.-H., 2018. Ni/SIRAL-30 as a heterogeneous catalyst for ethylene oligomerization. Appl. Catal. A Gen. 2018, 87–93.

Lee, K., Lee, M.-E., Kim, J.-K., Shin, B., Choi, M., 2019. Single-step hydroconversion of triglycerides into biojet fuel using CO-tolerant PtRe catalyst supported on USY. J. Catal. 379, 180–190.

Liu, G., Yan, B., Chen, G., 2013. Technical review on jet fuel production. Renew. Sustain. Energy Rev. 25, 59–70.

Luque, R., de la Osa, A.R., Campelo, J.M., Romero, A.A., Valverde, J.L., Sanchez, P., 2012. Design and development of catalysts for Biomass-To-Liquid-Fischer–Tropsch (BTL-FT) processes for biofuels production. Energy Environ. Sci. 5 1, 5186–5202.

Mahdaviani, S.H., Soudbar, D., Parvari, M., 2010. Selective ethylene dimerization toward 1-butene by a new highly efficient catalyst system and determination of its optimum operating conditions in a buchi reactor. Int. J. Chem. Eng. Appl. 1, 276–281.

Marsh, G., 2008. Biofuels: aviation alternative? Renew. Energy Focus 9 4, 48–51.

Martínez, A., Arribas, M.A., Concepción, P., Moussa, S., 2013. New bifunctional Ni–H-Beta catalysts for the heterogeneous oligomerization of ethylene. Appl. Catal. A Gen. 467, 509–518.

Matsumura, Y., Minowa, T., 2003. Fundamental design of a continuous biomass gasification process using a supercritical water fluidized bed. Int. J. Hydrogen Energy 29, 701–707.

Matsuoka, K., Hosokai, S., Kuramoto, K., Suzuki, Y., 2013. Enhancement of coal char gasification using a pyrolyzer–gasifier isolated circulating fluidized bed gasification system. Fuel Process. Technol. 109, 43–48.

Méndez, C.I., Ancheyta, J., 2020. Modeling and control of a Fischer-Tropsch synthesis fixed-bed reactor with a novel mechanistic kinetic approach. Chem. Eng. J. 390, 124489.

Mitchell, S.C., 1998. Hot Gas Cleanup of Sulphur, Nitrogen, Minor and Trace Elements. IEA Coal Research, London, UK.

Moses, C.A., 2008. Comparative Evaluation of Semi-Synthetic Jet Fuels Crc Project No. AV-2-04A.

Mountouris, A., Voutsas, E., Tassios, D., 2006. Solid waste plasma gasification: equilibrium model development and exergy analysis. Energy Convers. Manag. 47, 1723–1737.

Neuling, U., Kaltschmitt, M., 2018. Techno-economic and environmental analysis of aviation biofuels. Fuel Process. Technol. 171, 54–69.

Olgun, H., Ozdogan, S., Yinesor, G., 2011. Results with a bench scale downdraft biomass gasifier for agricultural and forestry residues. Biomass Bioenergy 35 1, 572–580.
Osada, M., Sato, T., Watanabe, M., Shirai, M., Arai, K., 2006. Catalytic gasification of wood biomass in subcritical and supercritical water. Combust. Sci. Technol. 178, 1–3.
Ostadi, M., Rytter, E., Hillestad, M., 2019. Boosting carbon efficiency of the biomass to liquid process with hydrogen from power: the effect of H_2/CO ratio to the Fischer-Tropsch reactors on the production and power consumption. Biomass Bioenergy 127, 105282.
Phung, T.K., Hernández, L.P., Lagazzo, A., Busca, G., 2015. Dehydration of ethanol over zeolites, silica alumina and alumina: Lewis acidity, Brønsted acidity and confinement effects. Appl. Catal. A Gen. 49, 77–89.
PNNL, 2018. PNNL and LanzaTech Team to Make New Jet Fuel. https://www.pnnl.gov/news/release.aspx?id=4527.
Pujan, R., Hauschild, S., Gröngröft, A., 2017. Process simulation of a fluidized-bed catalytic cracking process for the conversion of algae oil to biokerosene. Fuel Process. Technol. 167, 582–607.
Rabaev, M., Landau, M.V., Vidruk-Nehemya, R., Koukouliev, V., Zarchin, R., Herskowitz, M., 2015. Conversion of vegetable oils on Pt/Al2O3/SAPO-11 to diesel and jet fuels containing aromatics. Fuel 161, 287–294.
Richards, G.A., Casleton, K.H., 2010. Gasification technology to produce synthesis gas. In: Synthesis Gas Combustion Fundamentals and Applications. Taylor & Francis Group.
Samiran, N.F., Mohd Jaafar, M.N., Ng, J.-H., Lam, S.S., Chong, C.T., 2016. Progress in biomass gasification technique – with focus on Malaysian palm biomass for syngas production. Renew. Sustain. Energy Rev. 62, 1047–1062.
Santos, R.G.d., Alencar, A.C., 2020. Biomass-derived syngas production via gasification process and its catalytic conversion into fuels by Fischer Tropsch synthesis: a review. Int. J. Hydrogen Energy 45 (36), 18114–18132.
Shahabuddin, M., Alam, M.T., Krishna, B.B., Bhaskar, T., Perkins, G., 2020. A review on the production of renewable aviation fuels from the gasification of biomass and residual wastes. Bioresour. Technol. 312, 123596.
Shin, M., Jeong, H., Park, M.-J., Suh, Y.-W., 2020. Benefits of the SiO_2-supported nickel phosphide catalyst on ethylene oligomerization. Appl. Catal. A Gen. 591, 117376.
Silva Braz, D., Pinto Mariano, A., 2018. Jet fuel production in eucalyptus pulp mills: economics and carbon footprint of ethanol vs. butanol pathway. Bioresour. Technol. 268, 9–19.
SOLAR-JET, 2015. http://www.solar-jet.eu/. (Accessed 20 October 2020).
Steynberg, A.P., Espinoza, R.L., Jager, B., Vosloo, A.C., 1999. High temperature Fischer–Tropsch synthesis in commercial practice. Appl. Catal. A Gen. 186, 41–54.
SUN-to-LIQUID, 2016. https://www.sun-to-liquid.eu/. (Accessed 20 October 2020).
Swain, F., 2020. The Other Solar Power: How Scientists Are Making Fuel from Sunlight and Air. Discover Magazine. https://www.discovermagazine.com/environment/the-other-solar-power-how-scientists-are-making-fuel-from-sunlight-and-air. (Accessed 20 October 2020).
Tijmensen, M.J.A., Faaij, A.P.C., Hamelinck, C.N., van Hardeveld, M.R.M., 2002. Exploration of the possibilities for production of Fischer Tropsch liquids and power via biomass gasification. Biomass Bioenergy 23 2, 129–152.
Tomasek, S., Lónyi, F., Valyon, J., Hancsók, J., 2020. Fuel purpose hydrocracking of biomass based Fischer-Tropsch paraffin mixtures on bifunctional catalysts. Energy Convers. Manag. 213, 112775.

Tzanetis, K.F., Posada, J.A., Ramirez, A., 2017. Analysis of biomass hydrothermal liquefaction and biocrude-oil upgrading for renewable jet fuel production: the impact of reaction conditions on production costs and GHG emissions performance. Renew. Energy 113, 1388–1398.

Udomsirichakorn, J., Basu, P., Salam, P.A., Acharya, B., 2013. Effect of CaO on tar reforming to hydrogen-enriched gas with in-process CO_2 capture in a bubbling fluidized bed biomass steam gasifier. Int. J. Hydrogen Energy 38 34, 14495–14504.

Valmet, 2017. CFB's evolution from coal-only boiler to biomass-or anything in between. In: Valmet Technical Paper Series. https://www.valmet.com/globalassets/media/downloads/white-papers/power-and-recovery/cfb_evolution_coal_to_biomass_white-paper.pdf.

Vásquez, M.C., Silva, E.E., Castillo, E.F., 2017. Hydrotreatment of vegetable oils: a review of the technologies and its developments for jet biofuel production. Biomass Bioenergy 105, 197–206.

Vela-García, N., Bolonio, D., Mosquera, A.M., Ortega, M.F., García-Martínez, M.-J., Canoira, L., 2020. Techno-economic and life cycle assessment of triisobutane production and its suitability as biojet fuel. Appl. Energy 268, 114897.

Verma, D., Kumar, R., Rana, B.S., Sinha, A.K., 2011. Aviation fuel production from lipids by a single-step route using hierarchical mesoporous zeolites. Energy Environ. Sci. 4, 1667–1671.

Wang, W.-C., Tao, L., 2016. Bio-jet fuel conversion technologies. Renew. Sustain. Energy Rev. 53, 801–822.

Wang, W.C., Tao, L., Markham, J., Zhang, Y., Tan, E., Batan, L., et al., 2016. National Renewable Energy Laboratory, Review of Biojet Fuel Conversion Technologies, NREL/TP-5100-66291. http://www.nrel.gov/docs/fy16osti/66291.pdf.

Warsita, A., Al-attab, K.A., Zainal, Z.A., 2017. Effect of water addition in a microwave assisted thermal cracking of biomass tar models. Appl. Therm. Eng. 113, 722–730.

Wei, H., Liu, W., Chen, X., Yang, Q., Li, J., Chen, H., 2019. Renewable bio-jet fuel production for aviation: a review. Fuel 254, 115599.

Wright, M.E., Harvey, B.G., Quintana, R.L., 2008. Highly efficient zirconium-catalyzed batch conversion of 1-butene: a new route to jet fuels. Energy Fuels 22, 3299–3302.

Xie, Q., Borges, F.C., Cheng, Y., Wan, Y., Li, Y., Lin, X., et al., 2014. Fast microwave-assisted catalytic gasification of biomass for syngas production and tar removal. Bioresour. Technol. 156, 291–296.

Xu, R., Hou, C., Xia, G., Sun, X., Li, M., Nie, H., Li, D., 2020. Effects of Ag promotion for Co/Al_2O_3 catalyst in Fischer-Tropsch synthesis. Catal. Today 342, 111–114.

Yang, J., Xin, Z., He, Q., Corscadden, K., Niu, H., 2019. An overview on performance characteristics of bio-jet fuels. Fuel 237, 916–936.

Yilmaz, N., Atmanli, A., 2017. Sustainable alternative fuels in aviation. Energy 140, 1378–1386.

Yoo, G., Park, M.S., Yang, J.-W., 2015. Chemical pretreatment of algal biomass. In: Pretreatment of Biomass, pp. 227–258.

Zhang, X., Zhong, J., Wang, J., Zhang, L., Gao, J., Liu, A., 2009. Catalytic performance and characterization of Ni-doped HZSM-5 catalysts for selective trimerization of n-butene. Fuel Process. Technol. 90, 863–870.

Zhang, D., Al-Hajri, R., Barri, S.A.I., Chadwick, D., 2010a. One-step dehydration and isomerisation of n-butanol to iso-butene over zeolite catalysts. Chem. Commun. 46 23, 4088–4090.

Zhang, L., Jin, Q., Shan, L., Liu, Y., Wang, X., Huang, J., 2010b. $H_3PW_{12}O_{40}$ immobilized on silylated palygorskite and catalytic activity in esterification reactions. Appl. Clay Sci. 47 (3), 229–234.

Zhang, Q., Dor, L., Fenigshtein, D., Yang, W., Blasiak, W., 2012. Gasification of municipal solid waste in the Plasma Gasification Melting process. Appl. Energy 90, 106–112.

Zhang, C., Hui, X., Lin, Y., Sung, C.-J., 2016. Recent development in studies of alternative jet fuel combustion: progress, challenges, and opportunities. Renew. Sustain. Energy Rev. 54, 120–138.

Zhang, J., Tang, R., Shen, Z., Liang, S., Zhong, H., 2020. Catalytic Oligomerization of ethylene over nano-sized HZSM-5. J. Energy Inst. https://doi.org/10.1016/j.joei.2020.09.002.

Zhao, X., Sun, X., Cui, X., Liu, D., 2019. Production of biojet fuels from biomass. In: Sustainable Bioenergy, pp. 127–165.

Zhou, N., Zhou, J., Dai, L., Guo, F., Wang, Y., Li, H., et al., 2020. Syngas production from biomass pyrolysis in a continuous microwave assisted pyrolysis system. Bioresour. Technol. 314, 123756.

Zschocke, A., Scheuermann, S., Ortner, J., 2012. High Biofuel Blends in Aviation (HBBA), ENER/C2/2012/420-1 Final Report. https://ec.europa.eu/energy/sites/ener/files/documents/final_report_for_publication.pdf.

CHAPTER 3

Property specifications of alternative jet fuels

3.1 Introduction

Aviation jet fuel is a complex mixture of hydrocarbons designed to provide high performance for application in aircraft. The composition of the fuel varies depending on crude source and manufacturing process. Jet fuel is typically extracted from conventional crude oil in the middle distillate fractions, accounting for approximately 10% of the crude oil. The conventional hydrocarbon sources used for jet fuel production include crude oil, natural gas, liquid condensates, heavy oil, shale oil, and oil sands. Conventional jet fuel consists of a large variety of different species belonging primarily to four chemical families: long-chained unbranched alkanes (n-alkanes), long-chained, branched alkanes (iso-alkanes), cyclic alkanes (naphthenes), and aromatics. Table 3.1 shows the typical hydrocarbon components found in jet fuel and their corresponding characteristics. Jet fuel mainly consists of ~70%−85% paraffins (iso, normal and cyclic), of which the dominant straight-chain paraffin ensures high heat release and clean burning features. The cyclic isoparaffin has the characteristic of low temperature fluidity, which is essential to keep the jet fuel fluidic while operating at high altitude. Similarly, cycloparaffin contributes to the cold flow property owing to the nature of low freezing point, but the heat release rate is lower than n-paraffins due to lower hydrogen-to-carbon ratio. The aromatics content is approximately ~25%, with the main function of providing high heat release. There are some impurities in the jet fuel, such as less than 1% olefins and some trace amount of S, N, and O. The olefins are reactive and can lead to gum formation; hence, they are treated as contaminants. In addition, a small quantity of additives is usually added to improve the fuel properties. The jet fuel typically contains a carbon number distribution between 8 and 16 that falls in the gasoline (C_4-C_{12}) and diesel fuel (C_8-C_{23}) ranges. The supply of commercial jet

Biojet Fuel in Aviation Applications
ISBN 978-0-12-822854-8
https://doi.org/10.1016/B978-0-12-822854-8.00007-X

Table 3.1 Characteristics of the jet fuel components (Brooks et al., 2016).

Class	Structure	Characteristics	Composition (%wt)
n-Paraffins		• Clean burning with high heat release per unit weight	70%–85%
Isoparaffins	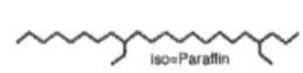	• Low fluid temperature fluidity	
Cycloparaffins (naphthenes)	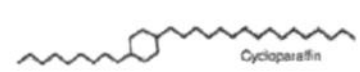	• Lower hydrogen-to-carbon ratio • Lower heat release per unit weight • Higher density and lower freeze point than n-paraffins	
Aromatics	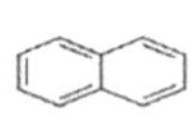	• Prone to form soot during combustion • High energy density	<25%
Olefins	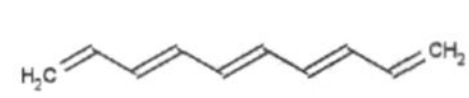	• Good combustion characteristics • Reactive and leads to gum formation	<1%

fuel needs to fulfill the stringent requirements as specified by the international agreed standards. The most widely used jet fuel nowadays is kerosene-based Jet A-1. This chapter discusses the property specifications that are adopted to regulate the quality of jet fuel derived from fossil and nonpetroleum sources. Furthermore, the additives that are present in jet fuel are discussed, followed by the relevant standards related to alternative jet fuel approval process.

3.2 Jet fuel specifications

The quality of the Jet A-1 fuel complies with the requirements of the International ASTM D1655 (ASTM D1655-19a, 2019) and UK Specification DEF STAN 91-091 (MOD, 2019), which are provided by the American Society for Testing Materials (ASTM) and the United Kingdom

Ministry of Defence, respectively. The ASTM D1655 specifies the standard specification requirements of conventional aviation turbine fuels from refinery to the aircraft. It also defines the minimum property requirements for Jet A and Jet A-1 aviation turbine fuels and lists acceptable additives for use in civil operated engines and aircrafts. Some of the main property requirements of jet fuel complying with ASTM D1655 are presented in Table 3.2. The Jet A-1 standard as specified in DEF STAN 91-091 is similar to the ASTM D1655, with DEF STAN 91-091 being more stringent in a small number of areas. Jet A is the standard jet fuel used widely in the United States for domestic flights and international flights originating from the United States. The Jet A is produced in accordance with the specifications of ASTM D1655 but with slight differences as compared with Jet A-1. The former is specified to a minimum flash point of 38°C and a freezing point of no greater than −40°C. The Jet A-1 typically contains more additives than Jet A, such as static dissipator, icing inhibitor, and antioxidant. The Jet A-1 has a lower maximum freezing point of −47°C, hence allowing the Jet A-1 to be used for long-haul international flight, especially on polar routes during the winter.

Due to the complex jet fuel supply arrangement and the need of storing different jet fuel products, the major jet fuel suppliers have jointly produced a document known as the Aviation Fuel Quality Requirements for Jointly Operated Systems (AFQRJOS-Issue 31) Check List to standardize the quality of the supplied Jet A-1 fuel (Joint Inspection Group, 2019). The Check List embodies the most stringent requirements of the DEF STAN 91-091 and ASTM D1655 for Jet A-1. The Check List is recognized by major aviation fuel suppliers including Agip, BP, Chevron Texaco, Equinor, ExxonMobil, Kuwait Petroleum, Shell, and Total. Under this approach, the supplied Jet A-1 is ensured to meet the specifications as required by the different standards, while allowing operators to utilize the same fuel distribution system and storage facility to handle the jet fuel. Jet fuel that meets the requirements of this Check List is referred as "Jet A-1 to Check List." The approach simplifies the jet fuel supply arrangement and reduces the cost of handling and storage.

JP fuels refer to jet fuel generally used for military aircraft with specific additives added to achieve the performance required by military aircraft engines. JP-8 jet fuel has similar properties as Jet A-1 fuel, but with the addition of more additives such as static dissipater additive, corrosion inhibitor/lubricity improver, antioxidant, metal deactivator, and fuel system

Table 3.2 Some Jet A/Jet A-1 requirements specified by ASTM D1655.

Fuel property	Jet A/Jet A-1 (ASTM D1655)	Test method (ASTM)
Composition		
Acidity, total mg KOH/g	≤ 0.10	D3242
Total aromatics, vol %	≤ 26.5	D6379
Sulfur, total mass%	≤ 0.3	D1266, D2622, D4294, D5453
Volatility		
Distillation range, °C		D86, D2887
10% recovery temperature	≤ 205	
50% recovery temperature	Report	
90% recovery temperature	Report	
Final boiling point, °C	≤ 300	
Distillation residue, vol %	≤ 1.5	
Distillation loss, vol %	≤ 1.5	
Flash point, °C	≥ 38	D56, D3828
Density, 15°C kg/m^3	775−840	D1298, D4052
Mobility		
Freezing point, °C	≤ −40 (Jet A), ≤ −47 (Jet A-1)	D5972, D7153, D7154, D2386
Kinematic viscosity at −20°C, mm^2/s	≤ 8	D445, D7042
Combustion		
Net heat of combustion, MJ/kg	≥ 42.8	D4529, D3338, D4809
(1) Smoke point, mm or	≥ 25	D1322
(2) Smoke point, mm, and	≥ 18	D1322
Napthalenes, % vol	≤ 3	D1840
Copper strip corrosion, 2 h, 100°C	No. 1	D130
Thermal stability		
Thermal stability filter pressure drop at 260°C, mm Hg	≤ 25	D3241
Cleanliness		
Existent gum, mg/100 mL	≤ 7	D381, D3948
Additive		
Antioxidant additive, mg/L	≤ 24	
Icing inhibitor additive range, vol%	0.07−0.15	

icing inhibitor (Brooks et al., 2016). JP-8 is produced in accordance with the requirements of the US Military Specification MIL-T-83188D, British military jet fuel specification DEF STAN 91-87 AVTUR/FSII, and NATO Code F-34 (CSG Network, 2013). It is also the dominant military jet fuel grade for the NATO-associated air fleet. JP-8 is specified to a maximum freezing point of −47°C and a flash point of 38°C. Jet B fuel is a distillate covering the naphtha and kerosene fractions, with a typical mixture of ∼30% kerosene and ∼70% gasoline. The ASTM D6615 specification defines the Jet B wide-cut aviation turbine fuel intended for use in aircraft that are certified to use such fuel (ASTM D6615-15a, 2019). Jet B is a relatively wide boiling range volatile distillate produced from blends of refined hydrocarbons derived from crude petroleum, natural gasoline, heavy oil, shale oil, or blends with synthetic hydrocarbons. The fuel contains a larger concentration of light hydrocarbons and naphthas than Jet A and hence weighs lesser. According to the standard specifications of ASTM D6615, the density range of Jet B is 751−802 kg/m^3 at 15°C. The vapor pressure of the fuel at 37.8°C is around 14−21 kPa, which is more volatile than Jet A and is more dangerous to handle. The total aromatics content in Jet B is limited to a maximum of 25 vol%, while the net heat of combustion needs to maintain above 42.8 MJ/kg; hence, the fuel combustion performance is similar to that of Jet A. Jet B fuel has the advantage for operations in very low temperature environments due to its enhanced cold weather capabilities, with a maximum freezing point of −50°C.

The main jet fuel grade used in Russia and the Commonwealth of Independent States is TS-1. It is a type of jet fuel specified by the Russia's latest industry standard edition (GOST 10227-2013), which is considered on par with the western Jet A-1 grade fuel. The TS-1 jet fuel is slightly more volatile with a minimum flash point of 28°C and has a lower freeze point (<−50°C) when compared with Jet A-1. The Russian standards emphasize on lower freezing point due to the colder climate in which the aircrafts are expected to operate. In Russia, the typical grade designation for jet fuel in Russia is T-1 to T-8, TS-1, or RT, governed by the State Standard (GOST) Number or a Technical Condition (TU) number. The TS-1 grade jet fuel also meets the specification specified by the Customs Union Technical Regulations (TR CU 013/2011, 2011). The RT-type fuel refers to the superior grade jet fuel that is not widely used.

The Chinese jet fuel specification covers five types of jet fuel. Previously, each grade was numbered with a prefix RP, but now has been renamed as No. 1 Jet Fuel, No. 2 Jet Fuel, and so on. Presently, the most widely used civilian jet fuel grade in China is No. 3 Jet Fuel (previously RP-3), of which the specification is comparable with the Jet A-1, in compliance with the local standard GB 6537 (GB 6537-2018). Table 3.3 shows the property specifications of No. 3 Jet Fuel as specified in the GB 6537-2018 standard. The No. 3 Jet Fuel contains low flash point (28°C minimum), similar to TS-1 jet fuel. The No. 1 and No. 2 Jet Fuels have lower freezing point of −60 and −50°C, respectively. The No. 4 Jet Fuel refers to wide-cut-type fuel similar to Jet B, whereas the No. 5 Jet Fuel is a high flash point kerosene.

3.3 Jet fuel from nonconventional sources

3.3.1 SASOL coal-based synthetic fuel

There are two synthetic jet fuels derived from nonconventional petroleum source that have been recognized as meeting the requirements of ASTM D1655 and DEF STAN 91-091. The approved coal-based synthetic fuels are (1) Sasol semisynthetic jet fuel and (2) Sasol fully synthetic jet fuel. The original Sasol approvals granted are detailed in the Annex B3 of DEF STAN 91-091 issue 11 (MOD, 2019). The Sasol semisynthetic Aviation Turbine Fuel manufactured from the Fischer−Tropsch (FT) process is defined as synthetic isoparaffinic kerosene (IPK). The synthetic component is derived from the FT process that has been polymerized and subsequently hydrogenated. The maximum allowable aromatic content for the Sasol semisynthetic jet fuels is 26.5%, while the minimum limit is 8.4% when measured using method IP436. The blending of IPK with conventional kerosene has been approved up to 50%. Sasol heavy naphtha #1 (HN1) produced from the FT process by fractionation and hydrogenation can be combined with IPK. Table 3.4 shows the specification requirements for HN1/IPK Blend as specified in DEF STAN 91-091 (Annex B3). The final synthetic blend shall contain at least 25% IPK by volume.

The Sasol fully synthetic kerosene is defined as the fuel blended from light distillate, heavy naphtha, and isoparaffinic kerosene streams. The batch certificate for the fuel shall state the fuel contains 100% synthetic components. The maximum flash point permitted is 50°C. The boiling point distribution shall have a minimum slope defined by $T50-T10 \geq 10°C$ and $T90-T10 \geq 40°C$. The aromatic content requirement for Sasol fully synthetic fuel is similar to those as specified for Sasol semisynthetic jet fuels,

Table 3.3 Comparison of the No. 3 Jet Fuel requirements specified by GB 6537-2018.

Fuel property	No. 3 Jet Fuel (GB 6537)	Jet A-1 (ASTM D1655)	Test method for GB 6537
Composition			
Acidity, total mg KOH/g	$\leq$ 0.015	$\leq$ 0.10	GB/T 12574
Total aromatics, vol %	$\leq$ 20.0	$\leq$ 26.5	GB/T 11132
Olefin, vol%	5.0	–	GB/T 1132
Sulfur, total mass%	$\leq$ 0.2	$\leq$ 0.3	GB/T 380, GB/T 11140
Volatility			
Distillation range, °C			GB/T 6536
10% recovery temperature	$\leq$ 205	$\leq$ 205	
50% recovery temperature	$\leq$ 232	Report	
90% recovery temperature	Report	Report	
Final boiling point, °C	$\leq$ 300	$\leq$ 300	
Distillation residue, %	$\leq$ 1.5	$\leq$ 1.5	
Amount of loss, % (v/v)	$\leq$ 1.5	$\leq$ 1.5	
Flash point, °C	$\geq$ 38	$\geq$ 38	GB/T 21789, GB/T 261
Density, 20°C kg/m^3	775–830	775–840	GB/T 1884
Mobility			
Freezing point, °C	$\leq$ −47	$\leq$ −47	GB/T 2430
Viscosity at 20°C, mm^2/s	$\geq$ 1.25	–	GB/T 30515
−20°C, mm^2/s	$\leq$ 8	$\leq$ 8	
Combustion			
Net heat of combustion, MJ/kg	$\geq$ 42.8	$\geq$ 42.8	GB/T 2429, GB/T 384
Smoke point, mm	$\geq$ 25	$\geq$ 25 or $\geq$ 18 and napthalenes $\leq$ 3 %vol	GB/T 382

Continued

Table 3.3 Comparison of the No. 3 Jet Fuel requirements specified by GB 6537-2018.—cont'd

Fuel property	No. 3 Jet Fuel (GB 6537)	Jet A-1 (ASTM D1655)	Test method for GB 6537
Thermal stability			
Filter pressure drop (260°C, 2.5 h)	$\leq$ 3.3 kPa	$\leq$ 25 mm Hg	GB/T 9169
Cleanliness			
Existent gum, mg/100 mL	$\leq$ 7	$\leq$ 7	GB/T 8019

i.e., minimum 8.4% and maximum 26.5%. Due to the near identical composition of Sasol synthetic fuels with conventional jet fuel, the performance in aircraft jet engine is expected to be similar.

3.3.2 Synthetic jet fuel from biofeedstocks

The call for a sustainable and clean alternative jet fuel over the past decade has led to considerable progress in the development of bio-based jet fuels. Bio-based aviation fuels obtained from sources other than fossil-based fuels, such as lignocellolusic biomass, hydrogenated fats, oils and waste fats have low carbon intensity. They could play an important role in mitigating the environment impact of the aviation industry. In order for the bio-based fuels to be used in aircraft jet engine, the fuels must undergo extensive tests to meet the stringent jet fuel requirements. The new fuel should have "drop-in" characteristic where no adaptation to the fuel distribution network or equipment engine fuel system is required. The fuel should ideally be used "as is" and can be blended with conventional jet fuel. At present, major international jet fuel standard has recognized the development of bio-based aviation fuel. The ASTM Specification D7566 (ASTM D7566-19b, 2019) was developed to provide quality control for fuels of novel compositions that include synthesized hydrocarbons from new sources. The ASTM D7566 was first issued in 2009 to provide supply control of synthesized paraffinic kerosene (SPK) derived from coal/natural gas through the FT process. The standard was revised in 2011 with approval of an annex covering SPK synthesized from esters and fatty acid in

Table 3.4 Batch requirements for HN1/IPK Blend as specified in DEF STAN 91-091 (Annex B3).

Property	Limits	Method
Thermal stability		
Jet fuel thermal oxidation test		IP 323/ASTM D3241
Test temperature, °C	$\geq$ 325	
Tube rating visual	Less than 3. No Peacock (P) or Abnormal deposit (A)	Use visual Tube rater within 120 min of completion of the test
Pressure differential, mm Hg	Maximum 25	
Fluidity		
Freezing point, °C	$\leq$ -40.0	IP 16/ASTM D2386
Combustion		
Specific energy, MJ/kg	$\geq$ 42.80	ASTM D3338 ASTM D4809
Composition		
Aromatics, % v/v Or total aromatics, % v/v	$\leq$ 7.0 $\leq$ 7.4	IP156/ASTM D1319 IP436/ASTM D6379

biofeedstock. Since then, four other types of bio-derived synthetic jet fuels have been certified for blending with conventional jet fuel under ASTM D7566, which are assigned by a specific annex as shown in Table 3.5. Recently, the Commercial Aviation Alternative Fuels Initiative (CAAFI) under International Civil Aviation Organization (ICAO) has announced the newly approved alternative jet fuel production pathway from algae-derived lipid, which is set to be included in the revised ASTM D7566 under Annex 7. The new pathway, termed as HH-SPK or HC-HEFA, describes the hydroprocessed hydrocarbons synthesized from the oils (fatty acids) yielded from the *Botryococcus braunii* algae. It is expected that 10% blending level is permitted for this synthetic jet fuel with conventional fossil jet fuel (CAAFI, 2020b). The DEF STAN 91-091 standard does not list out the approved synthetic jet fuel components from nonconventional sources. Annex B of DEF STAN 91-091 states the approval of synthetic jet fuel based on the specifications listed in ASTM D7566. This is to avoid

Table 3.5 Approved bio-based jet fuel production pathway under ASTM D7566.

Annex	Bio-based blendstock production pathway	Approval date	Blending limits (vol %)
A1	Fischer–Tropsch synthetic paraffinic kerosene	Sep 2009	50
A2	SPK from hydroprocessed fatty acid esters and free fatty acid	Jul 2011	50
A3	Hydroprocessing of fermented sugars—synthetic isoparaffinic kerosene	Jun 2014	10
A4	Synthesized kerosene with aromatics	Nov 2015	50
A5	Alcohol-to-jet synthetic paraffinic kerosene	Apr 2016	30
A6	Catalytic hydrothermolysis jet	Dec 2019	50
*A7	Hydroprocessed fatty acid from algae	May 2020	10

*, Recently approved, yet to be included in the ASTM D7566.

duplication and ensure harmonization between the two standards. At present, each of the synthetic jet fuel as approved in the ASTM D7566 is considered as a "batch" product that needs to be blended with ASTM D1655-approved jet fuel. Each synthetic jet fuel needs to fulfill the batch specification requirements as stated in the annexes of ASTM D7566, depending on the production pathway.

There are other emerging production pathways that are in the process of certification, such as the catalytic conversion of sugars by aqueous phase reforming, pyrolysis (hydrotreated depolymerized cellulosic jet), catalytic upgrading of alcohol intermediates, catalytic upgrading of ethanol, or direct use of a wider cut of HEFA with renewable diesel (US Dept of Energy, 2017). To be included in the ASTM standard as a certified alternative jet fuel, the fuel candidate has to undergo a rigorous certification process, which includes several test programs and reviews by manufacturers and flight authorities, which will be discussed in Section 3.7. At present, there are ongoing research focusing on developing potential feedstocks and technologies to produce jet fuel compatible blends, for example, using waste sludge or CO_2-rich waste gas stream from the industry to capitalize on the low-cost feedstocks, and catalytically converting different alcohol intermediates such as isobutanol or higher alcohol into jet fuel. Other new production methods such as using catalytic pyrolysis, syngas fermentation, and hydrotreating biooil derived from fast pyrolysis are currently being explored. Although still at early phase, these methods could eventually lead to fuel certification and commercial production as the technologies become mature.

3.4 Properties of synthetic jet fuel

3.4.1 Fischer–Tropsch hydroprocessed synthesized paraffinic kerosene

Fischer-Tropsh synthesized paraffinic kerosene (FT-SPK) is the first approved synthetic jet fuel production pathway from bioresource under ASTM D7566 in 2009. FT-SPK refers to synthetic paraffinic kerosene produced from biomass via the FT process. In this process, biomass is first gasified to produce synthesis gas, which is used to produce paraffins and olefins via the use of iron or cobalt catalysts. Subsequently, the hydrocarbon products undergo hydrotreating, hydrocracking, or hydroisomerization to be converted into jet fuel quality. Conventional refinery processes such as polymerization, isomerization, and fractionation may be included as part of the process. The jet fuel produced is expected to have similar properties as conventional jet fuel as defined by ASTM D1655. Table 3.6 shows the comparison of some of the batch property requirements for FT-SPK with Jet A-1. The FT-SPK can be produced with a freezing point of −40 or −47°C to meet the requirement of Jet A or Jet A-1, depending on the agreement between purchaser and producer. Other properties such as the flash point are expected to be similar as conventional jet fuel. The density of the FT-SPK is required to be in the range of 730–770 kg/m^3, which is slightly lower compared with the specifications of jet fuel (ASTM D1655). The lower density of the fuel is due to the limited aromatics content, which is capped at maximum 0.5 wt% mass.

3.4.2 Synthesized kerosene with aromatics derived by alkylation of light aromatics from nonpetroleum sources

The synthezised paraffinic kerosene plus aromatics (FT-SPK/A) approved in 2015 is an extension to the FT-SPK. It is produced using the same FT process, but the aromatics content is increased by alkylation of nonpetroleum derived light aromatics such as benzene with FT-derived olefins. The property specifications for FT-SPK/A are similar to FT-SPK, except

Table 3.6 Comparison of some of the batch property requirements for FT-SPK and FT-SPK/A (ASTM D7566) with Jet A-1 (ASTM D1655).

Property	Jet A-1 (ASTM D1655)	FT-SPK	FT-SPK/A	ASTM Test Methods
Flash point, °C	≥ 38	≥ 38	≥ 38	D56, D3828
Density at 15°C, kg/m^3	775–830	730–770	755–800	D1298, D4052
Freezing point, °C	≤ −47	≤ −40	≤ −40	D5972, D7153, D7154, D2386
Thermal stability (2.5 h at control temperature)				
Temperature, °C	≥ 260	≥ 325	≥ 325	D3241
Filter pressure drop, mm Hg	≤ 25	≤ 25	≤ 25	
Hydrocarbon composition				
Cycloparaffin, mass %	–	≤ 15	≤ 15	D2425
Aromatics, mass %	–	≤ 0.5	≤ 20	D2425
Aromatics, vol %	8.4–26.5	–	–	D6379

for the aromatics content and density. Even though the aromatics content in FT-SPK/A is increased to a maximum of 20 wt%, it still needs to be blended with conventional jet fuel to meet the required aromatics content as specified in ASTM D1655. It is expected that the batch property FT-SPK/A meets the requirements as stated in ASTM D7566, which is similar to Jet A-1 as shown in Table 3.6. The blending limit for FT-SPK and FT-SPK/A with jet fuel is 50% vol. The properties of SPK and jet fuel blend need to fulfill the extended requirements as specified in the ASTM D7566, as shown in Table 3.7.

3.4.3 Synthesized paraffinic kerosene from hydroprocessed esters and fatty acids

The SPK derived from hydroprocessed esters and fatty acids (HEFA) is another synthetic jet fuel that gained approval in July 2011 and included in ASTM D7566 under Annex 2. Lipid-based feedstocks, such as vegetable oils, used cooking oils, and tallow which contain mono-, di-, and triglycerides, free fatty acids, and fatty acid esters, have to undergo deoxygenation and hydrogenation processes to produce SPK. Annex 2 in ASTM D7566 specifies that carbon and hydrogen content needs to be at least

Table 3.7 Extended requirements applied to each batch of fuel containing a synthetic blending component.

Property	Limits	ASTM Test Methods
Composition		
Aromatics, vol% or	≥ 8	D1319
Aromatics, vol %	≥ 8.4	D6379
Volatility		
Distillation		D2887, D86, D7344, D7345
T50–T10, °C	≥ 15	
T90–T10, °C	≥ 40	
Lubricity		
Wear scar diameter, mm	≤ 0.85	D5001
Fluidity		
Viscosity −40°C, mm^2/s	≤ 12	D7945

99.5% by mass. It is well known that oxygen content in the fuel will reduce the heating value and affects the thermal stability of the fuel. The HEFA batch requirement specifies the limit of the cycloparaffin which is capped to maximum 15 wt%, while the aromatics content allowed is 0.5 wt%, as shown in Table 3.8. HEFA on its own does not fulfill the final jet fuel requirements of ASTM D1655 and hence will need to be blended with conventional jet fuel. The maximum blending limit allowed with jet fuel is 50% by volume. The synthetic jet blend containing HEFA-SPK is required to fulfill the additional jet fuel requirements related to fluidity under the ASTM D7566 standard, that is, the maximum viscosity allowed at −40°C is 12 mm^2/s.

3.4.4 Alcohol-to-jet synthetic paraffinic kerosene

The alcohol-to-jet synthetic paraffinic kerosene (ATJ-SPK) is defined as a type of synthetic jet fuel derived from ethanol and isobutanol. The SPK produced from alcohol undergoes the processes of dehydration, oligomerization, hydrogenation, and fractionation. The ATJ-SPK is the fifth synthetic jet fuel approved under ASTM D7566 in 2016, with property specifications listed in Annex 5 of ASTM D7566. The batch property

Table 3.8 Comparison of some of the batch property requirements for HEFA-SPK, ATJ-SPK and CHJ (ASTM D7566) with Jet A-1 (ASTM D1655).

Property	ATJ-SPK	HEFA-SPK	CHJ	ASTM Test Methods
Flash point, °C	≥ 38	≥ 38	≥ 38	D56, D3828
Density at 15°C, kg/m^3	730–770	730–772	775–840	D1298, D4052
Freezing point, °C	≤ −40	≤ −40	≤ −40	D5972, D7153, D7154, D2386
Thermal stability (2.5 h at control temperature)				
Temperature, °C	≥ 325	≥ 325	≥ 325	D3241
Filter pressure drop, mm Hg	≤ 25	≤ 25	≤ 25	
Hydrocarbon composition				
Cycloparaffin, mass %	≤ 15	≤ 15	Report	D2425
Aromatics, mass %	≤ 0.5	≤ 0.5	8.4–21.2	D2425

requirements are almost identical to FT-SPK and HEFA-SPK. Some of the batch property requirements for ATJ are shown in Table 3.8. For the hydrocarbon composition requirements, the cycloparaffins and aromatics are limited to a maximum of 15% and 0.5% by mass, respectively. Curiously, in spite of the similarity of batch fuel properties, the maximum blending limit with conventional jet fuel is only 30% by volume. It is envisaged that ATJ-SPK production will permit the use of all C2 to C5 alcohols as feedstocks, subject to the approval of the committee when data is available.

3.4.5 Synthesized kerosene from hydrothermal conversion of fatty acid esters and fatty acids

The sixth approved synthetic jet fuel production pathway is the catalytic hydrothermolysis jet (CHJ) as specified in Annex A6 under ASTM D7566. The process consists of hydrothermal conversion and hydrotreating operations that convert fatty acid esters and fatty acids such as waste fats, oils, and greases into jet fuel. The CHJ is a synthetic blending component that is comprised essentially of normal paraffin, cycloparaffin, isoparaffin, and aromatic compounds. Unlike HEFA-SPK which lacks aromatics compounds, it is required that a batch of CHJ fuel produced must contain a minimum of 8.4 wt% and a maximum of 21.2 wt% of aromatics, as shown in Table 3.8. The CHJ fuel needs to fulfill the density requirement of 775–840 kg/m^3, which is slightly denser as compared with HEFA-SPK due to the presence of aromatics. The blending limit for CHJ with conventional jet fuel is 50 vol%.

3.4.6 Synthesized isoparaffins from hydroprocessed fermented sugars

The permissible blending limits for synthetic fuels of FT-SPK, HEFA-SPK, SPK/A, and CHJ are 50 vol% with conventional jet fuel. The synthesized iso-paraffin (SIP) was approved by ASTM in 2014, but the blending limit is capped to 10% by volume. The synthetic blend component is composed of isoparaffins derived from farnesene ($C_{15}H_{24}$) produced from fermentable sugars. Farnesene is a type of branched alkene of C15 hydrocarbon molecule produced from fermentation process, where the biochemical conversion process utilizes modified yeast to convert sugar into hydrocarbon. The farnesene then undergoes the process of hydroprocessing where the farnesene is reacted with hydrogen, followed by the fractionation operation to separate gas and liquid to produce isoparaffin known as farnesane ($C_{15}H_{32}$). Table 3.9 shows some of the batch requirements for SIP as specified under ASTM D7566. The production pathway requires the produced hydrocarbons to have a minimum saturation level of 98 %wt, of which 97 %wt is farnesane. The permissible oxygenated compound of hexahydrofarnesol ($C_{15}H_{32}O$) in the blend is capped to 1.5 %wt. The olefins and aromatic compounds are limited to within 300 mgBr_2/100 g and 0.5 %wt, respectively. It is noted that the minimum flash point

Table 3.9 Some batch requirements for SIP (ASTM D7566).

Property	SIP	Test method
Flash point, °C	$\geq$ 100	D3828
Density at 15°C, kg/m^3	765–780	D1298, D4052
Freezing point, °C	≤ -60	D5972, D7153, D7154, D2386
Hydrocarbon composition		
Saturated hydrocarbons, mass %	$\geq$ 98	D7974
Farnesane, mass %	$\geq$ 97	D7974
Hexahydrofarnesol, mass %	$\leq$ 1.5	D7974
Olefins, mgBr_2/100 g	$\leq$ 300	D2710
Aromatics, mass %	$\leq$ 0.5	D2425

requirement is 100°C, which is significantly higher than those of other synthetic jet fuels. The density requirement of the fuel is 765–780 kg/m^3, while the freezing point is extended to a maximum of −60°C.

3.4.7 Coprocessing of biocrude

Another biojet fuel production pathway known as coprocessing has been approved. The coprocessing of mono-, di-, and triglycerides, free fatty acids, and fatty acid esters producing cohydroprocessed hydrocarbon synthetic kerosene is recognized as being acceptable for jet fuel manufacture. This means that vegetable oils, waste oils, and fats can be coprocessed along with conventional crude oil feedstocks in existing refining complexes, albeit in a small percentage of 5% by volume. There is no separate annex in ASTM D7566 for coprocessing, but rather, it is included in the amendment made to ASTM D1655. The DEF STAN 91-091 lists the coprocessing specifications in Annex B4. Similar to the ASTM counterpart, the permitted bio-feedstock used in coprocessing refinery unit is 5% by volume. The processes involved in coprocessing the lipid feedstocks are hydrocracking or hydrotreating and fractionation.

3.5 Performance characteristics of aviation turbine fuels

To ensure a safe and economic operation of aviation turbine, the fuel used needs to be essentially free from contamination and meets the specifications as stated in the ASTM D1655 standard. The following section discusses the key characteristics of jet fuel and their effects on the performance of aircraft engine systems.

3.5.1 Thermal stability

Thermal stability refers to the measurement of the amount of deposit formed in the engine fuel system by heating the fuel. Any jet fuel used in aviation engine needs to be thermally stable against oxidation and polymerization at a wide operating temperature range during flight. The oxidative thermal stability of jet fuel is determined via the D3241/IP 323 test method, in which a simple test run with tube temperature controlled at 260°C is conducted to ensure the minimum requirement is met. Two or more runs at different tube temperatures can be conducted to obtain the breakpoint, which is the highest tube temperature at which the fuel still passes the specification requirements of tube deposit color and pressure differential. The thermal stability test for the synthetic fuels is specified at the control temperature of 325°C, except for SIP from Annex 3 of ASTM D7566 where the control temperature is specified at 355°C to ensure the blend components are free of reactive species.

3.5.2 Combustion

The liquid jet fuel supplied to the combustion chamber is first atomized into tiny droplets before burning continuously with the stream of hot air. The fuel and air are mixed and burned at near stoichiometric conditions in the primary zone, where the heat is released. The burnt gases are diluted with the excess air, supplied via the dilution hole in the secondary zone of combustion chamber to lower the gas temperature to a safe level acceptable by the turbine. The effectiveness of the fuel combustion process and the level of soot emissions are strongly related to the fuel composition. Paraffins are straight-chain hydrocarbon, which offer the cleanest combustion characteristics. Olefins have good combustion characteristics, but their poor gum stability limits their use in aircraft turbine fuel to about 1% or less. Aromatics tend to produce soot that leads to undesirable thermal radiation in spite of their higher energy content. Within the aromatics group, naphthalene (a bicyclic aromatic) produces more soot than monocyclic aromatics and hence is the least desirable hydrocarbon class for aircraft fuel. In spite of the drawback, the study has shown that a minimal level of aromatics is needed to be present in the fuel to prevent the shrinkage of aged elastomer seals and fuel leakage (Chen and Liu, 2013). Therefore, the ASTM D7566 standard requires that the final blend for synthetic jet fuel with Jet A contains a minimum of 8 vol% of aromatics. To test the sooting tendency of the fuel, the smoke point test method (ASTM D1322) requires that a minimum smoke point of 25 mm to be met under the aviation jet fuel specification (ASTM D1655). Alternatively, a lower smoke point level of 18 mm is allowed provided that the total naphthalene is capped to a maximum of 3 vol%. Jet fuel with low smoke point tends to produce smoky flame, indicating high aromatic content.

3.5.3 Fuel metering and aircraft range

The density of jet fuel is an important fluid property needed for practical purposes, such as metering flow, and to determine the mass—volume relationship for commercial flights. A fuel with low density indicates low heating value per unit volume and would indicate reduced flight range for a given volume of fuel. The minimum requirement for density of jet fuel is 775 kg/m^3, as specified in ASTM D1655. The chemical energy in the jet fuel is converted into heat and mechanical energy. The aircraft flight range depends on the amount of fuel available and the energy obtainable from useful work. Therefore, an accurate determination of the net heat of combustion is vital so that the availability of a certain predetermined minimum amount of energy as heat is known for the flight operation. In the jet fuel specification, a minimum net heat of combustion requirement is specified to ensure the proper operation of the aircraft engine. The heat of combustion of jet fuel can be determined via the use of an aniline point and density relationship, as specified in the Test Method D4529. An alternative Test Method D3338, which is based on the correlations of aromatics content, gravity, volatility, and sulfur content, can be used to determine the net heat of combustion, thus avoiding the need to obtain the aniline points. In cases of dispute, the direct measurement method of Test Method D4809 or IP 12 can be used as the referee method.

3.5.4 Fuel atomization

Effective fuel atomization is important to maintain a consistent nozzle spray pattern and to ensure the breakup of spray into fine droplets for vaporization. The fuel physical properties are strongly related to the fuel volatility. For jet fuel, the 10% distilled temperature is limited to <205°C to ensure easy ignition. The final boiling point limit excludes heavier fractions that would be difficult to vaporize. Due to the extreme operating temperature range during flight, the fuel pumpability needs to be maintained to ensure a consistent nozzle spray pattern for continuous combustion. As such, the viscosity of aviation turbine fuel at −20°C is specified to exceed 5.5 mm^2/s for Jet A (−40°C freeze point) or 4.5 mm^2/s for Jet A-1 (−47°C freeze point). Another critical property that the jet fuel must possess is low freezing point. The temperature of fuel in an aircraft tank decreases as the outside temperature decreases with increasing altitude. A sufficiently low freezing point for the jet fuel is needed to ensure the fuel can flow through the filter screen to engine at extreme low temperature. The manual sampling method to determine freezing point, Test Method D2386/IP 16, is designated as the referee method.

3.5.5 Compatibility with elastomer and the metals in the fuel system and turbine

The jet fuel must not consist of elements that are reactive to any parts in the engine to ensure the durability and safe operation of the components, i.e., the elastomer and metals in the fuel system and turbine. The ASTM D1655 standard has specified the limits of some reactive components. The mercaptan gases are known to be reactive with certain elastomers. A limitation in the mercaptan content is specified to prevent such reactions and to minimize the unpleasant mercaptan odor. Another concern with the elastomer is the shrinkage of aged elastomer seals. A minimum level of aromatics is needed to keep the elastomer seals swollen to prevent fuel leakage (Chen and Liu, 2013). Synthetic fuels that contain little or no aromatics in batch form do not fulfill such requirement; hence, the final blend with conventional jet fuel needs to fulfill the minimum aromatics content requirement as stated in ASTM D1655. A maximum limit is imposed to prevent the formation of soot, which is considered as harmful pollutants.

Any deposition along the fuel delivery line must be avoided due to the inevitable contact between jet fuel and metal parts. The jet fuel has to undergo copper strip test to ensure the fuel does not contain any chemical species that is corrosive to the copper or copper-based alloys in various parts of the fuel system. Any impurities in the jet fuel, residual mineral acid, or caustic left as a result of the refining process need to be removed to avoid deposition in the fuel line. An acidity test is conducted to measure the presence of organic acids based on the Test Method D3242. The total acidity must be less than 0.1 mg KOH/g, as specified in the ASTM D1655 jet fuel standard. The sulfur content in the fuel needs to be kept minimum as the sulfur oxides formed during combustion are corrosive to the turbine metal parts. The maximum level of sulfur permitted in the ASTM D1655 jet fuel standard is 0.3% by mass.

3.5.6 Fuel storage stability and handling

The standardization of the jet fuel quality is integral to provide a guideline in handling and storage stability of jet fuel and to avoid contamination. The joint agreement developed by major jet fuel suppliers, i.e., Aviation Fuel Quality Requirements for Jointly Operated Systems (AFQRJOS) (Joint Inspection Group, 2019) allows the operators to use the same fuel distribution system and storage facility to handle jet fuel. Some of the properties requirements as per specified in the ASTM D1655 standard are related to the handling and storage of fuel. The flash point is an indication of the maximum temperature for fuel handling and storage without fire risk. For Jet A-1 fuel, the minimum flash point required is 38°C, which is the lowest temperature at which vapors above a volatile combustible substance will ignite in air when expose to flame. Thus, the surrounding temperature needs to be below 38°C to ensure safe handling of the fuel. Static electricity caused by the imbalance of charges can pose a significant fire and explosion risk during the handling of aviation fuel. This problem can be solved by adding electrical conductivity additives to dissipate the charges rapidly. The presence of nonvolatile residue (gum) after evaporation is indicative of jet fuel degradation or contamination by higher boiling oils or particulate matter, which could be attributed to long-term storage or poor handling practices. Constant monitoring of the gum level is needed to avoid problems with aircraft fuel system. The jet fuel specification requires the gum content to not exceed 7 mg/100 mL, measured based on the Test Method D381.

3.5.7 Fuel cleanliness and contamination

The cleanliness of aviation turbine fuel is important to ensure optimum performance. Contamination of jet fuel will result in the failure of filtration components in the engine system and put the flight operation at risk. Fuel cleanliness requires the relative absence of free water, solid particulates, and dirt. Undissolved (free) water in aviation fuel can lead to the growth of microorganism, corrosion of aircraft fuel tank, and icing to the filters in the fuel system. The presence of solid contaminants can potentially result in problems such as wear and plugging of filters in the engine system. The visual inspection of Test Method D4176 provides a qualitative method to determine suspended free water and contaminants in distillate fuel. The Procedure 1 provides a rapid pass/fail method for contamination, whereas the Procedure 2 provides a gross numerical rating of haze appearance. Quantitative testing to determine the presence of undissolved or free water in aviation turbine fuel can be performed via Test Method D3240 without exposing the sample fuel to the atmosphere. The test reading usually ranges from 1 to 60 ppm of free water, but dissolved waters in the fuel are undetectable. The control of free water in the fuel is exercised in ground fueling equipment by use of filter coalescers and water separators.

The presence of solid particulate contaminants such as dirt and rust may be detected by filtration of the jet fuel through membrane filters as prescribed by conditions under Test Method D2276. The test provides a gravimetric measurement of the particulate matter present in the sample of aviation turbine fuel by line sampling. The same standard also provides a qualitative assessment of particulate contaminants in fuel by filtering the fuel through a membrane. The color rating appearing on the membrane is compared against a standard color scale to indicate the level of contamination at a particular location. Apart from water and solid contaminants, microbial contamination in jet fuel is of concern as it can lead to fuel deterioration, leading to a variety of problems including corrosion, odor, filter plugging, and decreased fuel stability. The uncontrolled growth of microbes can even lead to the corrosion of aviation fuel storage and distribution network, thereby affecting the operation of the aircraft. There are two biocide additives approved by the airframe and engine manufacturers, namely Biobor JF and KATHON FP1.5, which are used in accordance with local regulations and agreement between the fuel supplier and end users. The ASTM D6469 lists the symptoms, occurrence, and consequences of chronic microbial contamination caused by microbes, as well as the guide on detection and control of microbes in fuels and fuel systems (ASTM D6469-20, 2020).

To prevent fuel contamination, the presence of surface-active substance known as "surfactants" has to be minimized. The presence of surfactants in the fuel affects the ability of filter separator to separate free water from the fuel, as the surfactants can disperse dirt and water, making them easier to pass through filters. Surfactants that are adsorbed on the surface of filters can interfere with the water droplet or particles and in some cases the solid level can be increased due to the lifting of rusts from piping surface. Test Method D3948 is used to detect the presence of surfactants in aviation turbine fuel. It is a common practice in the industry to conduct the test at the point of production and point of usage, in order to detect carryover traces of refinery treating residues and surface-active substances picked up by the fuel during transportation. A high rating indicates the fuel is free of surfactants, whereas a low rating indicates the presence of surfactants.

3.5.8 Fuel lubricity

Jet fuel plays an important side role of lubricating the sliding parts of engine fuel system components and fuel control units. However, jet fuel is also known to exhibit poor lubricity at high temperature and high load conditions. Lubricity improver additives can be added to improve the lubricity of fuel, but such additives could have adverse effects on fuel filtration systems and on fuel water separation characteristics due to the polar nature of the additives. Another concern is the depletion of the additives due to adsorption on tank and pipe surfaces. The Test Method D5001 is used for assessing fuel lubricity by evaluating the wear scar generated in the ball-on-cylinder lubricity evaluator (BOCLE). Low lubricity fuel will give a larger wear scar diameter. Synthesized hydrocarbons in batch form typically consist of pure hydrocarbons with low lubricity. The final blend of synthetic jet fuel with conventional jet fuel should meet the lubricity requirement of wear scar diameter up to 0.85 mm as specified for aviation turbine fuel under ASTM D1655.

3.6 Additives for alternative jet fuels

Any jet fuel produced must satisfy the stringent standards governing the physicochemical properties to ensure the fuel is fit for purpose (FFP) and does not impact the engine operation. The function of jet fuel is not only to provide the energy source for the engine, but also it plays an important role as hydraulic fluid and coolant for aircraft system. Therefore, the physical properties such as stability, lubricity, fluidity, volatility, noncorrosivity, and cleanliness are important to ensure the proper functioning of the system. In this context, fuel additives are used to maintain the desired properties in the base fuel to provide specific performance properties. Only additives approved by the aviation industry (including the aircraft certifying authority) are permitted to be added to the fuel. The approval process for additives is outlined in ASTM D4054 (ASTM D4054-20b, 2014). Some of the key considerations for jet fuel additives include being effective at below 1% (v/v), chemically stable, not having negative effects on the engine and fuel system components, having the required solubility in fuel, minimum environmental impact and cost-effective. For cases where the use of optional additives is desired, the supplier and purchaser must come to agreement on the types of additive to be used.

The additives approved in ASTM D1655 for fuel performance enhancing and their concentration limits are shown in Table 3.10. The typical additives used to ensure the optimum performance of jet fuel are antioxidant, metal deactivator and fuel system icing inhibitor. Oxidation of jet fuel is known to occur due to the oxidative degradation of hydrocarbon-based fuels during storage. The oxygen molecule reacts with hydrocarbon to form peroxides and hydroperoxides. Antioxidants are additives that are

Table 3.10 Fuel performance enhancing additives for aviation turbine fuels (ASTM D1655).

Additive	Category	Approved concentration
2,6-Ditertiary-butyl phenol	Antioxidant	≤ 24.0 mg/L
2,6-Ditertiary-butyl-4-methyl phenol	Antioxidant	≤ 24.0 mg/L
2,4-Dimethyl-6-tertiary-butyl-phenol	Antioxidant	≤ 24.0 mg/L
75% minimum, 2,6-Ditertiary-butyl phenol plus 25% maximum mixed tertiary and tritertiary butyl-phenols	Antioxidant	≤ 24.0 mg/L
55% minimum 2,4-Dimethyl-6-tertiary-butyl phenol plus 15% minimum 2,6-Ditertiary-butyl-4-methyl phenol, remainder as monomethyl and dimethyl tertiary-butyl phenols	Antioxidant	≤ 24.0 mg/L
72% minimum 2,4-Dimethyl-6-tertiary-butyl phenol plus 28% maximum monomethyl and dimethyl-tertiary-butyl-phenols	Antioxidant	≤ 24.0 mg/L
N,N'-Disalicylidene-1,2-propane diamine	Metal deactivator	≤ 2.0 mg/L (initial doping) ≤ 5.7 mg/L (cumulative concentration after field redoping)
Diethylene glycol monomethyl ether	Fuel system icing inhibitor	0.07%–0.15% by vol

added to improve fuel storage stability and inhibit the formation of peroxides, soluble gums or insoluble particulates produced from oxidation reaction. For synthetic fuels such as FT fuel and HEFA that are heavily hydroprocessed (including hydrotreatment, hydrofine, and hydrocracking), the lubricity of the fuel is typically low as the natural oxidants are removed. To prevent oxidation to the fuel, antioxidants in trace amount can be added. The list of approved phenol-based antioxidants is shown in Table 3.10, limited to maximum 24 mg/L in jet fuel.

Metal ions in jet fuel are considered as contaminants. Transition metals such as iron, nickel, manganese, cobalt, and copper are commonly found in the storage tanks or barrels, which can diffuse to the jet fuel in the distribution and transporting channels. The problem of metal contamination applies to both conventional and alternative jet fuels. The metal ions promote catalysis of oxidation reaction that subsequently lead to poor fuel thermal stability. Furthermore, these metals can react with organic substrates present in fuel such as naphthenates and stearates to form fuel-soluble organometallic complexes that are homogeneous oxidation catalysts. As such, metal deactivators are additives added as chelating agents to form stable water-soluble complexes with ions to prevent fuel degradation. At present, the approved metal deactivator additive for aviation fuel is N,N'-Disalicylidene-1,2-propane diamine. The concentration of the active material during initial doping of the fuel should not exceed 2.0 mg/L, while the cumulative addition of the additive is limited to a maximum of 5.7 mg/L when redoping the fuel.

The additives that are approved under ASTM D1655 with the aim to improve fuel handling and maintenance are shown in Table 3.11. They are mainly comprising the categories of lubricity improver, static dissipator, biocids, and electrical conductivity improver. Due to the need to store and distribute jet fuel, the exposure to the storage tank and pipelines subject the fuel to the risk of contamination. Corrosion inhibitors are added to the fuel to prevent contamination resulting from rust and corrosion caused by water and oxygen. It was later discovered that the corrosion inhibitor additives have the side benefit of improving fuel lubricity. Jet fuel needs to maintain a

Table 3.11 Additives approved for fuel handling and maintenance under ASTM D1655.

Additives	Category	Approved concentration
AvGuard SDA	Electrical conductivity improver	$\leq$ 3–5 mg/L
Stadis 450	Electrical conductivity improver	$\leq$ 3–5 mg/L
Tracer A (LDTA-A)	Leak detection	$\leq$ 1 mg/kg
HiTEC 580	Corrosion inhibitor/lubricity improver	$\leq$ 23 mg/L
Innospec DCI-4A	Corrosion inhibitor/lubricity improver	$\leq$ 23 mg/L
Nalco 5403	Corrosion inhibitor/lubricity improver	$\leq$ 23 mg/L
Biobor JF	Biocide	AMM*
Kathon FP1.5	Biocide	AMM*

*, Refer to manufacturer's Aircraft Maintenance Manuals.

certain level of lubricity as some components in the fuel system, such as pumps, rely on the fuel to lubricate moving parts. The current approved additives that function as corrosion inhibitor and lubricity improver are HiTEC 580, Innospec DCI-4A, and Nalco 5403 with a maximum permitted limit of 23 mg/L. An overdose in the treat rate could adversely impact the fuel filtration system and fuel water separation characteristics, owing to their polar nature and adsorption onto the metal surface. The military Qualified Products List (QPL-25017) states that a minimum effective concentration (MEC) must be maintained for the use of such additives, where HiTEC 580, Octel DCI-4A, and Nalco 5403 are required to have an MEC of 15, 9, and 12 g/m^3, respectively. Hydroprocessing of fuels removes trace components that provide the fuel with natural lubricating properties; hence, lubricity improvers need to be added. It has been pointed out that lubricity improvers are typically long-chain alkylphenol ethoxylates, carboxylic acids, or esters, such as linoleic acid derivatives (i.e., dilinoleic acid) (Black and Hardy, 1989).

Some fuel-handling procedures such as high-speed pumping and microfiltration can create charge separation at interfaces, which is capable of producing high voltage gradients inside fuel storage areas and increasing the risk of fire and explosion. Thus, static dissipator additive is widely used in jet fuel to eliminate the risk. Static dissipators are added to improve the naturally poor conductivity of jet fuel. Jet fuel has very low natural electrical conductivity of less than 5 pS/m, but the requirement in ASTM D1655 for the electric conductivity in jet fuel is 50–600 pS/m. As of now, Stadis 450 is the only static dissipator additive (SDA) approved for aviation turbine jet fuel. The concentration on first doping of the fuel is 3.0 mg/L, with a maximum concentration of 5.0 mg/L permitted.

Jet fuel produced in the refinery is sterilized due to high temperature, yet microorganisms such as bacteria and fungi can enter the fuel as soon as the fuel is in contact with air and water. Aviation jet fuel contains hydrocarbons, nitrogen, sulfur, phosphorous, oxygenated organic compounds, organometallic species, and other metal salts, which provides the essential nutrients for microbes to grow. The presence of microbes will accelerate fuel degradation and clog fuel filters. Biocides are additives that can be added to prevent the growth of microorganisms. Another effective method of retarding microbial activities is by controlling the water content in the fuel, since the lack of water will prohibit the growth of microorganisms.

3.7 Jet fuel certification process

Development of new alternative jet fuel must be certified to ensure the compatibility of the synthetic fuel with conventional jet fuel for safety reasons. The ASTM International Aviation Fuel Subcommittee (subcommittee J) was established to facilitate the development and deployment of alternative jet fuels. Two standards related to the specification criteria of alternative jet fuels have been issued by the subcommittee, which are ASTM D7566—*Standard Specification for Aviation Turbine Fuel Containing Synthesized Hydrocarbons*, and ASTM D4054—*Standard Practice for Qualification and Approval of New Aviation Turbine Fuels and Fuel Additives*. An alternative aviation fuel meeting the property and composition standards of ASTM D7566 can be considered as a "drop-in" replacement for the D1655 jet fuel, which can be integrated into the existing infrastructure designed to support the conventional jet fuel. The ASTM D4054 was developed to provide the guidance regarding testing and property targets necessary to evaluate a candidate alternative jet fuel.

Fig. 3.1 shows the process flow for testing fuel and additive as per ASTM D4054 (ASTM D4054-20b, 2014). The test program comprises specification properties (Tier 1), FFP properties (Tier 2), component or rig test (Tier 3), and engine test (Tier 4). Within each tier, there are several elements to be considered, as detailed in Table 3.12. Fuel producers will need to produce the Phase 1 research report based on the outcome of Tier 1 and 2 test programs. Next, the original engine manufacturer (OEM) will

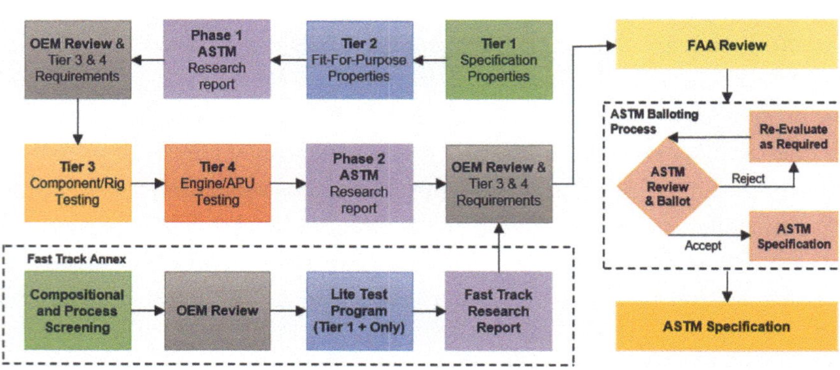

Figure 3.1 Overview of the testing protocols with four-tier programs for new aviation turbine fuel and additives as specified under ASTM D4054, *Standard Practice for Evaluation of New Aviation Turbine Fuels and Fuel Additives*. The Fast Track Annex for qualification and approval of new aviation fuel is highlighted in the box.

Table 3.12 Detail tests involved in each subtest program.

Tier 1: Fuel specification properties
• Relating to engine safety, performance, and durability
Tier 2: Fit for purpose properties
• **Chemistry**
- Hydrocarbon chemistry (carbon number, type and distribution)
- Trace materials
• **Bulk physical and performance properties**
- boiling point distribution, vapor pressure, thermal stability breakpoint
- Lubricity, response to lubricity improver, viscosity
- Specific heat, thermal conductivity
- Density, surface tension, bulk modulus
- Water solubility, air solubility (O_2/N_2)
• **Electrical properties**
- Dielectric constant, electrical conductivity and response to static dissipator
• **Ground handling/safety**
- Effect on clay filtration, filtration (coalescers and monitors)
- Storage stability (peroxides, potential gum), toxicity
- Flammability limits, autoignition temperature, hot surface ignition temperature
• **Compatibility**
- With other approved additives and fuels
- With engine and airframe seals, coatings and metallics
Tier 3: Component tests
• **Turbine hot section**
- Oxidative or corrosive attack on turbine blade metallurgy and coating
• **Fuel system**
- APU cold filter, fuel control, fuel pump, fuel nozzle
• **Combustor rig tests**
- Cold ignition (sea level to 10,000 feet)
- Lean blowout
- Aerial restarting
- Turbine inlet-temperature distribution
- Combustor efficiency
- Flow path carboning/plating
- Emissions
- Auxiliary power unit altitude starting
Tier 4: Engine endurance test
• Sea level endurance/durability
• Sea level performance and operability
• Altitude performance and operability
• Emissions

review the report and propose the requirements for Tier 3 and Tier 4 test programs. It is unlikely that all tests are to be performed. Hence, the OEM will provide the guidance on the type of tests to be run. Based on the test results, the fuel producer will revise the Phase 2 research report. The test results from the test program are then submitted to the OEM and flight authority for considerations. At this stage, additional information or tests may be requested. Even after a new fuel or additive passes the approval stage, a control test known as Controlled Service Introduction (CSI) is required to evaluate its long-term impact on the maintenance of the engine. The fuel or additive that passes the assessment will be included in the OEM specification or service bulletin. For the candidate fuel/additive to be included in the ASTM standards, the changes proposed by the OEM will need to be reviewed through the ASTM balloting process, which is a rigorous process that may require several iterations. Up until the time of writing, seven synthetic jet fuels have been approved in ASTM D7566 as shown in Table 3.5, while other production pathways are under evaluation to be included in the ASTM D7566 standard (DOE, 2020).

The ASTM D4054 was updated with a Fast Track Annex in early 2020 to expedite the approval of new jet fuel by reducing the set of property and compositional tests for evaluating new alternative jet fuels. The Fast Track Annex is limited to sustainable aviation fuel with conventional hydrocarbon fuel. Candidate fuel that meets the compositional and performance requirements of conventional jet fuel is allowed to blend up to 10% with Jet A or Jet A-1. The distillation point distribution of the blend must be similar to the slope of typical jet fuel, while the cycloparaffin and aromatic compositions must be less than 30 and 20 wt%, respectively. The Fast Track Annex for approving new aviation fuel with a limit of 10% blend is shown in Fig. 3.1.

Among the many local projects and initiatives to develop sustainable aviation fuel (ICAO, 2020), CAAFI is a coalition of stakeholders that seek to explore and facilitate the development and deployment of drop-in sustainable aviation fuel (CAAFI, 2020a). They have proposed a tool, known as the fuel readiness level (FRL) (CAAFI, 2016), which can be used to classify and track the progress on research, certification, and demonstration activities at difference milestones, as shown in Table 3.13. The tool provides a nine-level roadmap for fuel technology and certification. The FRL level of 1–5 indicates the progress of jet fuel development is at R&D stage. FRL 1 and FRL 2 focus on the fundamentals of the feedstock and process. At the proof-of-concept stage (FRL 3), the manufacturer needs to provide 500 mL of

Table 3.13 Fuel readiness level (FRL) developed by CAAFI.

FRL	Description	Milestone
Technology Development Phase		
1	Basic principles observed and reported	• Feedstock/process principles identified
2	Technology concept formulated	• Feedstock/complete process identified
3	Proof of concept	• Lab-scale fuel sample production feedstock • Energy balance analysis executed for initial environmental assessment • Basic fuel properties validated • Fuel quantity: 0.13 US gallons (500 mL)
4	Preliminary technical	• System performance and integration studies entry criteria/specification properties evaluated (MSDS/D1655/MIL 83133) • Fuel quantity: 10 US gallons (37.85 L)
5	Process validation	• Sequential scaling from laboratory to pilot plant • Fuel quantity: 80–225,000 US gallons (302.8–851,718 L)
Qualification Phase		
6	Full-scale technical evaluation	• Fitness, fuel properties, rig testing, and engine testing using ASTM-approved protocols • Fuel quantity: 80–225,000 US gallons (302.8–851,718 L)
7	Fuel approval	• Fuel class/type listed in international fuel standards
Deployment phase		
8	Commercialization validated	• Business model validated for production airline/military purchase agreements—facility-specific GHG assessment conducted to internationally accepted independent methodology
9	Production capability established	• Full-scale plant operational

lab-scale fuel samples for analysis, preliminary environmental assessment, and validation of basic fuel characteristics. Production of the fuel at pilot plant scale is said to reach FRL 5, making it ready to proceed to the certification stages (FRL 6–7). Under FRL 6, the manufacturer must be able to provide

80 to 225,000 US gallons of fuel for a full technical evaluation involving rig and engine tests. The alternative fuel that has been listed in the international standards is deemed to reach FRL7. Fuel production at Business and Economics levels are FRL 8–9, of which the fuel is said to reach FRL9 once full production capacity has been established. Preliminary assessments of environmental impacts are nested under FRL 3 and FRL 8.

Apart from the FRL, the feedstock readiness level (FSRL) tool is developed to track the development and availability of feedstock needed for the production of alternative jet fuels, which covers a broader aspect of production, market, program support, and regulatory compliance. These tools are designed to be used in conjunction with the FRL tool to assess the ready states of the fuel conversion and processing technology, fuel testing, and certification. The guides for commercialization of alternative jet fuel and environmental progression are also provided.

3.8 Summary

Commercial aviation jet fuel has to fulfill the stringent requirements as specified by the internationally agreed standard to ensure the safety of flight operation. Newly developed synthetic jet fuel needs to undergo stringent certification process to ensure the physicochemical properties of the fuel are met. Current practice employs the blending concept without modifying the engine; hence the developed synthetic jet fuel must be fit for purpose and does not impact the engine operation. At present, the ASTM D7566 is the widely referred standard for nonconventional jet fuel derived from new sources. The standard currently lists the property specifications of the approved alternative jet fuels in the Annexes of ASTM D7566, along with the limits allowed for blending with conventional jet fuel. In spite of the variation of batch properties of alternative jet fuel due to the use of different feedstocks and processing pathways, the jet fuel standard requires the final properties of the blended fuel to meet the requirements of conventional jet fuel. Fuel additives can be added to enhance the fuel performance of the blended fuel, such as to increase the oxidation resistance, inhibit the tendency for ice to form in the fuel system, deactivate the metal content in the fuel, and improve the electrical conductivity, among others. The process of certification for alternative jet fuel is discussed, based on the guidelines as specified in the ASTM D4054 standard. The test programs that an alternative jet fuel candidate needs to pass include the specification properties, FFP properties, component or rig test and engine test. The test reports are reviewed by the original equipment manufacturer and flight authority, before proceeding to the ASTM balloting process to be included in the ASTM standard. The Fast Track Annex was recently introduced in ASTM D4054 to expedite the qualification and approval of new alternative jet fuel with 10% blend limit. At present, there are seven approved alternative jet fuels from nonconventional sources as listed in the ASTM D7566 standard. More emerging synthetic jet fuels derived from different feedstocks and production pathways are expected to be certified in the future.

References

ASTM D1655-19a, 2019. ASTM International, Standard Specification for Aviation Turbine Fuels.

ASTM D4054-20b, 2020. ASTM International, Standard Practice for Qualification and Approval of New Aviation Turbine Fuels and Fuel Additives.

ASTM D6469-20, 2020. ASTM International, Standard Guide for Microbial Contamination in Fuels and Fuel Systems.

ASTM D6615-15a, 2019. Standard Specification for Jet B Wide-Cut Aviation Turbine Fuel. ASTM International.

ASTM D7566-19b, 2019. ASTM International, Standard Specification for Aviation Turbine Fuel Containing Synthesized Hydrocarbons.

Black, B.H., Hardy, D.R., 1989. The Lubricity Properties of Jet Fuel as a Function of Composition. Part 2. Application of Analysis Method.

Brooks, K.P., Snowden-Swan, L.J., Jones, S.B., Butcher, M.G., Lee, G.-S.J., Anderson, D.M., et al., 2016. Biofuels for Aviation: Chapter 6: Low-Carbon Aviation Fuel through the Alcohol to Jet Pathway. Elsevier.

CAAFI, 2016. Fuel Readiness Level. CAAFI. http://caafi.org/information/pdf/FRL_CAAFI_Jan_2010_V16.pdf.

CAAFI, 2020a. Commercial Aviation Alternative Fuels Initiative (CAAFI). http://www.caafi.org/.

CAAFI, 2020b. Two New Alternative Jet Fuel Production Pathways Approved. www.caafi.org/news/NewsItem.aspx?id=10502.

Chen, K., Liu, H., 2013. The impacts of aromatic contents in aviation jet fuel on the volume swell of the aircraft fuel tank sealants. SAE Int. J. Aerospace 6, 350–354 (2013-01-9001).

CSG Network, 2013. Aviation Turbine Fuel (Jet Fuel). CSG Network. www.csgnetwork.com/jetfuel.html.

DOE, 2020. Sustainable Aviation Fuel – Review of Technical Pathways. U.S. Department of Energy.

GB 6537-2018, 2018. No.3 Jet Fuel, State Administration for Industry and Commerce, People's Republic of China.

ICAO, 2020. Committee on Aviation Environmental Protection (CAEP). https://www.icao.int/ENVIRONMENTAL-PROTECTION/Pages/CAEP.aspx.

Joint Inspection Group, 2019. Bulletin 125, Aviation Fuel Quality Requirements for Jointly Operated Systems (AFQRJOS): Issue 31, November 2019, p. 1–9. Joint Inspection Group. Accessed at. www.jigonline.com.

MOD, 2019. Ministry of Defense, Defense Standard 91-091 Issue 11, Turbine Fuel, Kerosene Type, Jet A-1; Nato Code: F-35; Joint Service Designation. AVTUR.

TR CU 013/2011, 2011. Technical Regulations of the Customs Union, TR CU 013/2011, on Requirements to Automobile and Aviation Gasoline, Diesel and Marine Fuel, Jet Fuel and Heating Oil. Global Expert Group. http://globexpert.ru/en/.

US Dept of Energy, 2017. Alternative Aviation Fuels: Overview of Challenges, Opportunities, and Next Steps Alternative Aviation Fuels: Overview of Challenges, Opportunities, and Next Steps.

CHAPTER 4

Combustion performance of biojet fuels

4.1 Introduction

The growing demand for air traveling inevitably leads to more emissions as greater amount of jet fuels will be consumed. In the aviation turbine engine, the burning of conventional jet fuels typically produces carbon monoxide (CO), carbon dioxide (CO_2), water vapor, nitrogen oxides (NO_X), sulfur oxides (SO_X), unburned hydrocarbons (UHCs), and particulate matters (PMs), similar to those of ground transportations. These emissions are considered to be local air quality pollutants and greenhouse gases when emitted near the ground or at high altitude. Even though the aviation industry contributes to only 2%–3% of the global total greenhouse gases and 3% of the total NO_x, it is projected that the emissions will be significant in the near future, given the rapid growth of the industry (EASA, 2019). To achieve the goal negative carbon growth set by the International Civil Aviation Organization (ICAO), alternative jet fuels derived from nonconventional sources have been identified as one of the main strategies to achieve a sustainable and green aviation (ICAO, 2018). At present, only certified biojet fuel are allowed to be used in civilian aircraft. The ASTM D7566 standard established in 2009 specifies the property specifications of alternative jet fuels produced from nonconventional sources, placing emphasis on the "drop-in" characteristic so that the fuel is interchangeable with conventional jet fuel; hence, no modification to the engine or fuel distribution system is required. The certification process for alternative jet fuel, as described in the ASTM D4054 standard, requires a series of rigorous test programs and detailed reviews for fuel property certification, combustion, and flight tests. The whole process involves the participation of various stakeholders, including the fuel producer, original equipment manufacturers (OEMs), flight authority, and experts from the aviation community, in order for the developed alternative jet fuel to be accepted by

Biojet Fuel in Aviation Applications
ISBN 978-0-12-822854-8
https://doi.org/10.1016/B978-0-12-822854-8.00002-0

ASTM standard and subsequently included as one of the annexes in ASTM D7566. This chapter focuses on the combustion performance and characteristics of biojet fuel, specifically those related to the test program as per required by ASTM D4054 for jet fuel certification. Some of the fundamental combustion characteristics of alternative jet fuels conducted in research institution that are not related to jet fuel certification are also discussed, to provide a more comprehensive understanding on the combustion chemistry and properties of alternative jet fuel. It is to note that emphasis of the fuels is placed on the certified alternative jet fuel approved by ASTM D7566, while other noncertified alternative jet fuels such as liquid hydrogen, syngas, and biogas are beyond the scope of this chapter.

4.2 Principles of aircraft emissions

Aircraft operations in the vicinity of airports have raised concerns about local air quality and health-related impacts. The emissions produced from the gas turbine combustor largely depend on the fuel/air ratio, flame temperature, power setting, and jet fuel composition. In general, the main products emitted from the jet engines are CO_2 and H_2O, along with emissions in lesser quantity such as CO, NO_x, SO_x, UHCs, and PMs. Fig. 4.1 shows the estimated emissions produced from an aircraft with 150 passengers in an hour of flight. For every kilogram of jet kerosene burned, approximately 3.15 kg of carbon dioxide and 11.1 g of NO_x are produced. Other pollutants such as soot, CO, SO_2, and UHC are less than 0.1 g per kg of jet fuel burned. The formation of the emissions product is closely

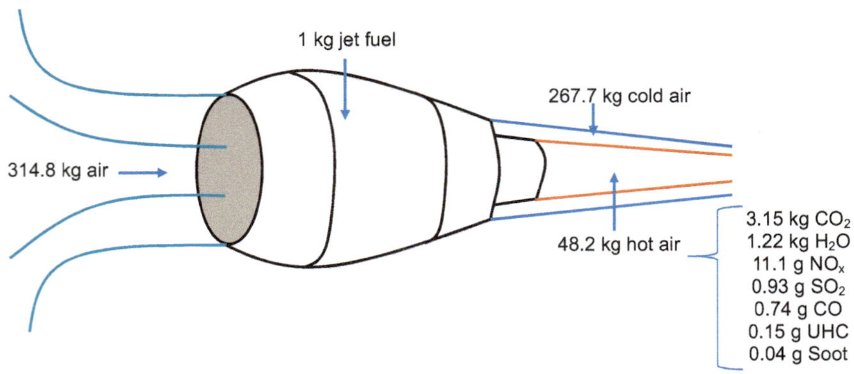

Figure 4.1 Amount of emissions produced from aviation turbine engine from the combustion of 1 kg of kerosene (Braun-Unkhoff et al., 2017).

related to the way the fuel is combusted. The generation of UHC, NO_x, and CO is strongly dependent on the combustion parameters in the combustor, such as temperature, pressure, turbulence level, and residence time. The generation of CO_2 and H_2O is proportional to the fuel combusted. Higher H/C ratio in the jet fuel will produce more H_2O. Some studies have shown that the water vapor produced will impact the formation of contrail, which can have greater impact as a heat trapping agent than the CO_2. Sulfur dioxide emission is proportional to the amount of sulfur in the fuel but is considered to be low due to the regulated amount as specified in the jet fuel standard.

The complete combustion of any hydrocarbon fuel will produce stable end products of carbon dioxide and water vapor. Such ideal fuel/air mixture with sufficient oxygen for complete combustion is known as a stoichiometric mixture. Thus, a stoichiometric fuel/air ratio in gas turbine term refers to the portion of fuel and air required for complete combustion. However, the use of stoichiometric mixture for operation is rarely adopted in actual gas turbine engine, as the flame temperature for a stoichiometric mixture is too high and will result in high portion of NO_x emissions. Very often, gas turbine operates based on lean combustion principle, in which more air is available than stoichiometric quantity. Such fuel-lean burning condition results in lower flame temperature and thus lowers NO_x. Furthermore, lean burning operation provides the conditions to oxidize unburned hydrocarbons as well as CO. For the aviation gas turbine, air entering the engine is partly used for combustion, while the excess air is used to cool the combustion products to within the turbine material's metallurgical limits. Additional air is introduced into the exhaust to tailor the temperature profile and to reduce the overall temperature of the exhaust gases entering the turbine.

Apart from the combustion aspect, the emissions are also strongly dependent on the flight operation, i.e., power settings, as shown in Fig. 4.2. The power settings used during the LTO cycle (landing-take off), taxi, approach, climb, and takeoff correspond to 7%, 30%, 85%, and 100% thrust, respectively. During aircraft idling and taxiing, the CO and UHC emissions tend to be high. Increasing the thrust during takeoff and climbing will elevate the flame temperature, thereby increasing the formation of NO_x and particulate matter, while the increase in combustion efficiency will reduce the emissions of CO and UHC. The emissions of NO_x, CO, UHC, and smoke produced during LTO cycle from aircraft engines are regulated by the ICAO.

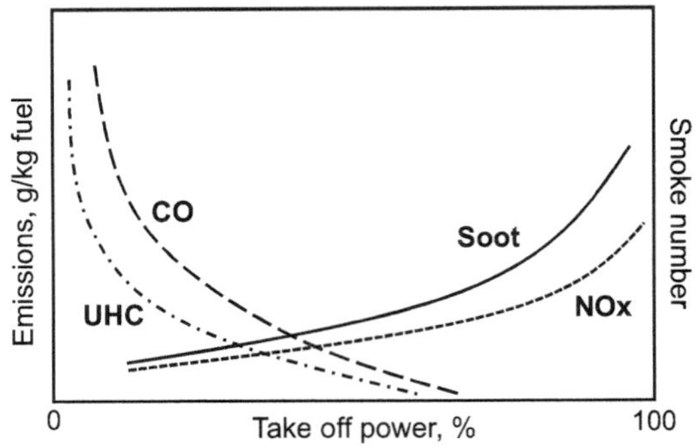

Figure 4.2 Emissions characteristics in an aircraft engine as a function of engine thrust.

4.2.1 Mechanism of aircraft pollutant formations

The oxides of nitrogen produced from the combustion in aviation turbine engine can lead to the formation of PMs, smog, and acid rain, which are detrimental to the environment and human health. The main routes responsible for the formation of NO have been identified as thermal NO, prompt NO, fuel NO, and nitrous oxide route. Thermal NO refers to the formation of NO driven by the high flame temperature, via the main reactions as prescribed in the Zel'dovich mechanism (Zel'dovich, 1946): $O + N_2 \Leftrightarrow NO + N$, $N + O_2 \Leftrightarrow NO + O$, and $N + OH \Leftrightarrow NO + H$. Thermal NO is found to be strongly dependent on the combustion temperature. At temperature around or above 1800K, which is the typical temperature for lean and stoichiometric burning in gas turbine combustor, thermal NO formation can be dominant. The prompt NO mechanism is typically referred as the Fenimore mechanism (Fenimore, 1971), which describes the formation of NO involving a rapid reaction of hydrocarbon radicals such as CH, CH_2, C_2 or C_2H with nitrogen that leads to the formation of hydrocyanic acid (HCN) and a nitrogen atom. The single nitrogen atom then reacts with the NO and O_2 to form NO, via the reactions in the Zel'dovich thermal NO mechanism. Bachmaier et al. (1973) showed that maximum prompt NO is attained at the fuel-rich region.

The fuel NO refers to the formation of NO due to the inherent presence of nitrogen compound in the fuel. The fuel NO production depends weakly on the local temperature as the reactions involved require low activation energy. The nitrogen compounds in the fuel provide the precursors necessary for the formation of fuel NO, such as HN_3, NH_2, NH,

HCN, and CN. It has been shown that the aromatic rings in fuel could contribute to the formation of fuel NO during the combustion process (Levy, 1982).

The formation of NO via the nitrous oxide mechanism requires the formation of N_2O species via the reaction $N_2 + O + M \Leftrightarrow N_2O + M$, after which the N_2O produced reacts with O atoms to form NO, via the reaction $N_2O + O \Leftrightarrow NO + NO$. These reactions become dominant under fuel-lean condition, during which the temperature is around 1500K. Such low temperature prohibits the formation of thermal NO and prompt NO due to the lack of CH radicals needed for the initiation of reaction (Kuo, 2005). In aviation turbine engine, thermal NO is expected to be the main NO formation route. The increase of the engine thrust results in the increase of flame temperature; therefore, the NO_x emission is increased.

The formation of CO is inversely related to the NO in gas turbine combustor. The CO emissions are seen to peak at low aircraft engine thrust, while the NO emissions are correspondingly low. The formation of CO at high temperature oxidation involves the formation of HCO via the oxidation of methyl radical, where $CH_3 + O_2 \Leftrightarrow HCO + H_2O$. The HCO produced contributes to the formation of CO via the reactions $HCO + OH \Leftrightarrow CO + H_2O$ and $HCO + M \Leftrightarrow H + CO + M$. The emissions of CO become less important during cruising as the emissions of CO become insignificant under high engine thrust.

4.2.2 Emission index calculation

Aircraft emissions are typically expressed in the form of emission index, which is a method of normalizing the emission species mass flow with fuel mass flow to have a better comparability of the emissions for different power settings and fuel mass flow. The unit for emission index is in g/kg fuel (Riebl et al., 2017). For flight emission measurement, the emission index of nitrogen oxides (NO_x) can be calculated from the emission index of carbon dioxide (CO_2) and the carbon content in the fuel. Eq. (4.1) shows an example of an emission index calculated from measurement results.

$$EI_n = \frac{E_n}{E_{CO_2}} \times \frac{M_n}{M_{CO_2}} \times EI_{CO_2} \tag{4.1}$$

Aircraft engine certification values are in mass of pollutant per unit mass of thrust, which is the pure emission index referenced on the ICAO thrust levels. The ICAO published the boundary values in Annex 16 Volume II which limits the emissions of NO_x, CO, and hydrocarbons. The limits imposed are dependent on the maximum rated thrust of the engine at sea-level static conditions without water injection, engine age, and engine pressure ratio (ICAO, 2017). The ICAO Aircraft Engine Emissions Databank contains information on exhaust emissions of production aircraft engines, submitted by engine manufacturers to the certification authority as part of the certification process.

4.3 Component or rig test for alternative jet fuel

The approval of alternative jet fuel for application in aircraft requires the alternative jet fuel to meet the fuel specification properties. The fuel needs to be proven as "fit for purpose" by meeting the specified properties and characteristics including the bulk thermodynamic and transport properties, compatibility with existing fuels, additives, handling and storage, and combustion process. The ASTM D4054 defines the approval process for alternative fuels as commercial fuels (refer to Chapter 3 for detailed description). Within the ASTM D4054, the test program of Tier 3, also known as "Component or Rig Test," requires the fuel candidate to undergo a series of tests in different combustor components to assess the fuel performance and the impact on other components. The combustor rig testing is defined in terms of the desired engine performance to be evaluated, such as cold ignition at sea level and high altitude (10,000 ft), flame extinction, lean blowout (LBO), aerial restarting after a flame-out event, turbine inlet temperature distribution, combustor efficiency, emissions, and auxiliary power unit altitude starting. The ASTM D4054 does not specify any standardized criteria for the test rigs and testing conditions, but rather, the list of tests provided is merely a guide. Not all tests will be conducted, but the OEM will provide advice on the tests to be conducted, depending on the availability of testing components and facilities as appropriate.

The combustion performance of alternative jet fuels that are considered most important from the standpoint of fuel effects on safety and operability includes cold ignition limits, LBO limits, and altitude relight limits. These combustion performance issues have been termed as figures of merit (FOMs).The secondary FOMs include temperature field (including flame structure, pattern and profile factors, radiation), combustion efficiency and emissions (including CO, HC, NO_x, and smoke), and combustor coking (Edwards et al., 2010). The following sections discuss some of the component tests and results of alternative jet fuels conducted at research institutions or OEM facilities.

4.3.1 Spray atomization

Fuel atomization is the process that occurs before evaporation and mixing takes place. It is directly related to the fuel combustion efficiency and emissions performance. An effective liquid fuel combustion in a combustor requires a consistent nozzle spray pattern and fine jet fuel breakup. The spray quality of alternative jet fuel needs to be thoroughly investigated to ensure the breakup of fuel into fine droplets, so that effective droplet vaporization can take place to ensure a good mixing with air and effective combustion. The atomization process is known to strongly relate to the fuel physical properties, such as fuel viscosity, density, volatility, and surface tension. Poor atomization can lead to detrimental effect to the ignition and altitude relight, LBO, and emissions performance. Hence, the atomization tests at pressure are part of the test program to provide accurate data for the combustion models, as current fundamental models for atomization are not adequate (Edwards et al., 2010). Table 4.1 shows the type of alternative jet fuels that have been tested in different test spray facilities.

In the test program conducted by OEM to evaluate the spray characteristics and combustion performance of 50/50 blend of alcohol-to-jet (ATJ) fuel, FAA, (2016) a pilot injector was used to test the spray quality at ambient temperature with ambient and cold fuels (−40°C) using laser diagnostic methods. The test results show that the ATJ fuel blend has similar mean droplet sizes as baseline JP-8 fuel at four different injection pressures vary between 100 and 400 psi. Measurement of the spray cone angle, radial, and circumferential droplet flow distribution via patternation test indicates the similarity of cone angles between the tested fuels. The data generated formed part of the research report submitted to the American Society for Testing and Materials (ASTM) for the evaluation and approval process of the ATJ-SKA fuel. Another ATJ fuel spray characteristics test was conducted by Bokhart et al. (2017) as part of the National Jet Fuel Combustion Program (NJFCP). The spray measurement was conducted using a pressure swirl nozzle with the pressure differentials of 172−517 kPa across the nozzle, while maintaining the ambient condition of the pressure vessel at 204 kPa and 394K. The droplet Sauter mean diameter (SMD) distribution shows that ATJ fuel droplets are smaller at the locations that shift radially outward from the center of the spray, and the spray cone angle was also found to vary slightly between ATJ and Jet A-1. The difference could be due to the variation in fuel physical properties, as the ATJ fuel has lower surface tension and density

Table 4.1 Types of alternative jet fuels tested in nonreacting spray facilities.

Year	Tested fuel	Spray injector conditions	References
2012	Coal-derived FT fuel	• Pressure-swirl injector • P_{inj} = 0.1−0.9 MPa, T_{fuel} = 10°C	Lin et al. (2012)
2014	Gas-derived FT fuel	• Pressure swirl injector • P_{inj} = 0.3−0.9 MPa	Kannaiyan and Sadr (2014a,b)
2015	Camelina HEFA (UOP)	• Simplex swirl injector • P_{inj} = up to 160 psi	Sivakumar et al. (2015)
2016	20/80 and 70/30 Jatropha HEFA/Jet A-1	• Hollow cone pressure swirl injector • P_{inj} = 100−300 kPa	Sivakumar et al. (2016)
2017	FT fuel (Yitai Petrochemical)	• Pressure swirl injector • P_{inj} = 0.05−0.85 MPa, T_{fuel} = 25°C	Zhao et al. (2017)
2016	ATJ (Swedish Biofuel)	• Pilot pressure injector • T_{fuel} = 25 and −40°C, P_{inj} = 100−400 psi	FAA (2016)
2020	GTL FT fuel	• Pressure swirl injector • P_{inj} = 300 kPa, T_{fuel} = 288K • P_{amb} = 100−1300 kPa, T_{amb} = 300−375K	Kannaiyan and Sadr (2020)
2017	ATJ (Gevo)	• Pressure swirl injector • P_{inj} = 1.7−5.2 bar, T_{fuel} = 322K • T_{amb} = 394K	Bokhart et al. (2017)

P_{amb}, ambient gas pressure; P_{inj}, injection pressure; T_{amb}, ambient gas temperature; T_{fuel}, fuel temperature.

compared with Jet A-1. The droplet diameter is also sensitive to the pressure differential across the injector swirler, as an increase in the pressure drop will result in the decrease in droplet diameter.

There have been a number of spray studies conducted using FT-based fuels. Lin et al. (2012) evaluated the spray characteristics of FT-based jet fuel derived from coal. The test was conducted using a pressure swirl atomizer under atmospheric condition and at the operating temperature of 10°C. The injection pressure of the fuel was varied between 0.1 and 0.9 MPa. The generated spray for FT fuel at different tested pressure shows uniform and similar droplet size distribution as conventional jet fuel (RP-3), except at low pressure of 0.1 MPa where FT fuel droplets are slightly smaller. In another separate atomization test using FT fuel at ambient temperature of 25°C, the SMD of the synthetic jet fuel was found to be consistently smaller and more uniform in size than those of RP-3 jet fuel for injection pressures below 0.8 MPa. The spray cone angle for the FT fuel was also found to be larger, indicating a better spray quality was achieved for the FT fuels. The lower SMD for FT fuel was attributed to the lower surface tension of the fuel. Fig. 4.3 shows the effect of injection pressure on the spray cone angle for FT fuels and conventional RP-3 jet fuels (Zhao et al., 2017).

Kannaiyan and Sadr (2014a,b) compared the spray characteristics of GTL synthetic jet fuels using a pilot-scale pressure swirl nozzle. The GTL and Jet A-1 fuels were shown to have similar global spray parameters, such as the effective spray cone angle and global SMD, while the differences in local SMD for radial profiles were minor at atmospheric condition. The difference was more prominent at high injection pressure due to the influence of inertial force and surface tension of the fuels. The lower kinematic viscosity and surface tension of the GTL fuel resulted in faster disintegration and dispersion of the droplets in the core region of the spray as compared with the Jet A-1 fuel (Kannaiyan and Sadr, 2014b). At elevated ambient conditions, the liquid sheet breakup for GTL was slightly longer than Jet A-1 (Kannaiyan and Sadr, 2020). Such discrepancy is attributed to the differences in fuel evaporation characteristics, which is more pronounced at elevated ambient conditions compared with atmospheric

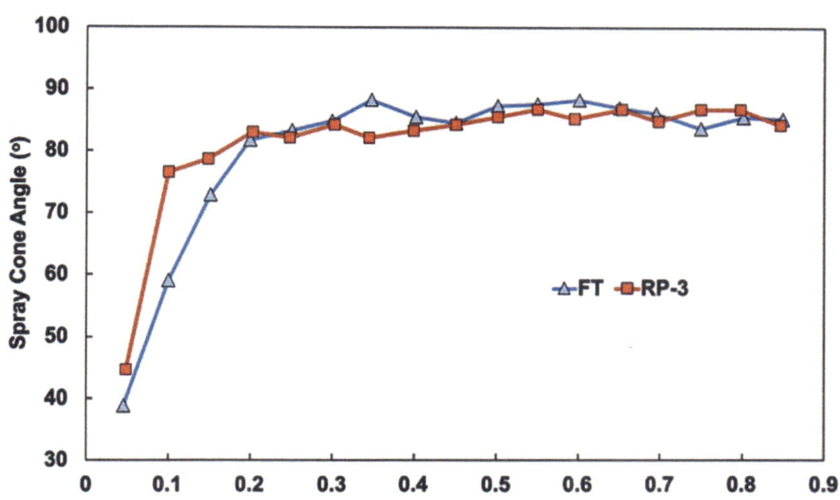

Figure 4.3 Effects of oil supply pressure on spray cone angles (Zhao et al., 2017).

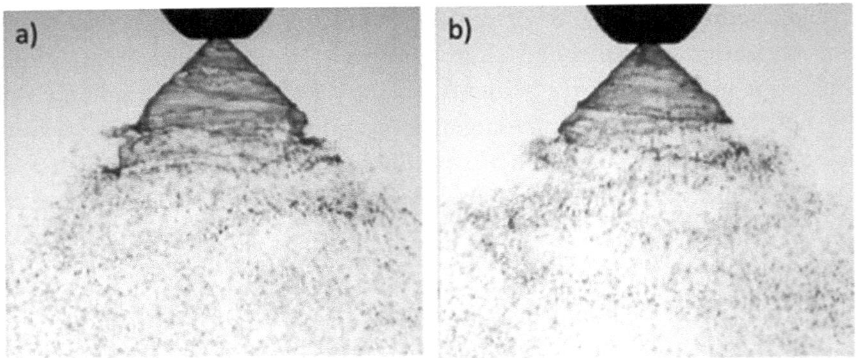

Figure 4.4 Images of hollow cone fuel sprays discharging from a simplex swirl atomizer at 300 kPa for (left) 70/30 Jatropha HRJ/Jet A-1 and (right) Jet A-1 (Sivakumar et al., 2016).

condition. It was also reported that the effect of ambient gas pressure on the spray characteristics is more significant when compared with that of ambient gas temperature.

There have been some studies on the spray characteristics of hydrogenated fatty acid-based biojet fuels. Sivakumar et al. (2015) studied the atomization characteristics of camelina biojet fuel using a simplex swirl atomizer. The measured breakup length of biojet fuel sheet and wavelength discharged from the atomizer agrees well with the prediction using film breakup model. The measured SMD for biojet fuel concurs with the empirical correlations derived from conventional hydrocarbon fuel sprays. They extended the experiments to compare the spray characteristic of jatropha-derived jet fuel with Jet A-1. Only minor differences were recorded in the spray characteristics, for instance, a marginal decrease in SMD along the spray axis was observed for jatropha biojet fuel compared with Jet A-1 owing to the lower boiling point for the former that facilitates rapid evaporation (Sivakumar et al., 2016). Fig. 4.4 shows the comparison of the hollow cone fuel sprays discharged from a simplex swirl atomizer for biojet fuel blend and Jet A-1.

4.3.2 Ignition

The ignition characteristic of jet fuel is important as it relates to the cold starting and high altitude relighting. The ignitibility of a mixture depends strongly upon the equivalence ratio in the vicinity of the igniter. Other factors that govern the ignition characteristics are fuel volatility, droplet size and distribution, mixing of the fuel and oxidizer, ignition energy, fuel-air temperature, air velocity, and various other parameters (Mosbach et al., 2011). It is to note that the volatility and the droplet size are dependent on the physical properties of the fuel. Although the final blend of the alternative jet fuel with conventional jet fuel has to meet the ASTM D1655 standard specification, the varied chemical properties for alternative jet fuel mean that the combustion chemistry may differ; thus, investigation of the combustion properties including ignition needs to be conducted. Table 4.2 shows the ignition and extinction performances of alternative jet fuels conducted by different groups. Hermann et al. (2005) compared the ignition characteristics of FT-SPK developed from syngas with Jet A-1 fuel. The FT-SPK was found to ignite at slightly fuel-rich conditions than Jet A-1, due to the slightly higher flash point and viscosity for the FT-SPK than Jet A-1. The latter produces finer droplets, which enable faster evaporation and subsequently facilitate ignition. The extinction point was found to be similar for both fuels. The ignition and high altitude relight performance of GTL FT fuels was investigated at simulated high altitude conditions of 25,000–30,000 ft using a subatmospheric sectorial rig (Fyffe et al., 2011). The combustor inlet pressures were set at 6 and 8 psi with the inlet air temperatures of 265 and 278K. The FT fuels tested showed similar ignition performance as the regular Jet A-1. Lower iso-to-normal paraffin ratio fuels were found to exhibit better ignition performance.

From the comparative study of ignition performance using two combustors, it was found that the FT GTL fuel exhibited similar ignition behavior as Jet A-1 fuel in the can-annular combustor equipped with an airblast-type atomizer (Rye and Wilson, 2012). However, the annular combustor fitted with two lean airblast atomizers showed a consistent improved ignition probability for FT fuel compared with Jet A-1, attributable to the sufficient presence of light hydrocarbons in the vicinity of spark kernel to vaporize, thus enabling ignition at leaner condition in the primary zone. Lin et al. (2012) showed that the ignition performance for FT fuel is slightly better than those of RP-3, in particular at low liner pressure drop conditions, which is attributable to the lower viscosity of the fuel and finer atomized droplets that improve droplet vaporization. Burger et al. (2014) compared the ignition and extinction characteristics of FT fuels with Jet A-1. The tests were conducted in a rich-burn, quick-mix, lean-burn (RQL) combustor equipped with an airblast atomizer, under a fixed combustor inlet pressure of 41.4 kPa and 265K. The tests showed that FT fuels have similar ignition characteristic as Jet A-1 fuel, but the lean

Table 4.2 Ignition and extinction performances of alternative jet fuels.

Year	Test component (OEM)	Fuel	Test conditions	References
2005	Single can-type combustor (Volvo Aero)	FT-SPK	• Equipped with air-assisted nozzle • Atmospheric, ambient temperature 20°C • Airflow rates were varied from 10 g/s up until ignition was not possible	Hermann et al. (2005)
2011	Subatmospheric relight sector rig (Rolls Royce)	GTL FT-SPK	• Coaxially staging lean burn fuel injector • Simulated altitude ignition at combustor inlet pressure 6–8 psi, air temperature 265–278K, fuel temperature 288K	Fyffe et al. (2011)
2011	Twin-sector combustors	FT-SPK	• Air pressure 5.9–7.9 psia, air temperature 265–278K • Fuel temperature 288–290K • Fuel/air ratio of 0.08 and 0.055	Mosbach et al. (2011)
2012	Single-cup rectangular combustor	CTL FT-SPK	• Pressure swirl atomizer and counter-rotating swirl cup • Atmospheric inlet pressure (0.1–0.11 MPa) and inlet air temperatures of 275–278K	Lin et al. (2012)
2012	Annular and can-annular combustor	GTL FT	• Inlet air temperature ~310K, fuel inlet temperature 290K, inlet air flow rate (0–0.47 kg/s) • Lean airblast atomizer and airblast atomizer	Rye and Wilson (2012)
2014	RQL combustor	SPK	• Air supplied at 41.4 kPa and 265K, fuel temperature 288K	Burger et al. (2014)
2016	APU combustor rig (Pratt & Whitney AeroPower)	ATJ	• Fuel supplied at ambient and −40°C • Simulate altitude of sea level—41,000 ft	FAA (2016)

extinction behaviors show higher variability with no significant trend observed for low air flow rate. Mosbach et al. (2011) studied the temporal behavior of flame ignition process at high altitude with FT fuel in a twin-sector combustor. It was reported that the temporal radiation emitted after the spark ignition and failed ignition events for FT fuel and Jet A-1 were similar. To ensure a high probability of high altitude relight, a fuel-rich, low velocity region between the ignitor and fuel injector is desirable to ensure the initiation of flame kernel and subsequent flame propagation.

Ignition tests were conducted in an APU combustor rig at three simulated altitudes using 50/50 ATJ-SKA/JP-8 fuel blend supplied at both ambient and cold temperatures (−40°C) (FAA, 2016). During the test, combustor inlet airflow was set to simulate low-speed engine airflow condition. The ignition boundary that separates light off and no-light data was determined. It was reported that the ignition boundary of 50/50 ATJ-SKA/JP-8 blend fuel is comparable with the baseline fuel of JP-8 and Jet A, implying similar ignition capabilities among the fuels tested.

4.3.3 Lean blowout

Alternative fuel developed for aviation purpose must be operationally fit and compatible with existing engine, apart from meeting the fit for purpose requirements. The lean blowout behavior of the fuel is one of the important characteristics that needs to be assessed to ensure the safe flight operation. In a liquid fuel spray flame system, the process governing blowoff can be rather complex, as a number of processes can affect blowoff, including kinetics, atomization, vaporization, mixing, and heat transfer. As part of the alternative jet fuel approval process, LBO test has been included as part of the component test. Burger et al. (2012) investigated the LBO behavior of 26 types of liquid fuels, including Sasol fully synthetic jet fuel, SPK, Jet A-1, and other fuel blends in a heterogenous combustor. The flame stability limits were determined over a range of air flow rates that correspond with liner pressure drops ranging from 1% to 6%, while maintaining the inlet air condition of 310K and 1 bar. The LBO limit was determined by reducing the fuel flow rate while keeping the air flow rate constant until the flame went off. It has been found that density and viscosity exhibited a positive correlation with blowout equivalence ratio. A decrease in the density and viscosity resulted in lower LBO and led to the increase of stability limit of the flames. The fuel physical property is related to the atomization of the fuel, as blowout in rich primary zone combustor is governed by evaporation and mixing. Reaction rates were shown to have a lesser influence on the stability limits. Lin et al. (2012) showed that CTL FT fuel possesses better LBO performance than RP-3 at low liner pressure drop conditions, which could be due to finer droplets for the former that facilitates vaporization and hence extends the flame stability limit. However, the LBO result needs to be interpreted with care, as the result is geometry specific and highly dependent on the operating conditions. Rock et al. (2017) examined the blowoff phenomena of 10 different liquid fuels at two air inlet temperatures, i.e., 450 and 300K. It was concluded that blowoff event is limited by fuel vaporization temperature at 300K, where fuels with higher volatility tend to resist blowoff, but the physics that govern blowoff at higher inlet temperature of 450K remained unclear. Given that a flight operation involves a wide range of conditions, rig component tests should cover the possible range of flight including extreme operating conditions.

4.3.4 Emissions of alternative jet fuels

4.3.4.1 Gaseous emissions

Different test setups have been utilized to assess the emission performance of alternative jet fuels at flight conditions, such as lab-scaled gas turbine combustor, APU unit, sectorial combustor, and full engine test, as shown in Table 4.3. These emission tests are typically designed based on the flight operation including ground idle, landing and takeoff, and cruising conditions, while the emissions of interest include the gaseous emissions and particulate matters. This section discusses the gaseous pollutants emitted from the tests of alternative jet fuel, followed by the emissions of particulate matters in the subsequent section. Earlier jet engine study on the emissions of alternative jet fuel can be traced back to 2007 conducted by Corporan et al. (2007). The FT fuel was tested in a T63 turboshaft engine and swirl-stabilized research combustor at idle and cruise conditions. Analysis of the gaseous emissions of FT fuel and blends shows only minor effects on the CO, CO_2, and NO_x species. Water vapor was shown to increase with FT blend fraction due to the higher H/C ratio. The sulfur-free FT fuel results in a linear decrease in SO_2 with increasing FT concentration in the blend. Lobo et al. (2012) evaluated the emissions of FT synthetic jet fuel in an APU unit at idle and full power conditions. The report showed 5% and 5%–10% reductions in NO_x and CO emissions for the alternative fuels and blend tested, respectively. The UHC was shown to increase by 7% at idle condition for CTL FT relative to Jet A-1.

Table 4.3 Emissions of alternative jet fuels produced from different engine conditions.

Year	Aircraft engine/combustor	Fuel	Operating conditions	References
2007	T63 turboshaft engine, swirl-stabilized research combustor	FT	• Engine was operated at idle and cruise power • Injector is a generic swirl-cup liquid-fuel nozzle fitted with a pressure swirl atomizer	Corporan et al. (2007)
2011	Combustion chamber sector rig	80/20 Jet A-1/FAME, 50/50 Jet A-1/GTL FT, GTL FT	• Mass airflow 0.465 kg/s, inlet temperature 315K • Inlet pressure 16 kPa, equivalence ratio 0.23	Pucher et al. (2011b)
2012	Artouste Mk113 APU	CTL FT, GTL FT, 50/50 GTL/Jet A-1	• Idle and full power	Lobo et al. (2012)
2012	T63-A-700 Allison turbine engine	Camelina HRJ, 50/50 (by vol) HRJ/JP-8, tallow-derived HRJ, 16% trimethylbenzene/ Tallow HRJ	• Idle and cruise conditions, 150 h duration test	Klingshirn et al. (2012)
2013	Combustion chamber sector rig	GTL FT, 50/50 Jet A-1/FT, Algae HRJ, 50/50 Jet A-1/camelina blend	• Air mass flow 0.465 kg/s, inlet temperature 315K • Pressure drop across the inlet 16 kPa, equivalence ratio of 0.23	Purcher et al. (2013)
2013	GE CFM-56-2C1 engine aboard of a DC-8 aircraft	HRJ, FT, 50/50 HRJ/JP-8	• Engine set to six different power settings between 4% and 100% of max rated thrust	Huang and Vander Wal (2013)

Continued

Table 4.3 Emissions of alternative jet fuels produced from different engine conditions.—cont'd

Year	Aircraft engine/combustor	Fuel	Operating conditions	References
2014	Artouste MK113 APU gas turbine engine	GTL, 50/50 HEFA/Jet A-1, 75/25 HEFA/Jet A-1	• Idle: EGT 445°C, AFR 80 • Full power: EGT 460°C, AFR 76	Altaher et al. (2014)
2015	GTCP85 aircraft APU	UCO−HEFA	• No load, environmental control systems, main engine start	Lobo et al. (2015a)
2016	GE CF-700-2D-2 engine core	CH-SKA, FT-SPK, 50/50 HEFA-SPK/Jet A-1	• Ground idle, 80%, and 95% of rated power 20 kN	Chan et al. (2016)
2018	Rolls-Royce Tay gas turbine combustor	GTJ, ATJ, HEFA, SPK	• Air mass flow rate 200 g/s, • Fuel flow rate 1.8 g/s for stable burning and 0.5 g/s for lean burning conditions	Zheng et al. (2018)
2018	Turbojet engine (GTM 140 series)	48% camelina HEFA blended with Jet A-1	• 33,000−120,000 rpm, maximum thrust is 140 N	Gawron and Bialecki (2018)
2019	CFM 56-7B26 turbofan engine	32% HEFA/Jet A-1	• Ground idle (34% engine thrust) and • Climb-out engine thrust (~85% engine thrust)	Liati et al. (2019)
2019	CFM56-5C4 jet engine	ATJ-SPK (Gevo), HEFA-SKA (Readijet)	• Idle to takeoff conditions (20%−96% of rated power)	Schripp et al. (2019)
2019	Single-can combustor	SPK (UOP), jatropha HEFA,	• Inlet pressure 3.66−4.46 bar, inlet temperature 375 and 500K • Combustor pressure loss 1.3−1.4, fuel−air ratio 0.0115−0.0132	Sundararaj et al. (2019)
2020	J-85 engine	30% and 70% camelina HEFA/Jet A-1	• Idle to full power	Kumal et al. (2020)

In the turbine engine emissions test using HEFA conducted by Klingshirn et al. (2012), a reduction of CO emissions by ~20% was shown by neat HEFAs compared with JP-8 in Fig. 4.5A, which could be attributed to the lower overall carbon content in the renewable fuels. No significant difference was observed in NO_x emissions for the operating conditions tested, indicating the similar combustion temperatures for the fuels tested. Fig. 4.5B shows alternative fuels emitted lower UHC at idle engine condition due to the low aromatics contents. Likewise, the emissions of benzene, toluene, ethylbenzene, and xylene (BTEX) were also shown to reduce due to similar reason. No UHC was detected during cruise engine condition as the combustion efficiency is high. The negligible amount of sulfur in the HEFA is the reason for the low sulfur oxide emissions detected in the alternative fuels.

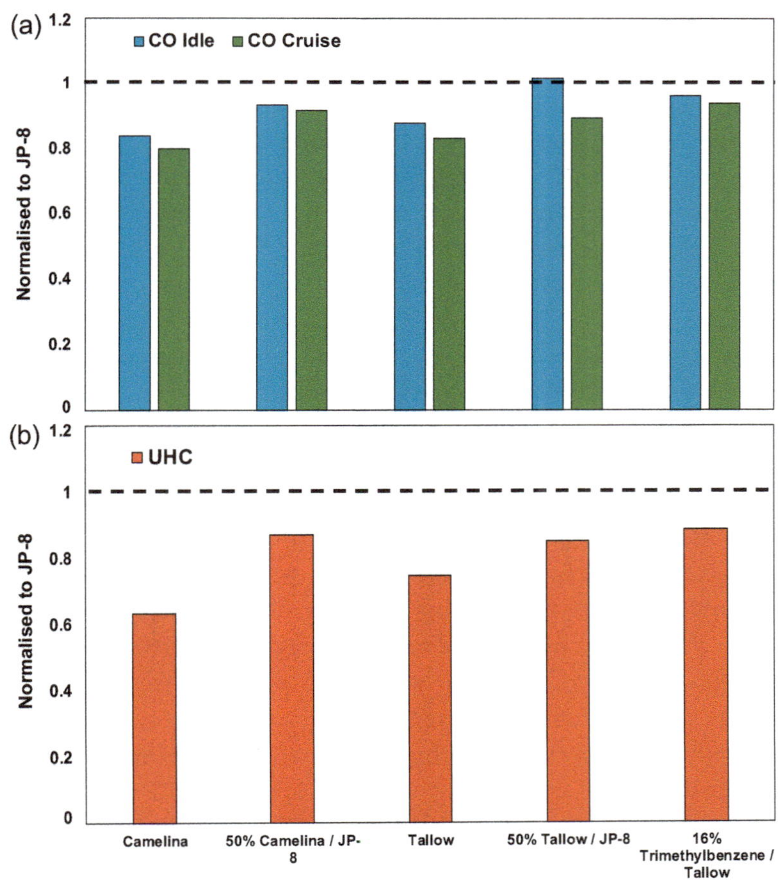

Figure 4.5 Comparison of emissions of (A) carbon monoxide and (B) unburned hydrocarbons relative to JP-8 at different engine conditions (Klingshirn et al., 2012).

Altaher et al. (2014) investigated the emissions of unburned nonmethane hydrocarbons (NMHCs) and oxygenated VOCs (carbonyl compounds) in the exhaust gas from an APU gas turbine engine using renewable fuels such as GTL and HEFA blends. The emissions of NMHC were most significant at idle condition, which was about four to nine times higher compared with full power condition. The content of aromatics in the fuel is the most pronounced factor that affects the emissions of NMHC. Fuels with low aromatics lead to reduced aromatics hydrocarbon emissions including benzene and toluene. GTL fuel showed a significant reduction in aldehyde emissions, but HEFA blends and Jet A-1 showed similar carbonyls emissions level. The CO_2 emitted from Jet A-1 is highest, followed by the 100% CH-SKA, 50% HEFA-SPK, and then 100% FT-SPK. NO_x emissions are strongly related to the flame temperature; hence, higher engine load leads to higher NO_x emissions. The FT-SPK and 50% HEFA-SPK showed lower NO_x emissions than Jet A-1 at high engine load conditions, whereas the CH-SKA showed comparable NO_x emissions level as the latter (Chan et al., 2016). The emission performance of HEFA/Jet A-1 blend has been tested in a miniature turbojet engine (Gawron and Bialecki, 2018). The HEFA blend showed a reduced thrust and thrust specific fuel consumption compared with the baseline Jet A-1. Unsurprisingly, the measured turbine inlet temperature for HEFA blend is also lower across all engine rotational speeds tested. The lower flame temperature could explain the lower NO_x emissions measured for HEFA blend. The CO_2 and CO emissions were found to be slightly lower by approximately 1% and 4%, respectively, at certain conditions.

Sundararaj et al. (2019) reported that increasing blend ratio of camelina-based biojet fuel showed an increase in NO_x emissions, but drops in CO, UHC, and soot were observed. Increase in flame temperature is the primary reason for higher NO_x and reduced CO and UHC emissions, but the lack of aromatic content in biojet fuel is the main reason for reduced soot. Jatropha-based biojet fuel has lower viscosity, which affects the spray quality at low power setting, thus affecting the emissions, as observed with higher CO and UHC due to incomplete mixing. Schripp et al. (2019) showed that the emission indices of CO and NO_x per fuel burned for ATJ and CHJ fuels were not significantly different, as shown in Fig. 4.6. The obtained emission indices show a good correlation to the values recorded in the ICAO database for CFM56-5C4 engine. However, Wang et al. (2020) demonstrated that the emissions of NO and CO for HEFA-SPK and blends can be reduced by using distributed combustion conditions, which was

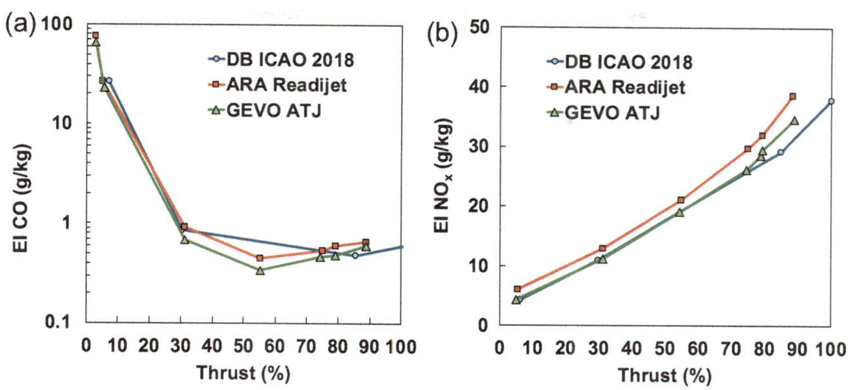

Figure 4.6 Emission indices of (A) carbon monoxide and (B) nitrogen oxides for different alternative jet fuels at different fuel flow settings and fuels. Reference values for the CFM56-5C4 engine (blue) were derived from the ICAO database 2018 (Schripp et al., 2019).

achieved by reducing the oxygen concentration in the oxidizer through adding diluent gases of N_2 and CO_2, thereby simulating the internal entrainment of hot reactive gases into the combustor.

4.3.4.2 Particulate matters

Solid particulate matter (PM) comprises mainly soot and, to a small extent, ash (metal particles). Soot generated from combustion is known to have adverse effects on human health. Soot particles generated from the aviation gas turbine engine depend on the fuel types and engine operating conditions. At present, the aircraft emissions of PM during landing and takeoff are regulated under the CAEP/10 standard introduced by ICAO (Jacob and Rindlisbacher, 2019). The regulatory limit concentration of nonvolatile particulate matters (nvPM) applies to all in production engine types with rated thrust of greater than 26.7 kN or after January 1, 2020, based on the equation $nvPM_{mass} = 10(3 + 2.9F_{oo} - 0.274)$, where F_{oo} refers to the rated thrust (Jacob and Rindlisbacher, 2019). The PM emissions standard proposed under the CAEP/11 are expectedly more stringent, prompting future engine development to consider the full interdependencies between pollutant formation and fuel burn.

The interest in decarbonizing the aviation industry via the use of alternative jet fuels has prompted the detailed investigation of the combustion characteristics and emissions performance of the substitute fuels. In earlier studies, the effects of the fuel properties of a natural gas–derived synthetic jet fuel on the PM and gaseous emissions produced from a T63 engine and a swirl-stabilized research combustor were examined. FT fuel without aromatics showed a reduction of 90% and 80% in particle number and smoke number, respectively, compared with JP-8. The absence of sulfur and high hydrogen-to-carbon ratio of the FT fuel resulted in the reduction in sulfur oxide and increased in water vapor, respectively (Corporan et al., 2007). In a CFM56-7B engine test simulating the LTO cycle, Lobo et al. (2011) showed a reduction of 62% of PM emissions was achieved for the neat FT fuel compared with Jet A-1 fuel. The reduction of PM emissions was attributed to the absence of aromatics content in FT fuel, as opposed to the 18.5 vol% of aromatics in the Jet A-1. Similar conclusion was reached by Lobo et al. (2012), in which the reduction of PM emissions were found to be pronounced for FT fuels due to the lack of aromatics, with up to 90%, 72%, and 65% of reduction by mass recorded for GTL, 50:50 GTL: Jet A-1, and CTL at idle and full power conditions. Although lower aromatics content in synthetic fuel blend results in lower soot, other problem such as insufficient swelling of the engine seal was reported (Link et al., 2008; Jürgens et al., 2019). Hence, it has been specified in the ASTM D7566 standard that a minimum of 8.4 vol% aromatics content for synthetic jet fuel blend needs to be retained.

Klingshirn et al. (2012) evaluated the PM emissions of two types of HEFA derived from camelina and tallow in a T63 turbine engine. The lack of aromatics in the neat HEFA fuel showed the least soot with 90%–98% reduction in particle number. The tallow-HEFA blended with 16% trimethylbenzene (aromatics) shows the highest soot emissions among all alternative fuels tested, but a reduction of ~30% in smoke number was achieved in cruising condition compared to jet fuel. For the duration test of 150 h using 50/50 HRJ/JP-8 fuel, it was reported that similar heating and sooting patterns were achieved compared with JP-8 fuels. Huang and Vander Wal (2013) studied the characteristics of particulate matters emitted from a CFM56-2C1 engine on a DC-9 aircraft operating with FT fuel and 50/50 HRJ/JP-8 in the power range of 4%–100%, covering ground-idle, intermediate, cruise, and takeoff levels. Measurements showed that the soot particle size varies as a function of engine power, albeit in a different trend depending on the type of fuel. The primary soot particle size for the renewable fuels decreases with increasing power, but the conventional JP-8 shows a reverse trend, as shown in Fig. 4.7. The reduced soot particle size

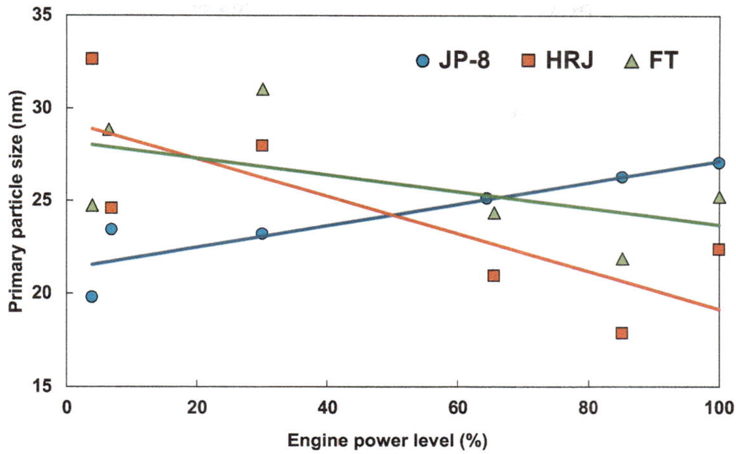

Figure 4.7 Primary particle distributions of JP-8-, HRJ-, and FT-derived soot across engine power levels from 4% to 100% (Huang and Vander Wal, 2013).

for renewable fuels indicates less time and low concentration of growth species, as higher temperature with higher power setting accelerates fuel pyrolysis reactions, increases fuel–air mixing, and lowers local equivalence ratio. Conversely, the high aromatic content in JP-8 directly contributes to the inception of soot growth. Furthermore, the paraffinic and cyclo-paraffinic compounds pyrolyzed at high power could have contributed aromatics for soot growth. It is to note that the blended HRJ fuel and FT fuel contain 9.8 and 1.7 vol% of aromatics, respectively, as opposed to JP-8 which contains 21.8 vol% of aromatics. The soot nanostructure for renewable fuels and baseline jet fuel are different, suggesting that the soot formation path is dependent on the chemical composition of the fuel and burning conditions in the combustor.

Purcher et al. (2013) evaluated the emissions and deposits of several synthetic jet fuels, including 100% GTL FT, 50/50 FT/Jet A-1, 100% HEFA and 50/50 HEFA/Jet A-1 blends. Overall, the smoke emission and deposit levels for all synthetic jet fuels were substantially lower than Jet A-1. The camelina/Jet A-1 blend showed a reduction of 70% in smoke relative to neat Jet A-1, despite the presence of 50% conventional jet fuel in the blend. A significant reduction of mass deposit of 72% was reported for the camelina–HEFA blend relative to Jet A-1 over the course of testing conducted on the rig. The 100% algae-based HEFA showed a remarkable

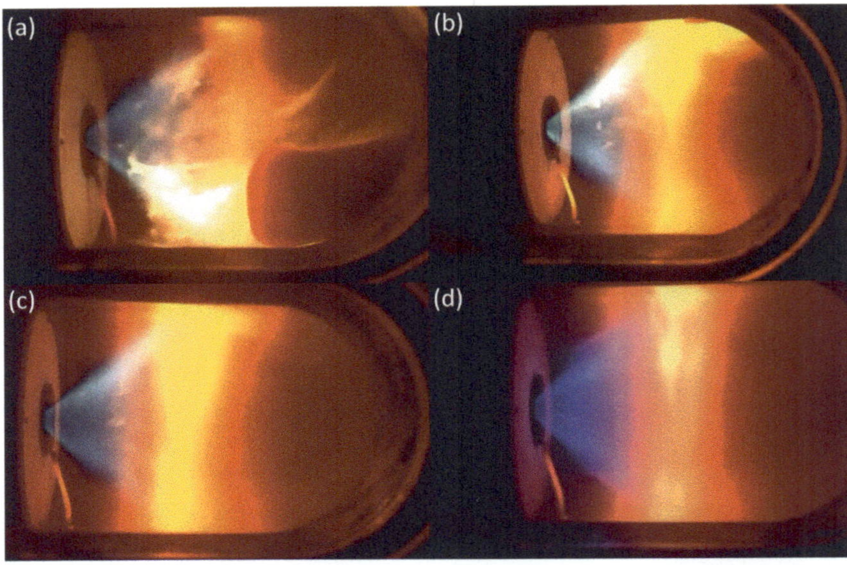

Figure 4.8 High pressure optical chamber axial view at pressure of 1 bar and equivalence ratio of 0.9 for blended fuels of (A) neat Jet A-1 (B) Jet A-1:HEFA at 80/20% vol (C) Jet A-1/HEFA at 50/50% vol and (D) neat HEFA (Buffi et al., 2017).

96% reduction in smoke number compared with Jet A-1. Significant amount of deposits builtup on the surfaces of the combustor can be observed for Jet A-1. FT synthetic fuel was shown to be exhibit clean burning characteristics, with no clear carbon deposited on the injector and combustor wall based on visual inspection (Pucher et al., 2011a). They further reported 20% fatty acid methyl esters (FAME) blend with Jet A-1 resulted in lower soot emission, although FAME is not an ASTM-certified biojet fuel. Buffi et al. (2017) further reinforced this in an emissions characterisation test for comparing various blends of the conventional Jet A-1 fuel with HEFA. Fig. 4.8 shows the presence of soot in the blended fuel flames using a pressurised swirl combustor in a high pressure optical chamber. It is clear that the HEFA component in the blend has a higher oxygen consumption rate, hence producing a higher temperature flame which goes through a more complete combustion reaction.

The particulate matters emitted from used cooking oil (UCO)—based HEFA biojet fuel in different blends were investigated using an aircraft APU. The reductions of PM were shown to increase with increasing HEFA content, owing to the increase of hydrogen content. The PM size distributions were found to narrow and shift to smaller sizes as the UCO–HEFA component of the fuel blend increased (Lobo et al., 2015a). The ASTM-approved 50:50 blend of UCO-HEFA and Jet A-1 showed a reduction of 60% by mass compared with baseline Jet A-1 (Lobo et al., 2015a). Chan et al. (2016) evaluated the gaseous emissions and PM emitted from a GE CF-700-2D-2 turbofan engine (20 kN) for three alternative fuels. The fuels tested were neat CH-SPK, derived from *Brassica carinata* plant oil via catalytic hydrothermolysis process, 50% camelina HEFA-SPK blended with Jet A-1, and neat FT-SPK fuels. The particle number emissions for CH-SKA showed a reduction of 7%–25% over the range of engine load conditions, in spite of the similar aromatics level between CH-SKA and Jet A-1. The HEFA-SPK and FT-SPK showed a significant reduction of 40%–60% and 70%–95%, respectively, mainly due to the reduced level of aromatics compared to Jet A-1.

Zheng et al. (2018) systematically evaluated the sooting tendency of 16 different types of jet fuels including ATJ, HEFA, GTL, and others. They concluded that aromatics content is the main determinant for soot formation; in particular, di- and cycloaromatics are more prone to produce soot than alkylbenzenes. In addition, fuels with higher density, cycloparaffin content, and surface tension were also found to have higher soot formation tendency. In the engine test study of the soot reactivity conducted by Liati et al. (2019), it was reported that 32% HEFA/Jet A-1 blend showed reduced soot reactivity during ground idle conditions as compared with Jet A-1, an important improvement to local air quality in airport area during ground-idle condition. The HEFA blend shows higher soot reactivity during climb-out conditions, but the overall lower soot emitted by HEFA blend could result in net positive effects.

Schripp et al. (2019) evaluated the performance of catalytic hydrothermolysis jet fuel (ReadiJet) and an unblended ATJ fuel in a CFM56-5C4 engine. Measurements of the gaseous emissions and particulate matters were performed at idle and takeoff conditions, which translated to roughly 20% and 96% of the total engine thrusts. Measurements of the PM emissions were conducted at 25 m behind the engine exit plane rather than in-plane of the engine exhaust as prescribed in AIR 6241 (AIR6241, 2013). The tested ATJ-SPK fuel consists of a few different

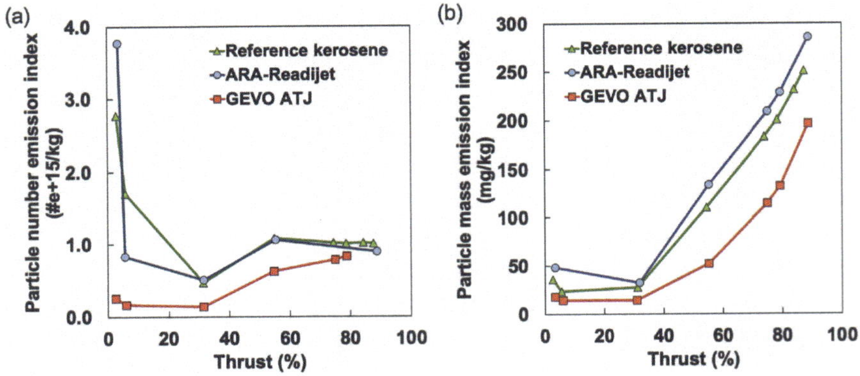

Figure 4.9 (A) Total particle number and (B) particle mass emission index for different alternative jet fuels and thrust settings. (Schripp et al., 2019).

branched aliphatic compounds (primarily iso-$C_{12}H_{26}$ and iso-$C_{16}H_{34}$), while the CHJ-SPK (ReadiJet) contains high level of aromatics and cycloparaffins that are preserved during the production process. As expected, the higher content of aromatics (20.9 vol%) in CHJ-SPK led to higher emissions of particulate matters, whereas the unblended neat AJT-SPK with <1% aromatics content produced lower soot than the baseline Jet A-1, as shown in PM emission index presented in Fig. 4.9. Calculation of the emission index (EI) for particulate emissions and combustion gases follows the proposed method in Chapter 7 of AIR 6241 (AIR6241, 2013).

$$EI\left(\#/_{kg}\right) = PN\left(\#/_{cm^3}\right) \times 10^6 \times \frac{0.082 \times T(K)}{p(atm)\left[M_c\left(g/_{mol}\right) + \alpha \times M_H\left(g/_{mol}\right)\right] \times [(CO_2) - (CO_2)_{BG}]} \tag{4.2}$$

$$EI\left(mg/_{kg}\right) = PM\left(mg/_{m^3}\right) \times \frac{0.082 \times T(K)}{P(atm)\left[M_C\left(g/_{mol}\right) + \alpha \times M_H\left(g/_{mol}\right)\right] \times [(CO_2) - (CO_2)_{BG}]} \tag{4.3}$$

where PN is the particle number concentration or PM particle mass concentration, CO_2 is the measured concentration of carbon dioxide in air, $(CO_2)_{BG}$ is the background carbon dioxide concentration, P is the atmospheric pressure, α is the molar ratio of hydrogen to carbon in the fuel, and the M_C and M_H are the molar masses of carbon and hydrogen, respectively (Schripp et al., 2019).

In the MERMOSE project (Smith et al., 2017) that focuses on the characterization of particulate matters emissions of aircraft engine, measurement at downstream of the exhaust plane of a turbofan engine at ground level revealed that the size of PM increases with engine thrust, concurring with the findings shown in other engines (Lobo et al., 2015b). In spite of the differences in particle sizes, the soot morphology within the primary particles does not vary across the test regimes. The organic content of the emitted particles was found to significantly decrease at 30% and 70% engine thrust levels (Smith et al., 2017). Lobo et al. (2015b) compared the PM sampled from different in-service engines. It was concluded that the PM emissions indices vary with engine size, type, and technology, with older technology engines showing higher PM emissions in mass. The particulate matters emitted from aircraft turbine engines are composed of branched-chain fractal aggregates of multiple primary particles (Saffaripour et al., 2019). The primary diameter, aggregate size, and density strongly depend on the engine power setting and fuel type. Kumal et al. (2020) studied the morphology and aggregate size of the particulate matters emitted from 30% to 70% camelina-based biofuel/Jet A-1 blends under various engine thrust conditions. The camelina-based biofuel consists primarily of n-paraffins and isoparaffins, produced via the catalytic hydroprocessing pathway. It was reported that the size of the PM increases with engine power (Fig. 4.10). The increase in paraffinic and hydrogen contents with increasing biofuel fraction results in the significant reduction in PM size. The low tendency of soot formation for biojet fuel is attributable to the reduced aromatics required for soot nucleation, as well as limited by the

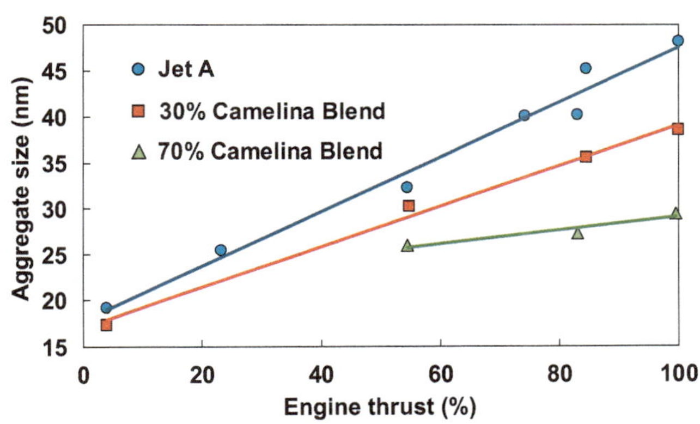

Figure 4.10 Variation of aggregate size as a function of engine thrust for Jet A and camelina/Jet A blend fuels (Kumal et al., 2020).

kinetic delay in biojet fuel fraction. In addition, the highly turbulent environment in the combustor dilutes the fuel-rich pockets, thus lowering the local equivalence ratio in the soot-forming region.

4.4 Flight test

The successful application of aviation biofuels in commercial airline was first demonstrated by Virgin Atlantic in a B747 flight using the blend of Jet A-1 with 20% biofuels derived from coconut and babassu oil (Greenair, 2008). Since then, the frequency of flight trials with alternative fuels has been increasing over the years. The alternative jet fuels used have largely focused on the use of drop-in sustainable aviation fuel (SAF), which is defined by the IATA to be (1) jet fuel production from alternative feedstock in an alternative manner, (2) proven sustainable in the context consistent with economic, social, and environment, and (3) jet fuel that meets the technical and certification requirements for use in commercial aircraft (IATA, 2020). Fig. 4.11 shows the timeline of the flight demonstrations using SAF by different airlines, aircraft OEMs, and fuel manufacturers. Air New Zealand demonstrated the flight test with blend of 50% SAF derived from Jatropha in a B747-400 fitted with Rolls Royce engines, marking the first flight test using nonedible feedstock (Express, 2008). The SAF used by Continental Airlines in 2009 was derived from algae and

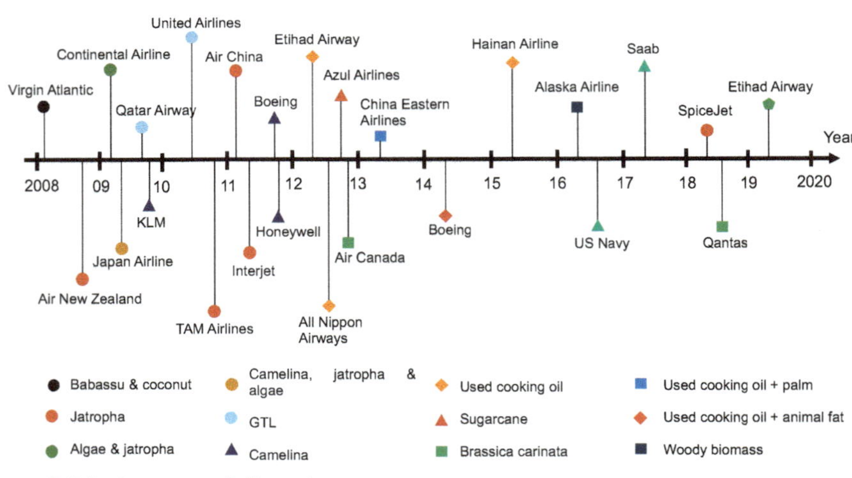

Figure 4.11 Alternative fuel-powered flights between 2008 and 2020 by various airlines (IATA, 2015).

jatropha, implying a blend of drop-in SAF is feasible regardless of the feedstock type (GE Aviation, 2009). Qatar Airways first demonstrated the use synthetic jet fuel derived from natural gas in aircraft engine in 2009. The gas-to-liquid synthetic fuel was blended with jet fuel and used in an Airbus A340-600 flight (Greenair, 2009). In the following year, Sasol, the South Africa fuel manufacturer demonstrated the world's first passenger aircraft flight using 100% synthetic jet fuel (Sasol, 2010).

After HEFA's approval in 2011, various airlines and fuel producers have demonstrated flight trials using blends of SAF with regular jet fuel to demonstrate their environmental commitment and market interest. Air China tested a Boeing 747-400 flight powered by Pratt & Whitney engines using blends of 50% HEFA derived from jatropha with conventional jet fuel. The SAF was the first biojet fuel produced in China by China National Petroleum Corp (China Daily, 2011). Interjet and Airbus jointly completed a flight trial in 2011 using biojet fuel blend. The flight test was conducted using an Airbus A320 fueled with 27% HEFA-SPK produced by Honeywell's UOP with jatropha seeds (Biofuels Digest, 2011). Honeywell, being an SAF producer, further conducted flight test in 2011 with 50% HEFA-SPK derived from camelina in a Gulfsream G450 powered with Rolls Royce engines (Marketwatch, 2011). In 2012, Etihad Airways completed a 14-hour flight with a Boeing 777-300 ER aircraft from Seattle to Abu Dhabi using waste cooking oil biojet fuel supplied by SkyNRG (Airportwatch, 2012). The same fuel was fueled in a Boeing 787 by All Nippon Airways to complete a transpacific flight in the same year (ANA, 2012). In fact, KLM has already started using waste cooking oil biojet fuel in their regular commercial flight since 2011 (WIRED, 2011).

In Brazil, Azul Airlines demonstrated the use of sugarcane-based biojet fuel, produced by Amyris, in the Embraer 195 aircraft powered by GE's CF34-10E engines (Digest, 2012). The first flight test using 100% biojet fuel was conducted by Air Canada in 2012. The biojet fuel (ReadiJet), produced from Brassica Carinata oil seeds, was developed by Applied Research Associates and Chevron Lumnus Global (Newatlas, 2012). In 2013, China's locally produced biojet fuel (CBF-1) by Sinopec from used cooking oil and palm oil was tested in an Airbus A320 by China Eastern Airline (China Daily, 2013), paving way for the biojet fuel certification process by Civil Aviation of Administration of China for commercial use in the following year. This later enabled Hainan Airlines to complete the first commercial flight operating with 50% biojet fuel made from waste cooking oil on a Boeing 737 flight (Breaking Travel News, 2015). In 2014,

The ecoDemonstrator program by Boeing demonstrated the feasibility of flying with 15% biofuels derived from fatty acids of different sources in a B787 aircraft. The program later showed a successful flight with a Boeing 777 freighter fueled with 100% HEFA biofuels produced from animal fat and beef tallow waste (Greenaironline, 2014). In 2018, India's first biofuel test flight was conducted by Spicejet with a Bombardier Q400 aircraft fueled with 25% jatropha biojet fuel (Indiatimes, 2018). In 2019, Etihad Airways conducted a flight trial with a Boeing 787 operated with biojet fuel derived from *Salicornia* plant, which is a type of seed plant grown in desert with seawater (Gulfnews, 2019). To cater for the demand of SAF, Norway's Oslo airport became the first airport in the world to regularly offer SAF to all departures in 2016, followed by Los Angeles, Stockholm and Bergen, Norway (Boeing, 2019). As test flights are part of the rigorous process in achieving biojet fuel certification, it is expected that more flight tests will be conducted in the near future given that different alternative jet fuel methods are under development. At the same time, the commitment shown by commercial airlines in using SAF indicates the growing acceptance of biofuels, which is important in the long run to achieve decarbonization goal in the aviation section.

4.5 Fundamental combustion properties

Since the fuel composition in alternative fuel may vary significantly compared with conventional jet fuel, understanding the fundamental combustion properties is important to gain insights into the impact of the fuel properties on the combustion performance. Significant effort has been devoted to investigate the chemistry of alternative jet fuels under reacting conditions. The fundamental combustion properties of interest include ignition characteristics, flame speed, flame blowout and extinction characteristics, and chemical reaction pathway, which are closely linked to the fuel composition and molecular structure. The fuel chemistry forms the building blocks of flames that has direct impact to the combustion performance in engine such as cold startup, high altitude relight, combustion efficiency, and emissions. Furthermore, the understanding of the combustion properties plays an important role in facilitating the development of fuel and engine technology. For example, the development of the detailed reaction mechanism requires the identification of suitable surrogate fuels and valid kinetic targets (such as ignition delay time, laminar flame speed, and temporal combustion speciation, among others) to test, refine, and validate the mechanism. The chemical kinetic models developed can be used to predict emissions and combustion performances, which can lead to significant cost savings in the development and testing programs. The fundamental combustion data can also facilitate the fuel developing process by screening out the unacceptable fuel prior to testing in costly component or engine tests. The fundamental combustion tests using alternative jet fuels are presented in the following sections.

4.5.1 Ignition delay time

The study of the ignition, pyrolysis, and oxidation behavior of practical jet fuels has been conducted using shock tube and rapid compression machine (RMC). Measurements of the ignition delay time using these devices are performed with homogenous fuel/oxidizer mixtures, so that the reaction is solely controlled by the chemical kinetics that depends on the pressure and temperature histories of the fuel/oxidizer mixture, and without the complication of physical processes such as mixing, atomization, vaporization, and fluid dynamics encountered in jet engine. The ignition delay time data derived at elevated pressure and temperature are typically used as global kinetic target to validate the chemical kinetic models (Flora et al., 2017). There is a slight difference in the definition of ignition delay time between shock tube and RMC. For RMC, the ignition delay is defined as the occurrence of an inflection point in the pressure trace during the pressure rise due to ignition that is shown after the end of compression. Fig. 4.12A shows the ignition delay measurement derived from an RCM experiment with S-8/air mixture, $\varphi = 1.15$, and at 15 bar (Kumar and Sung, 2010). The end of compression is termed as the starting point of the

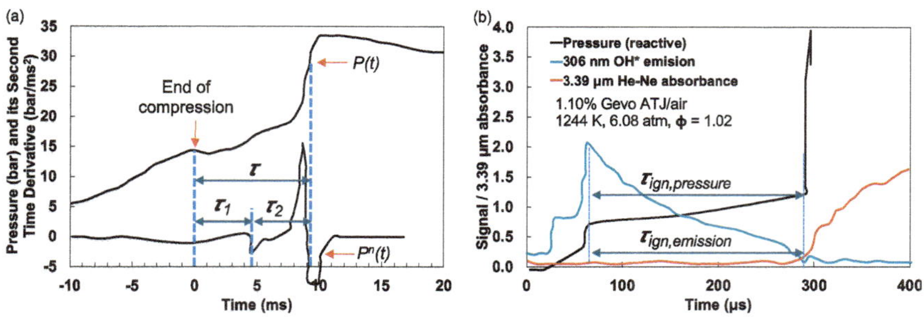

Figure 4.12 Ignition delay time measurements derived from a (A) rapid compression machine with S-8/air, oxidizer-to-fuel mass ratios of = 13, $\varphi = 1.15$, 15 bar (Kumar and Sung, 2010) and (B) shock tube with 1.1% Gevo ATJ/air, $\varphi = 1.02$, 1244 K, 6.08 atm (Zhu et al., 2015).

ignition. The first stage of ignition delay (s1) corresponds to the time interval between the end of compression and the onset of the first-stage ignition, while the duration of the second-stage ignition is termed as s2. The sum of the two intervals $s = s1 + s2$ is the overall ignition delay. For shock tube, the pressure traces derived from the ignition experiment of a shock tube conducted with 1.1% ATJ/air at $\varphi = 1.02$, 1244 K and 6.08 atm are shown in Fig. 4.12B (Zhu et al., 2015). They used laser absorption method to trace the OH* emissions in conjunction with the pressure traces for reacting mixture. The ignition delay time is defined as the time interval between the arrival of the reflected shock and the onset of ignition. The point of ignition can be determined by extrapolating the maximum slope of the pressure or OH* signal back to the baseline. The operating range between shock tube and RCM differs quite significantly. Shock tube has been used to determine the short ignition delay measurements in the high temperature range, while RCM is generally used in the low-to-intermediate temperature range for relatively longer ignition delay measurements. In practice, measurement data from shock tube and RCM can overlap, although discrepancies in data have been observed. Therefore, the ignition delay results obtained from different facilities have to be interpreted with care (Zhang et al., 2016).

Comparison on the autoignition delay times for conventional and alternative jet fuels has been performed by several groups, as shown in Table 4.4. Kahandawala et al. (2008) measured the ignition delay time of FT fuel and JP-8 behind the reflected shock wave under the conditions of $\phi = 0.5$, pressure of 21 atm, and the preignition temperature of 1100−1600K. The FT fuel and JP-8 fuel show similar ignition delays for the range of temperature tested, although the FT fuel contains two times more cycloparaffin than JP-8 with no aromatics. Kumar and Sung (2010) extended the study of autoignition delay time of FT fuel by using a rapid compression machine at lower and intermediate temperature of 615−933K, covering an extended pressure range of 7−30 bar and equivalence ratio of 0.43−2.29. The FT fuel was found to have the shortest overall ignition delay time, followed by Jet A-1 and JP-8, as shown in Fig. 4.13. The mixture's equivalence ratio is reported to have a strong influence on the ignition propensity. Gokulakrishnan et al. (2008) utilized a pressure flow reactor to measure the ignition delay time of FT fuel at 900−1200K, $\phi = 0.5−1.5$, and atmospheric condition. Result showed that the FT fuel has lower ignition time compared with JP-8, which is attributable to the lower aromatic content and higher activation energy.

Table 4.4 Ignition delay measured for alternative jet fuels.

Year	Fuel	Operating condition	References
2008	FT (Syntroleum)	Shock tube, $\phi = 0.5$, pressure of 21 atm and the pre-ignition temperature of 1100–1600K	Kahandawala et al. (2008)
2008	FT (S8)	Atmospheric pressure flow reactor, 900 and 1200K, $\phi = 0.5–1.5$	Gokulakrishnan et al. (2008)
2010	FT (Syntroleum)	Rapid compression machine, initial pressure of 7, 15, and 30 bar, temperature 615–933K, $\phi = 0.43–2.29$	Kumar and Sung (2010)
2012	GTL FT fuel (S-8, Shell), CTL FT, Sasol IPK	Shock tube, 8–39 atm, 651–1381K, $\phi = 0.25–1.5$	Wang and Oehlschlaeger (2012)
2014	2,6,10-trimethyl dodecane (farnesane)	Initial pressure 20 atm	Won et al. (2014)
2015	FT (Sasol, Shell), HRJ (tallow, camelina), ATJ (Gevo), Swedish BioJet, hydro-refined algal oil	Initial temperature 1047–1520K, and $\phi = 0.25–2.2$, in two pressure and mixture regimes: for fuel/air mixtures at 2.07–8.27 atm, and for fuel/4% O_2/Ar mixtures at 15.9–44.0 atm	Zhu et al. (2015)
2015	ATJ (Gevo), sugar-to-hydrocarbon (Amyris Farnesane)	Rapid compression machine, compressed pressure 20 bar, initial temperature between 600–700K, $\phi = 1.0$, 0.5, 0.25	Min et al. (2015)
2017	GTL, GTL surrogate of 32% isooctane, 25% n-decane, and 43% n-dodecane	Spherical chamber, $\phi = 0.8–1.2$, initial pressure 8.6, 10, and 12 atm at initial temperature of 450K.	Askari et al. (2017)
2017	CTL FT (Sasol), DSHC (Amyris), ATJ (Gevo), HEFA based biojet (corn, canola and soy)	Shock tube, preignition temperature range of 980–1800K at a pressure of 16 ± 0.8 atm, $\phi = 0.5$ and argon as the diluent (93% by vol)	Flora et al. (2017)
2018	50/50 BTL FT/Jet A-1 blend	700–1200K at 20 atm, $\phi = 1.0$	Han et al. (2018)

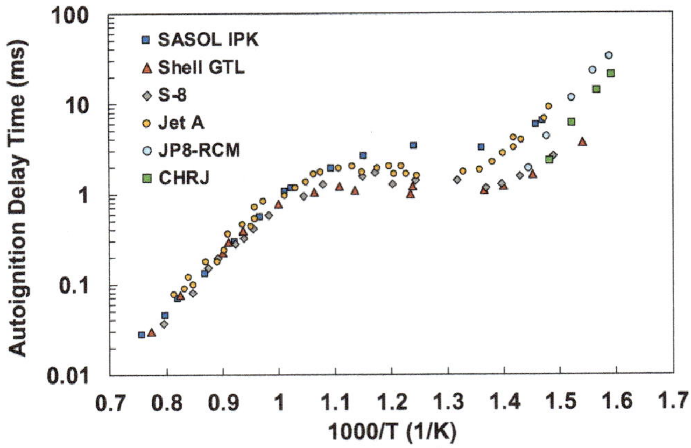

Figure 4.13 Autoignition delay times for different alternative jet fuels measured at stoichiometric and pressure of 20 atm using shock tubes and rapid compression machine (Zhang et al., 2016; Vasu et al., 2008).

Wang and Oehlschlaeger (2012) measured the ignition delay times of four FT fuels using a shock tube at the condition of 8–39 atm, 651–1381K, and $\phi = 0.25-1.5$. Comparison of ignition delay for FT fuels and Jet A-1 showed no discernible difference in reactivity at T > 1000K. However, the negative temperature coefficient (NTC) and low temperature region showed an inverse correlation between the fuel's derived cetane number and ignition time. The compositional variation between the FT fuels influenced the reactivity of the fuel. For example, Shell GTL exhibits slightly shorter ignition delay times in the NTC regime than S-8 due to its larger n-alkane fraction, while the 90% isoalkanes and 10% cycloalkanes have relatively low reactivity. A comparison of the ignition delay time data obtained with shock tube (Wang and Oehlschlaeger, 2012) is compared with those derived with RMC (Allen et al., 2012). Both set of data show similar trends, albeit the temperature range for shock tube is significantly extended compared to RMC. Han et al. (2018) measured the ignition delay of 50/50 BTL FT/Jet A-1 blend in a shock tube at the preignition temperature range of 700–1200K and pressure of 20 atm. The ignition delay of biojet fuel blend is similar to Jet A-1, except at the condition below 1000K where the former shows a reduced ignition delay time by ~50%, which could be due to the higher derived cetane number contributed by the biojet fuel component.

Won et al. (2014) compared the reflected shock ignition delay characteristic of 2,6,10-trimethyl dodecane (farnesane) with FT (S-8) surrogate fuel (n-dodecane/isooctane, 51.9/48.1 mol%) at 20 atm. The farnesane exhibited a notable difference in ignition delay at temperature lower than 870K when compared with S-8 surrogate fuel. The latter shows a faster ignition delay time by a factor of 2 at the low temperature kinetic regime. Min et al. (2015) examined the ignition delay characteristics of ATJ (GEVO) and direct sugar to hydrocarbon fuel (Amyris Farnesane) in a rapid compression machine. The ATJ fuel has a low derived cetane number (DCN) of 15 and shows the longest ignition delay time and prominent multistage ignition behavior. The farnesane fuel contains the highest DCN among the tested fuels, with the shortest ignition delay time except for $\phi = 0.25$. The ATJ fuel does not ignite at $\phi = 0.25$ due to low reactivity and was observed to enter the NTC region faster than Jet A-1 fuel. Measurement of the ignition delay of alternative jet fuels of ATJ fuel (GEVO), direct sugar to hydrocarbon (farnesane), and CTL-FT fuel via a single-pulse shock tube was performed under lean, high pressure conditions (Flora et al., 2017). The tests were conducted at 980−1800K, at a pressure of 16 atm and an equivalence ratio of 0.5 in the presence of 93 vol% dilution with argon. ATJ fuel showed slightly longer ignition delay time at higher temperature. Kinetic modeling results show that predominant reactions are oxidation of C1−C4 fuel fragments; hence the chain branching of large n-paraffins hardly has any impact on the ignition delay (Flora et al., 2017). The ignition delay characteristics of nine neat alternative fuels and six blends were investigated using a high-pressure shock tube under the preignition temperatures of 1047−1520K, compressed pressures of 2.07−8.27 atm, and equivalence ratios of 0.42−2.19 (Zhu et al., 2015). It was shown that the fuels tested showed similar orders of magnitude of ignition delay data, with all isoparaffin fuels and conventional jet fuels showing similar ignition delay times. However, the ATJ fuels display lower reactivity than JP-8 at 3 atm, as shown in Fig. 4.14A. The GEVO ATJ is less reactive than Swedish BioJet fuel at higher temperature. At elevated pressure of 6 atm, the differences in reactivity among the alternative fuels and JP-8 are less pronounced, as shown in Fig. 4.14B.

4.5.2 Derived cetane number

The indicator used to describe the ignition characteristics of diesel fuel in compression ignition engine is known as cetane number (CN). Higher value of CN indicates higher reactivity and autoignition tendency under diesel-relevant conditions, which implies better combustion performance.

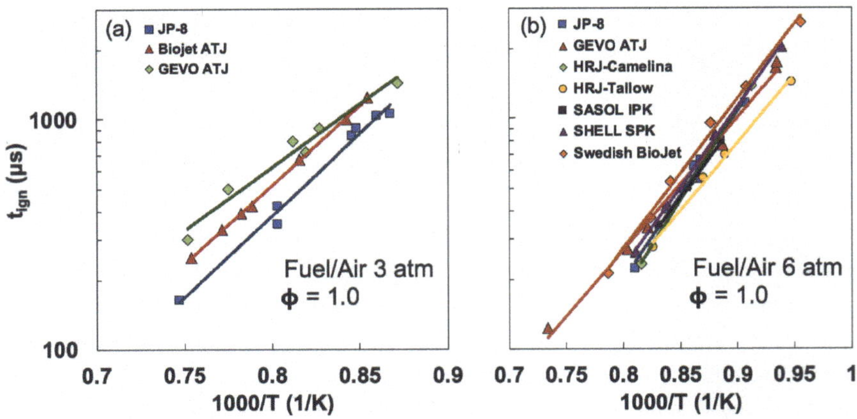

Figure 4.14 Comparison of ignition delay time for alternative fuels in air at (A) 3 atm and (B) 6 atm and $\phi = 1.0$ (Zhu et al., 2015).

The DCN is another method that has been developed to characterize the autoignition reactivity of liquid fuel based on pressure trace. At present, there are three ASTM-approved devices and procedures to derive the DCN values, which are the fuel ignition tester (FIT), ignition quality tester (IQT), and cetane ignition delay (CID) based on the respective approved ASTM standard of D7170 (ASTM, 2016b), D6890 (ASTM, 2016a) and D7668 (ASTM, 2017). These methods employ a constant volume combustion chamber to measure the ignition delay time of the fuel injected at designated standard conditions of pressure (2.4 MPa for FIT, 2.137 MPa for IQT) and temperature (545K for IQT, 510K for FIT). The ignition delay time measured is defined as the time interval from injection of the fuel to the start of ignition, taking into account the process of spray injection, vaporization, mixing with oxidizer, and reaction. The DCN value is then calculated based on an empirical correlation using the ignition delay time (IDT), via the equation $DCN = 4.460 + 186.6/IDT$ (ASTM, 2016a). The DCN is inversely proportional to ignition delay time. Similar to CN, higher DCN value implies shorter ignition delay time and better autoignition propensity.

Table 4.5 shows the DCN values of conventional jet fuel and alternative jet fuels measured using different standards. It can be seen that the measured DCN shows a wide spectrum of ignition reactivity for the fuels. Fossil-based jet fuel (Jet A) shows minor difference in the DCN values (47–49) measured using different methods. One should note that the DCN values

Table 4.5 Derived cetane number for jet fuel and alternative jet fuels.

Fuel	FIT, ASTM D7170	IQT, ASTM D6890	CID, ASTM D7668
Jet A	49.35	47.1	47.01
Syntroleum S-8s	66.50	58.7	–
Shell GTLs	64.69	59.1	–
Shell SPKs	–	58.4	62.43
Sasol IPKs	33.46	31.28	31.71
Camelina HRJ	60.70	53.94	59.77
Tallow HRJ	65.85	58.1	–
Gevo ATJ	–	15.5, 18	18.24
Farnesane (2,6,10-trimethyl dodecane)	–	59.1	–

Data are compiled from Hui et al. (2012), Dooley et al. (2012a), Bessee et al. (2011), Won et al. (2013, 2014), Zhu et al. (2015), Dickerson et al. (2015), Kang et al. (2019b).

vary according to the method used. FT fuels generally show higher DCN value than conventional jet fuels except for Sasol IPK. The FT fuels of S-8, Shell GTL, and Shell SPK exhibit a DCN value between 58 and 67, but the Sasol IPK shows a significantly lower DCN value of 31–33 compared with Jet A due to the presence of branched alkanes, which are less reactive. In spite of the lower DCN value, the measured ignition delay time is comparable to conventional jet fuel (Zhu et al., 2015). Farnesane (2,6,10-trimethyl dodecane) shows a comparable DCN value (59.1) as those of FT-SPK fuels (Won et al., 2014). Interestingly, HEFA-based synthetic jet fuels show similar DCN values as FT fuels, which is within the range of 58–67, but the ATJ-SPK developed by GEVO showed a significantly lower DCN value of 15–19. One might expect the ignition delay characteristic to be impacted due to the stark difference in DCN, but experimental measurements have shown that GEVO ATJ-SPK exhibited similar ignition delay trend as regular jet fuel (Flora et al., 2017; Zhu et al., 2015). The similar ignition delay characteristics of GEVO ATJ and Sasol IPK with conventional jet fuel despite having lower DCN values indicates that DCN value alone is insufficient to characterize the reactivity of the fuels. The DCN value for alternative jet fuel blends can be estimated based on the percentage of blend fraction of synthetic jet fuel, as previous studies have shown a fairly linear relationship between the DCN values and synthetic jet fuel volume fraction in the binary fuel blend (Zhu et al., 2015).

4.5.3 Laminar flame speed

Laminar flame speed is defined as the propagation rate of the normal flame front relative to the unburned mixture. It is an important property for a premixed flame as it embodies the fundamental information of diffusivity, reactivity, and exothermicity of the combustible hydrocarbon mixture. Laminar flame speeds are also practical building blocks for understanding fuel behavior in devices that operate via mixture deflagration. On a practical level, laminar flame speed is related to the burning rate in the combustor, which can affect the combustion efficiency and exhaust emissions. Values for laminar flame speeds can be used directly in turbulent combustion models, or indirectly as validation targets for chemical kinetic models.

Measurement of the laminar flame speeds of FT fuel (S-8) and conventional Jet A has been performed by Kumar et al. (2011) via the use of a counterflow twin-flame configuration at elevated preheated temperatures of 400—470K and $\phi = 0.7-1.4$. From the plot of stretched laminar flame speed as a function of stretch rate, they applied a linear extrapolation technique to derive the unstretched laminar flame speed. Results indicate that the Jet A and FT fuel exhibit similar flame speeds across the equivalence ratios tested. The same group later extended the study to investigate the laminar flame speeds at elevated pressure of 1—3 atm, with result showing that S-8 has the same laminar flame speed characteristics as Jet A across fuel-lean and fuel-rich regions (Hui and Sung, 2013). Mze-Ahmed et al. (2012) measured the laminar flame speeds of CTL-FT fuel and Jet A-1 using the cone angle method. The measurements were performed at 473K and pressures of 1 and 3 atm at the equivalence ratios of 0.95—1.30. The measured flame speeds for CTL-FT fuel are close to Jet A-1 at atmospheric pressure.

Wang et al. (2018) studied the laminar flame speed of GTL-FT fuel/air/diluent in a spherical vessel over a wide range of temperatures of 490—610K, pressures of 0.5—3.2 atm, $\phi = 0.7-1.2$, and two different diluent concentrations of 5% and 10%. A mixture of 32% isooctane, 25% n-decane, and 43% n-dodecane by volume was used as a surrogate to represent GTL fuel. The laminar flame speed decreases with the increase of pressure but increases with increasing preheated temperatures. Hwang et al. (2020) measured the laminar flame speeds of two types of alternative jet fuel using the Bunsen flame method at elevated temperature of 550K and atmospheric pressure. The tested synthetic jet fuels were synthesized cyclohydrocarbons derived via chemical reactions from crude oil—derived chemicals. It is noted that the fuels tested were not conforming to the ASTM D7566 standards, as the properties (e.g., density) deviate from the batch property specification as required by the synthetic jet fuel. The measured peak laminar burning velocities for the alternative jet fuels are higher than Jet A-1, but the flame speeds at fuel-lean region are comparatively lower.

Vukadinovic et al. (2012) measured the laminar burning velocity of GTL-FT fuel using a spherical bomb method. By imaging the spherical expanding flame in a constant volume chamber, information of the propagation rate of the flame kernel with different stretch rate can be obtained. The unstretched burning velocity can be obtained by extrapolating back to zero stretch rate. Result shows that the laminar flame speed of GTL and GTL blend with aromatics have similar values as Jet A-1. Kick et al. (2012) utilized the cone-angle method of Bunsen flame to derive the laminar flame speed of FT-SPK, CTL-FT, and blends. Fig. 4.15A shows the example of premixed conical flame established using GTL fuel at $\phi = 1.2$. The flame speed is derived from the cone angle and the velocity of the unburned gas based on the nozzle diameter and volumetric flow rate, $S_u = V_u \sin \alpha$, as illustrated in Fig. 4.15B. Among the fuels tested, no discernible difference in flame speed was observed, but the comparison with Jet A-1 shows some deviation at $\phi < 1.1$, where the synthetic jet fuel velocities are lower than Jet A-1.

Munzar et al. (2014) showed the laminar flame speed is sensitive to the blend ratio of HEFA with Jet A-1. In their study, they utilized a jet stagnation flame method to derive the flame speed for HEFA/Jet fuel blends of

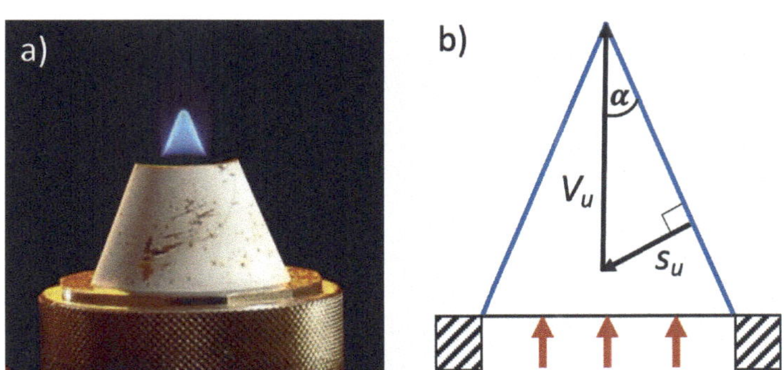

Figure 4.15 (A) Premixed conical flame used to obtain the laminar flame speed of GTL/air established at $\phi = 1.2$. (B) Cone angle method used to derive the flame speed (Kick et al., 2012).

different ratios at preheat temperature of 400K and atmospheric pressure. The 20% camelina–HEFA blend with Jet A-1 showed similar flame speed trend as Jet A-1, but 50% camelina–HEFA blend showed slightly higher reactivity across all equivalence ratios. Similar blend fractions of 20% and 50% of Jatropha HEFA with Jet A-1 showed lower flame speed at fuel-lean and fuel-rich regions, but the stoichiometric flame speed for the blends is slightly higher. Hui et al. (2012) compared the laminar flame speeds of FT fuels and HEFA with Jet A derived at 400 and 470K as a function of equivalence ratios. The flame speeds of alternative jet fuels are quite similar to Jet A, regardless of the preheating temperatures and equivalence ratios tested owing to the similar heat of combustion. The increase in preheat temperature results in the increase of laminar flame speeds by about 30% for all fuels.

Fig. 4.16 shows the comparison of atmospheric laminar flame speeds of several alternative jet fuels derived at the preheat temperature of 470–473K with conventional jet fuel. The laminar flame speeds for Jet A-1 are found to be quite different, which can be attributed to the differences in experimental techniques and the associated uncertainties in measurements and extrapolation methods used by different research group. Larger discrepancies in flame speed can be observed on the fuel-rich side for Jet A-1, but most alternative jet fuels are measured within the jet fuel band. Within each

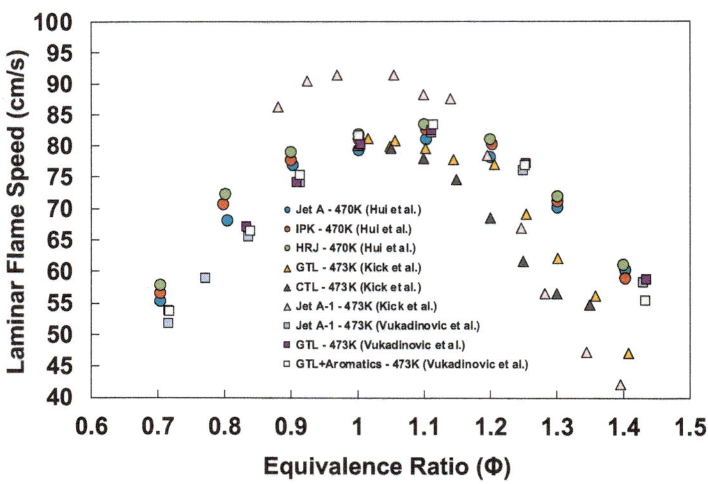

Figure 4.16 Laminar flame speeds of conventional and alternative jet fuels derived at elevated temperature ~470-473K as a function of equivalence ratios (Hui et al., 2012; Kick et al., 2012; Vukadinovic et al., 2012).

group of synthetic jet fuel, the measured values are similar and close to conventional jet fuel, which implies similar reactivity. Laminar flame speed is known to be strongly dependent on adiabatic flame temperature rather than fuel composition; thus, the Arrhenius kinetics and heat of combustion are important factors that influence flame speed. This explains the general similarity in laminar flame speed characteristic for different alternative jet fuels due to similar heat of combustion. The peaking flame speed at near stoichiometric condition for all fuels indicates that heat release is highest at this condition.

4.5.4 Extinction strain rate

Another combustion property of interest is the flame extinction limit, a phenomenon that is induced by the incomplete reaction in the flame, or the nonequidiffusivity of heat and mass in conjunction with the flame stretch effect manifested by the flow nonuniformity, flame curvature, and flow-flame unsteadiness. A typical flame configuration used to measure the flame extinction limit is counterflow twin-flame configuration. Under this setup, a flat flame is established between two nozzles under a stretched condition. The extinction of the flame is induced by gradually increasing the flow rates through the burner nozzles until the flame blows off abruptly. The maximum axial velocity gradient of the flame just before the flame blows out is defined as the extinction strain rate, which characterizes the interaction between characteristic flame/flow time and chemical reaction time. This parameter is used to describe flame stability and blowout event under engine operating conditions. There have been some studies conducted on measuring the extinction strain rates of alternative jet fuels.

Hui et al. (2012) compared the extinction characteristics of FT S-8, Sasol IPK, and camelina HRJ with Jet A. The fuels were first vaporized and established as flat flame using a twin-flame counterflow burner. Fig. 4.17 shows the result of the extinction stretch rate for the fuels at different equivalence ratios. It can be seen that the extinction characteristics are comparable at fuel-lean condition, but some deviation can be noticed at fuel-rich regions where the HEFA exhibits a higher stretched extinction than Sasol IPK and S-8, while Jet A shows the least resistant to extinction with the lowest extinction stretch values. Ji et al. (2011) also showed that synthetic jet fuels of S-8, Shell GTL, and R-8 are more resistant to extinction compared with JP-8 for nonpremixed flames established under opposed flame at 403K and atmospheric pressure. No distinct difference is

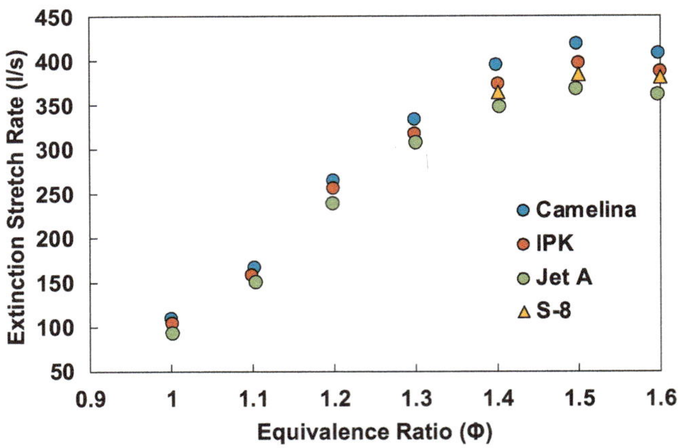

Figure 4.17 Extinction stretch rates of Jet A, S-8, IPK, and camelina—HEFA at 400K as a function of equivalence ratios (Hui et al., 2012).

shown in the extinction strain rates for S-8, Shell GTL, and R-8 fuel under nonpremixed flame condition. The same study also showed that premixed flames for synthetic jet fuels are more resistant to flame extinction than conventional jet fuel. In general, the lower resistance to flame extinction is attributable to the presence of aromatics, as the disintegration of the aromatic ring results in lower reactivity compared with aliphatics (Hui et al., 2012). Higher amount of aromatics will lead to lower extinction strain rate.

4.5.5 Sooting propensity

Evaluation of the sooting tendency of jet fuels can be performed via the smoke point (SP) measurement, which is a standard specification for aviation fuels as stated in ASTM D7566. Differing from the PM exhausted from the engines, the SP measurement represents the fundamental sooting tendency of a fuel indicative of the balance between soot formation and oxidation in a nonpremixed flame without the complexity in combustor. The SP of a fuel is defined as the maximum flame height in milimeters at which the fuel burns without smoking, which can be measured via a wick-fed lamp in accordance with the ASTM D1322 method (ASTM, 2019). In general, sooting tendency is inversely proportional to SP, where low SP indicates high sooting tendency and vice versa. The current ASTM D7566 specifies the minimum SP required for the final blend is 25 or 18 mm with a maximum 3 vol% of naphthalenes, although there is no batch requirement for synthetic jet fuel in neat form. Still, the SP values of alternative jet fuels or blends are of practical and scientific interests. Won et al. (2016) compared the derived smoke point with those of conventional jet fuels, as shown in Table 4.6. Since the alternative jet fuels contain little or no aromatic components, the H/C ratio and measured smoke point are considerably higher than those petroleum-derived jet fuels. Llamas et al. (2012) showed that the measured smoke point values for biokerosene/Jet A-1 blends are linearly related with biokerosene fuel fraction up to 20 vol%.

Another method that is commonly used to assess the sooting tendency of alternative jet fuel is via the use of threshold sooting index (TSI), which is related to smoke point and molecular structure as defined by Calcote and Manos (1983) via Eq. (4.4):

$$\mathrm{TSI} = \mathrm{a}\left(\frac{\mathrm{MW}}{\mathrm{SP}}\right) + \mathrm{b} \tag{4.4}$$

where a (mol mm/g) and b (dimensionless) are experimentally determined constants based on the fuel tested. Larger smoke point indicates lower sooting propensity, which is reflected in lower TSI due to their inverse relationship, thus making the latter a useful parameter to predict the sooting tendency of alternative jet fuel in practical combustor. The TSI of alternative jet fuels has been determined by some research groups, as shown in Table 4.6. It can be seen that the TSI may differ greatly for the same fuel, as the TSI depends upon an accurate determination of the molecular weight of the fuels, while the coefficients of a and b used are experiment specific. As expected, conventional jet fuels exhibit highest TSI values, while the alternative jet fuels show significantly lower TSI values. The discrepancy in TSI is largely due to the presence of aromatic content in

Table 4.6 Hydrogen/carbon ratio, smoke point and threshold sooting index for conventional and alternative jet fuels (Won et al., 2013,2016; Dooley et al., 2012a).

Fuel	H/C ratio	Smoke point	TSI
JP-8	2.02	24.4	19.3
Jet A	1.96	22.1	21.4
S-8	2.14	79.2	–
SPK	2.24	84.4	9.11
IPK	2.19	42.5	17.3
ATJ	2.17	35.2	–
Camelina HRJ	2.20	59.2	12.0
Tallow HRJ	2.18	62.1	11.6

the fuel. Synthetic jet fuel with low aromatic content generally exhibits low sooting tendency compared with jet fuel (Saffaripour et al., 2011). Han et al. (2018) investigated sooting propensity of 50% FT-derived biojet fuel blend via smoke point measurement. The soot emissions were reportedly reduced by half compared with conventional jet fuel. Among the alternative jet fuels, Sasol IPK has the highest TSI value due to its heavily branched alkane composition.

The IPK and ATJ show relatively higher sooting tendencies due to the presence of isoalkane and higher aromatic contents compared with straight-chain alkanes such as S8, Shell SPK, and HRJ8 (Kang et al., 2019b). The soot propensities for different hydrocarbon classes can generally be ranked as aromatics > cyclic alkanes > branched alkanes > linear alkanes (Calcote and Manos, 1983). Due to the limitation of smoke point to <45 mm, Won et al. (2013) utilized a "virtual" smoke point to derive the TSI values of alternative jet fuels, by extracting the smoke point values that vary in the blending ratio between a fuel and another fixed chemical component. Xue et al. (2017) investigated the soot volume fraction of biojet fuel including FT-SPK, camelina–HEFA, and ATJ fuels in the nonpremixed flame configuration at atmospheric condition using laser-induced incandescence technique. The TSI shows an approximate linear correlation with the maximum soot volume fraction for all the fuel blends. The sooting propensity of the jet fuels is ranked largely in accordance to the aromatic contents, where conventional jet fuel > ATJ > FT-SPK > HEFA (Xue et al., 2019), except ATJ which has lower aromatic content than FT-SPK. The ATJ fuel has higher sooting propensity due to the presence of heavier hydrocarbons.

4.5.6 Formulation of surrogates for alternative jet fuels

Practical liquid fuels for use in transportation engines consist of hundreds of hydrocarbons of different molecular classes. Such complexity and variation in the fuel composition makes it almost impossible to simulate the fuel performance using the exact fuel composition, as the computational power required would be extremely high, not to mention the formidable challenge in developing the exact fuel chemical kinetic mechanism. In view of the extensive use of computational engine modeling by engine manufacturers to aid in design and test of new engine component from system level prior to prototype fabrication stage, the ability to predict the engine's efficiency and combustion emissions with high level of confidence becomes highly critical, as it is directly related to the cost and time of production. The ability to predict the fuel performance relies largely on the accuracy of the fuel models, which should be of high fidelity so that the fuel properties and behaviors can be accurately reflected in different engine operating conditions. In the case of a newly developed alternative jet fuel, the combustion performance and emissions must be predicted with high reliability when applied in aviation turbine engine. To overcome the difficulty of overcomplex fuel components, a surrogate fuel model approach can be adopted to mimic the behaviors of the target fuel. A surrogate typically consists of a limited number of hydrocarbons, which are formulated to emulate the real practical fuel's thermophysical and chemical properties. A single component surrogate may not be sufficient to reproduce the reactive characteristics and is certainly insufficient to represent the varied molecular classes and carbon distribution of real fuels. Thus, fuel modelers adopt the multicomponent surrogate approach to capture a broad range of characteristics, be it under engine operating conditions or fundamental combustion aspects for a wide range of application. Such approach simplifies the components required and reduces the reactions to manageable size, so that the computational time frame required for simulation can be reasonably achievable.

There have been ongoing efforts to design and optimize the surrogates for conventional jet fuel, as they play an important role in aero engine development. The ability to predict the engine's performance and combustion characteristics accurately can significantly cut down the cost and production time. This can be extended to the development of alternative jet fuel, where surrogates can provide a screening process by reducing the number of large-scale rig or engine tests required for testing and certifying a candidate jet fuel. Much effort has been devoted to study the surrogates for alternative jet fuels since the introduction of drop-in alternative jet fuels. Some of the recent developments of surrogates for alternative jet fuels are shown in Table 4.7. It can be seen that the surrogates developed varied in terms of composition and validation targets, even for the same target fuel. This is due to the different strategies adopted to develop fuel surrogates, including the use of distillation curve analysis, chemical structure-based modeling method, and the physical and combustion properties matching method. Although the primary aim of jet fuel surrogate is to simulate the fuel oxidation chemistry to reflect the combustion kinetics, the ability of

Table 4.7 List of surrogates developed for alternative jet fuels.

Target fuel	Surrogate fuels	Validation targets	References
S-8	n-Nonane/2,6-dimethyloctane/3-methyldecane/n-tridecane/n-tetradecane/n-pentadecane/n-hexadecane 0.03/0.28/0.34/0.13/0.20/0.015/0.005 (by mol)	Distillation curve	Huber et al. (2008)
S-8	n-Decane/i-octane 0.6/0.4 (by vol)	Autoignition delay time and species profile	Mawid (2007)
S-8	Isooctane/n-decane 80/20 (by vol)	Ignition delay time	Gokulakrishnan et al. (2008)
S-8	4-Methyloctane/2,5-dimethylnonane/2,3,5-trimethyldecane/n-tridecane/n-pentadecane 0.105/0.281/0.164/0.227/0.223 (by mol)	Distillation curve	Huber et al. (2011)
Shell GTL	Isooctane/n-decane/n-dodecane 28/61/11 (by mol)	Autoignition temperature, laminar flame speed, extinction strain rate, NO_x emissions	Naik et al. (2011)
S-8	Isooctane/n-decane/n-dodecane 32/25/43 (by mol)		
S-8	n-Dodecane/isooctane 51.9/48.1 (by mol)	Species profiles, ignition delay time, extinction limit of diffusion flame	Dooley et al. (2012b)
GTL FT	n-Decane/isooctane/n-propylcyclohexane 0.58/0.33/0.09 (by mol)	Species profiles, ignition delay time, laminar flame speed	Dagaut et al. (2014)
CTL FT	n-Decane/n-dodecane/n-tetradecane/isooctane/methylcyclohexane 0.026/0.603/0.229/0.117/0.025 (by mol)	Droplet SMD, CO, and NO_x emissions	Xu et al. (2017)

Table 4.7 List of surrogates developed for alternative jet fuels.—cont'd

Target fuel	Surrogate fuels	Validation targets	References
IPK	n-Dodecane/isocetane/ isooctane/decalin 0.1416/0.3141/0.4016/ 0.1427 (by vol)	Ignition delay time	Kim et al. (2017)
S-8	n-Dodecane/n-decane/ isocetane/isooctane 0.3073/0.4234/0.2309/ 0.0384 (by vol)		
CHCJ	n-Butylcyclohexane n-Butylbenzene/n-dodecane 0.64/0.36 (by mol) n-Butylcyclohexane/n-butylbenzene/n-dodecane 0.104/0.582/0.314 (by mol), 0.246/0.508/0.246 (by mol), 0.453/0.396/0.151 (by mol)	Burn duration in crank angle degrees, crank angle degree location where 50% of the fuel has burned, maximum rate of heat release, engine thermal efficiency, ignition delay	Prak et al., 2017
IPK	n-Dodecane/isocetane/ isooctane/decalin/2,2,4,6,6-pentamethylheptane 16.91/ 7.38/23.2/10.89/41.63 (by vol) n-Dodecane/isooctane/ decalin/2,2,4,6,6-pentamethylheptane 18.03/ 19.77/11.71/50.49 (by vol)	Density, distillation curve, ignition delay time	Kim and Violi (2018)
IPK	n-Dodecane/isocetane/ isooctane/decalin 0.1416/0.3141/0.4016/ 0.1427 (by vol)	CO emissions, heat release rate, ignition delay, threshold sooting index	Kang et al. (2019a)
S-8	n-Dodecane/n-decane/ isocetane/isooctane 0.3073/ 0.4234/0.2309/0.0384 (by vol)		

the surrogate to reflect the physical properties is also important as it is related to the spray atomization process in the aviation turbine engine (Slavinskaya et al., 2010).

The synthetic jet fuel of derived from natural gas (S-8) is the most investigated target fuel for surrogate fuel formulation. Huber et al. (2008) developed a seven-component surrogate to represent the thermophysical properties of natural gas—derived synthetic jet fuel, S-8. The surrogate is able to match the experimental data to within 1%. The overall shape of the distillation curve is primarily governed by the four major components, i.e., 2,6-dimethyloctane, 3-methyldecane, n-tridecane, and n-tetradecane, while the presence of n-nonane and n-hexadecane in small amount reflects the initial boiling behavior and the tail of the distillation curve. They later demonstrated the concept of dynamic data evaluation within the NIST ThermoData Engine to generate equations of state (EOS) on demand to develop a surrogate for synthetic fuel derived from biomass (Huber et al., 2011). A five-component S-8 surrogate was shown to provide a good match with the distillation curve of S-8 to within 0.1% accuracy. Gokulakrishnan et al. (2008) developed a surrogate for S8 that consists of 80% vol isooctane and 20% n-decane to represent the content of isoparaffins and n-paraffins, respectively. The surrogate model was able to predict the ignition characteristic of the fuel reasonably well, partly due to the simplicity of the component in the synthetic fuel that could be sufficiently represented with a two-component surrogate. The result also showed that the NTC region between 700 and 900K is sensitive to the presence of isoparaffins to within two orders of magnitude. From the ignition characteristic evaluation, the two-component surrogate for S-8 (n-decane and isooctane) developed by Mawid (2007) was shown to have higher reactivity compared with the surrogates of JP-8. Naik et al. (2011) developed a high-temperature (>1000K) reaction mechanism for the surrogates of Shell GTL and S-8 using three basic components of isooctane/n-decane/n-dodecane of different ratios. The surrogates were intended to emulate not only the combustion and emissions characteristics for high-temperature chemical kinetics relevant to jet engine combustion but also the physical and chemical properties of the target fuels. The method used to construct the mechanism was done in steps, where the single mechanism containing the surrogate components was first assembled using a surrogate blend optimizer software, followed by the addition of NO_x and PAH submechanisms. Finally, the accuracy of the mechanism was optimized by applying the rate rules. Validation of the mechanism of surrogates against fundamental properties including laminar flame speeds and extinction rates with good agreement was achieved. Furthermore, the mechanisms were validated against premixed stretched flames and was used to predict the NO_x emissions, but the sooting characteristics have not been validated.

Dooley et al. (2012b) proposed a surrogate for S-8 that consists of n-dodecane and isooctane. The surrogate fuel matches the target fuel of S-8 from the aspects of H/C ratio and DCN, but the mixture averaged molecular weight number is slightly mismatched due to the heavier n-dodecane. The surrogate fuel is validated via the measurements of flow reactor oxidation, shock tube ignition delay, and diffusion flame strained extinction with good agreement, indicating the reactivity of the S-8 is sufficiently emulated. Dagaut et al. (2014) proposed a three-component surrogate for GTL-FT that consists of n-decane, isooctane (2,2,4-trimethyl pentane), and n-propylcyclohexane. The surrogate was formulated in accordance with the chemical composition of the fuel determined quantitatively and was used to simulate the oxidation kinetics, laminar flame speed, and autoignition delay time. Comparison against the experimental data shows that the surrogate is able to predict the species profiles and laminar flame speed accurately, but the calculated ignition delay time is slightly larger than the measurements. The measured ignition delay time for GTL-surrogate showed similar trend as those of GTL fuel. Kim et al. (2017) used the surrogate optimizing method to derive the surrogate fuels for FT fuels of Sasol IPK and S-8. The method uses models and correlations to estimate the various chemical and physical properties of the HC mixtures to determine the surrogate compositions that best fit the target fuel properties. The cycloalkane contents in IPK and S-8 are emulated using their respective surrogates. Apart from emulating the chemical and physical properties critical to spray and ignition behavior, the DCN of the surrogates matched the values determined experimentally. The surrogates were able to capture the trends shown experimentally, especially in the high-temperature regime. Kim and Violi (2018) utilized the surrogate optimizer model to identify the suitable components for IPK synthetic fuel. Two surrogates (four- and five-component) with substantial amount of branched alkane (2,2,4,6,6-pentamethylheptane) was shown to provide a good match with the densities and distillation curves. The ignition characteristics were also better represented, especially in the low temperature region.

Xu et al. (2017) utilized hybrid mixing model based on explicit equations and artificial neural network (ANN) to develop a surrogate jet fuel for CTL-FT fuel. The ANN model is used to predict the physical properties of the surrogate mixtures, while the thermo and chemical properties are estimated using three linear equations. The formulated five-component surrogate was able to match the target physicochemical properties to within 4.6%. Measurements of the spray droplet size show that the surrogate fuel was able to reflect the droplet SMD of FT fuel. The emissions of CO between the surrogate and FT fuel are quite similar, but the NO_x emissions for surrogate are slightly higher. Prak et al., 2017 formulated five different surrogates with one to three components to represent the properties of catalytic hydrothermal conversion jet (CHCJ) fuel. The formulation was done by matching the physical properties with CHCJ, including the density, kinematic viscosity, speed of sound, bulk modulus, surface tension, flash point, and derived cetane number. They validated the surrogates using data obtained from engine operating conditions. It is clear that validation of the surrogates requires extensive validations, from fuel's physicochemical properties, fundamental combustion properties to system-level engine data. To develop reliable surrogates for alternative jet fuels, more experimental efforts are needed to provide data for target validations.

4.6 Summary

The rigorous jet fuel certification process consists of fuel tests of physicochemical properties, rig and component tests, engine tests, and flight tests. This chapter reviews some of the combustion tests of alternative jet fuels needed to fulfill the requirements for certification. Due to regulatory requirement, the produced alternative jet fuel must be "drop-in" in nature, so that it can be blended in certain limits with conventional fuel and used as operating fuel without compromising flight safety. Since the establishment of ASTM D7566, more emerging technologies have shown the capability to produce alternative jet fuels with different compositions. Substitution of the jet fuel with alternative jet fuels requires comprehensive knowledge of the physicochemical properties and combustion characteristics of alternative jet fuel. This leads to the interest in continuously refining the fuel certification program to streamline the required tests as to better understand the impact of fuel properties on combustor performance. Component and rig tests of alterative jet fuels are usually carried out in facilities dedicated for specific tests such as spray atomization, ignition, flame blowout, and emissions tests, even though there are no standardized practices and procedures. The component tests can be complemented with fundamental combustion data set to gain better understanding on the fuel chemistry effects. Fundamental flame studies including ignition delay time, laminar flame speeds, extinction limits, and fuel chemical kinetics study are related to engine performance such as altitude relight, flame propagation, lean blowout, and engine emissions. The fundamental combustion data can serve as validation targets for surrogate fuel models. There are several surrogate fuel models developed for alternative jet fuels with reasonably well validation achieved. The high fidelity of the surrogate fuel model is important to ensure the accuracy of combustion modeling and development of predictive tool. Thus, further studies on alternative jet fuel combustion are required for a better understanding of fuel oxidation chemistry and comprehensive surrogate models with a wider range of predictability. To achieve the goal of sustainable and green aviation, development of low-carbon alternative jet fuel requires thorough screening, characterization, and rigorous testings, as well as the validated fuel models that can be used for modeling and predicting the engine performance. The combination of fundamental combustion research and rig tests is essential to facilitate the development of alternative jet fuel and aid in the design of advanced aero engine and related aviation infrastructure.

References

AIR6241, 2013. Procedure for the Continuous Sampling and Measurement of Nonvolatile Particle Emissions from Aircraft Turbine Engines, SAE Aerospace Information Report (AIR).

Airportwatch, 2012. Etihad Airways Operates First Biofuel (Recycled Cooking Oil) Powered Delivery Flight. https://www.airportwatch.org.uk/2012/01/next-stop-for-long-haul-airlines-set-a-course-to-the-east/.

Allen, C., Toulson, E., Edwards, T., Lee, T., 2012. Application of a novel charge preparation approach to testing the autoignition characteristics of JP-8 and camelina hydroprocessed renewable jet fuel in a rapid compression machine. Combust. Flame 159, 2780–2788.

Altaher, M.A., Li, H., Blakey, S., Chung, W., 2014. NMHC and VOC speciation of the exhaust gas from a gas turbine engine using alternative, renewable and conventional jet A-1 aviation fuels. In: Proceedings of ASME Turbo Expo 2014: Turbine Technical Conference and Exposition, Düsseldorf, Germany, GT2014–25445.

ANA, 2012. The World's First Trans-atlantic Flight Using Biofuels. https://www.ana.co.jp/eng/aboutana/corporate/csr/stakeholders/environment/biofuel.html.

Askari, O., Elia, M., Ferrari, M., Metghalchi, H., 2017. Auto-ignition characteristics study of gas-to-liquid fuel at high pressures and low temperatures. J. Energy Resour. Technol. 139, 012204−1−6.

ASTM, 2016. ASTM D7170−16. Standard Test Method for Determination of Derived Cetane Number (DCN) of Diesel Fuel Oils—Fixed Range Injection Period, Constant Volume Combustion Chamber Method. ASTM International.

ASTM, 2018. ASTM D6890−18. Standard Test Method for Determination of Ignition Delay and Derived Cetane Number (DCN) of Diesel Fuel Oils by Combustion in a Constant Volume Chamber. ASTM International.

ASTM, 2017. ASTM D7668-17. Standard Test Method for Determination of Derived Cetane Number (DCN) of Diesel Fuel Oils—Ignition Delay and Combustion Delay Using a Constant Volume Combustion Chamber Method. ASTM International.

ASTM, 2019. ASTM D1322−19, Standard Test Method for Smoke Point of Kerosene and Aviation Turbine Fuel. ASTM International.

GE Aviation, 2009. Continental Airlines Flight Demonstrates Use of Sustainable Biofuels as Energy Source for Jet Travel. https://www.geaviation.com/press-release/jv-archive/continental-airlines-flight-demonstrates-use-sustainable-biofuels-energy.

Bachmaier, F., Eberius, K.H., Just, T., 1973. The formation of nitric oxide and the detection of HCN in premixed hydrocarbon-air flames at 1 atmosphere. Combust. Sci. Technol. 7 (2), 77−84.

Bessee, G.B., Hutzler, S.A., Wilson, G.R., 2011. Propulsion and Power Rapid Response Research and Development (R&D) Support. Delivery Order 0011: Analysis of Synthetic Aviation Fuels. DTIC Document.

Biofuels Digest, 2011. Interjet, Airbus Complete Jatropha Aviation Biofuels Trial in Mexico. https://www.biofuelsdigest.com/bdigest/2011/04/04/interjet-airbus-complete-jatropha-aviation-biofuels-trial-in-mexico/.

Boeing, 2019. Boeing Pilots the Birth of Biofuel for Aviation's Sustainable Growth. https://www.boeing.com/features/innovation-quarterly/2019_q3/btj-sustainability.page.

Bokhart, A.J., Shin, D., Gejji, R., Buschhagen, T., Naik, S.V., Lucht, R.P., 2017. Spray measurements at elevated pressures and temperatures using phase Doppler anemometry. In: AIAA SciTech Forum, 55th AIAA Aerospace Sciences Meeting, Grapevine, Texas, 2017. https://doi.org/10.2514/6.2017-0828.

Braun-Unkhoff, M., Riedel, U., Wahl, C., 2017. About the emissions of alternative jet fuels. CEAS Aeronaut. J. 8, 167−180.

Breaking Travel News, 2015. Hainan Airlines brings first commercial biofuel flight to China. https://www.breakingtravelnews.com/news/article/hainan-airlines-brings-first-commercial-biofuel-flight-to-china.

Burger, V., Yates, A., Viljoen, C., 2012. Influence of fuel physical properties and reaction rate on threshold heterogeneous gas turbine combustion. In: Proceedings of the ASME Turbo expo. Copenhagen, GT2012−68153.

Buffi, M., Valera-Medina, A., Marsh, R., Pugh, D., Giles, A., Runyon, J., Chiaramonti, D., 2017. Emissions characterization tests for hydrotreated renewavle jet fuel from used cooking oil and its blends. Appl. Energy 201, 84−93.

Burger, V., Mosbach, T., Yates, A., Gunasekaran, B., 2014. Fuel influence on targeted gas turbine combustion properties part II: detailed results. In: Proceedings of ASME Turbo Expo. Düsseldorf, GT2014−25105.

Calcote, H., Manos, D., 1983. Effect of molecular structure on incipient soot formation. Combust. Flame 49, 289−304.

Chan, T.W., Chishty, W.A., Canteenwalla, P., Buote, D., Davison, C.R., 2016. Characterization of emissions from the use of alternative aviation fuels. J. Eng. Gas Turbines Power 138, 011506−1−9.

China Daily, 2011. Air China Conducts First Biofuel Test Flight. www.chinadaily.com.cn/china/2011-10/28/content_13998217.htm.
China Daily, 2013. China Eastern Tests Flight Using Biofuel. http://www.chinadaily.com.cn/business/2013-04/25/content_16447065.htm.
Corporan, E., DeWitt, M.J., Belovich, V., Pawlik, R., Lynch, A.C., Gord, J.R., Meyer, T.R., 2007. Emissions characteristics of a turbine engine and research combustor burning a Fischer-Tropsch jet fuel. Energy Fuels 21, 2615–2626.
Dagaut, P., Karsenty, F., Dayma, G., Diévart, P., Hadj-Ali, K., Mzé-Ahmed, A., et al., 2014. Experimental and detailed kinetic model for the oxidation of a Gas to Liquid (GtL) jet fuel. Combust. Flame 161, 835–847.
Dickerson, T., McDaniel, A., Williams, S., Luning-Prak, D., Hamilton, L., Bermudez, E., Cowart, J., 2015. Start-up and steady-state performance of a new renewable alcohol-to-jet (ATJ) fuel in multiple diesel engines. In: SAE Pap. 2015, Paper No. 2015-01-0901.
Digest, B., 2012. The Airlines: Who's Doing what in Aviation Biofuels? https://www.biofuelsdigest.com/bdigest/2015/10/19/the-airlines-whos-doing-what-in-aviation-biofuels/4/.
Dooley, S., Won, S.H., Heyne, J., Farouk, T.I., Ju, Y., Dryer, F., et al., 2012a. The experimental evaluation of a methodology for surrogate fuel formulation to emulate gas phase combustion kinetic phenomena. Combust. Flame 159, 1444–1466.
Dooley, S., Won, S.H., Jahangirian, S., Ju, Y., Dryer, F., Wang, H., Oehlschlaeger, M.A., 2012b. The combustion kinetics of a synthetic paraffinic jet aviation fuel and a fundamentally formulated, experimentally validated surrogate fuel. Combust. Flame 159, 3014–3020.
EASA, 2019. European Aviation Environmental Report. https://www.easa.europa.eu/eaer/.
Edwards, T., Moses, C., Dryer, F., 2010. Evaluation of combustion performance of alternative aviation fuels. In: 46th AIAA/ASME/SAE/ASEE Joint Propulsion Conference & Exhibit, Nashville, TN, 2010, AIAA 2010-7155.
Express, 2008. Air New Zealand Biofuel Flight Test. https://www.express.co.uk/news/world/77677/Air-New-Zealand-biofuel-flight-test.
FAA, 2016. Federal Aviation Administration, Swedish Biofuel Performance Evaluation, Report no.: DOT/FAA/AEE/2016-05, 2016.
Fenimore, C.P., 1971. Formation of nitric oxide in premixed hydrocarbon flames. Symp. (Intl.) Combust. 13, 373–380.
Flora, G., Balagurunathan, J., Saxena, S., Cain, J.P., Kahandawala, M.S.P., DeWitt, M.J., et al., 2017. Chemical ignition delay of candidate drop-in replacement jet fuels under fuel-lean conditions: a shock tube study. Fuel 209, 457–472.
Fyffe, D., Moran, J., Kannaiyan, K., Al-Sharshani, A., Sadr, R., 2011. Effect of GTL-like jet fuel composition on gt engine altitude ignition performance: Part I—combustor operability. In: Proceedings of the ASME Turbo Expo. Vancouver, GT2011–45487.
Gawron, B., Bialecki, T., 2018. Impact of a Jet A-1/HEFA blend on the performance and emission characteristics of a miniature turbojet engine. Int. J. Environ. Sci. Technol. 15, 1501–1508.
Gokulakrishnan, P., Klassen, M.S., Roby, R.J., 2008. Ignition characteristics of a fischer-tropsch synthetic jet fuel. In: Proceedings of ASME Turbo Expo 2008: Power for Land, Sea and Air, Berlin, Germany, GT2008–51211, pp. 1–9.
Greenair, 2008. Biofuelled Virgin Boeing 747 Takes to the Skies on Pioneering First Flight. https://www.greenaironline.com/news.php?viewStory=116.
Greenair, 2009. Qatar Airways Undertakes First Commercial Passenger Flight Powered by a Natural Gas Blended Jet Fuel. https://www.greenaironline.com/news.php?viewStory=624.

Greenaironline, 2014. Boeing's Latest 777F ecoDemonstrator Flight Programme Tests 100% Biofuel Use and Turbulence Detection on GreenAir Online. https://www.greenaironline.com/news.php?viewStory=2462.

Gulfnews, 2019. Etihad Flies World's First Commercial Flight Using UAE-Made Fuel from Plants. https://gulfnews.com/business/aviation/etihad-flies-worlds-first-commercial-flight-using-uae-made-fuel-from-plants-1.1547634775232.

Han, H.S., Kim, C.J., Cho, C.H., Sohn, C.H., Han, J., 2018. Ignition delay time and sooting propensity of a kerosene aviation jet fuel and its derivative blended with a bio-jet fuel. Fuel 232, 724–728.

Hermann, F., Hedemalm, P., Orbay, R., Gabrielsson, R., Klingmann, J., 2005. Comparison of combustion properties between a synthetic jet fuel and conventional Jet A-1. In: Proceedings of ASME Turbo Expo. Nevada, GT2005–68540.

Huang, C.-S., Vander Wal, R.L., 2013. Effect of soot structure evolution from commercial jet engine burning petroleum based JP-8 and synthetic HRJ and FT fuels. Energy Fuels 27, 4946–4958.

Huber, M.L., Smith, B.L., Ott, S., Bruno, T.J., 2008. Surrogate mixture model for the thermophysical properties of synthetic aviation fuel S-8: explicit application of the advanced distillation curve. Energy Fuels 22, 1104–1114.

Huber, M.L., Bruno, T.J., Chirico, R.D., Diky, V., Kazakov, A.F., Lemmon, E.W., et al., 2011. Equations of state on demand: application for surrogate fuel development. Int. J. Thermophys. 32, 596–613.

Hui, X., Sung, C.-J., 2013. Laminar flame speeds of transportation-relevant hydrocarbons and jet fuels at elevated temperatures and pressures. Fuel 109, 191–200.

Hui, X., Kumar, K., Sung, C.-J., Edwards, T., Gardner, D., 2012. Experimental studies on the combustion characteristics of alternative jet fuels. Fuel 98, 176–182.

Hwang, B.J., Kang, S., Lee, H.J., Min, S., 2020. Measurement of laminar burning velocity of high performance alternative aviation fuels. Fuel 261, 116466.

IATA, 2015. IATA Sustainable Aviation Fuel Roadmap. IATA.

IATA, 2020. What Is SAF. IATA. https://www.iata.org/contentassets/d13875e9ed784f75bac90f000760e998/saf-what-is-saf.pdf.

ICAO, 2017. Local Air Quality. https://www.icao.int/environmental-protection/Pages/local-air-quality.aspx.

ICAO, 2018. Sustainable Aviation Fuels Guide. https://www.icao.int/environmental-protection/Documents/Sustainable%20Aviation%20Fuels%20Guide_100519.pdf.

Indiatimes, 2018. SpiceJet Operates India's First Biofuel-Powered Flight from Dehradun to Delhi. https://indianexpress.com/article/business/aviation/spicejet-operates-indias-first-biofuel-powered-flight-5326913/.

Jacob, S.D., Rindlisbacher, T., 2019. The Landing and Take-Off Particulate Matter Standards for Aircraft Gas Turbine Engines, Chapter 3, Local Air Quality. https://www.icao.int/environmental-protection/Documents/EnvironmentalReports/2019/ENVReport2019_pg100-105.pdf.

Ji, C., Wang, Y.L., Egolfopoulos, F.N., 2011. Flame studies of conventional and alternative jet fuels. J. Propul. Power 27 (4), 856–863.

Jürgens, S., Oßwald, P., Selink, M., Piermartini, P., Schwab, J., Pfeifer, P., et al., 2019. Assessment of combustion properties of non-hydroprocessed Fischer-Tropsch fuels for aviation. Fuel Process. Technol. 193, 232–243.

Kahandawala, M.S.P., DeWitt, M.J., Corporan, E., Sidhu, S.S., 2008. Ignition and emission characteristics of surrogate and practical jet fuels. Energy Fuels 22, 3673–3679.

Kang, D., Kim, D., Kalaskar, V., Boehman, A., Violi, A., 2019a. Experimental characterization of jet fuels under engine relevant conditions – Part 2: insights on optimization approach for surrogate formulation. Fuel 239, 1405–1416.

Kang, D., Kim, D., Kalaskar, V., Violi, A., Boehman, A.L., 2019b. Experimental characterization of jet fuels under engine relevant conditions – Part 1: effect of chemical composition on autoignition of conventional and alternative jet fuels. Fuel 239, 1388–1404.

Kannaiyan, K., Sadr, R., 2014a. Effect of fuel properties on spray characteristics of alternative jet fuels using global sizing velocimetry. Atomization Sprays 24 (7), 575–597.

Kannaiyan, K., Sadr, R., 2014b. Experimental investigation of spray characteristics of alternative aviation fuels. Energy Convers. Manag. 88, 1060–1069.

Kannaiyan, K., Sadr, R., 2020. Macroscopic spray performance of alternative and conventional jet fuels at non-reacting, elevated ambient conditions. Fuel 266, 117023.

Kick, T., Herbst, J., Kathrotia, T., Marquetand, J., Braun-Unkhoff, M., Naumann, C., Riedel, U., 2012. An experimental and modeling study of burning velocities of possible future synthetic jet fuels. Energy 43, 111–123.

Kim, D., Violi, A., 2018. Hydrocarbons for the next generation of jet fuel surrogates. Fuel 228, 438–444.

Kim, D., Martz, J., Abdul-Nour, A., Yu, X., Jansons, M., Violi, A., 2017. A six-component surrogate for emulating the physical and chemical characteristics of conventional and alternative jet fuels and their blends. Combust. Flame 179, 86–94.

Klingshirn, C.D., Dewitt, M., Striebich, R., Anneken, D., Shafer, L., Corporan, E., et al., 2012. Hydroprocessed renewable jet fuel evaluation, performance, and emissions in a T63 turbine engine. J. Eng. Gas Turbines Power 134, 051506–1–8.

Kumal, R.R., Liu, J., Gharpure, A., Wal, R.L.V., Kinsey, J.S., Giannelli, B., et al., 2020. Impact of biofuel blends on black carbon emissions from a gas turbine engine. Energy Fuels 34, 4958–4966.

Kumar, K., Sung, C.-J., 2010. A comparative experimental study of the autoignition characteristics of alternative and conventional jet fuel/oxidizer mixtures. Fuel 89, 2853–2863.

Kumar, K., Sung, C.-J., Hui, X., 2011. Laminar flame speeds and extinction limits of conventional and alternative jet fuels. Fuel 90, 1004–1011.

Kuo, K.K., 2005. Principles of Combustion. John Wiley & Sons, Inc.

Levy, A., 1982. Unresolved problems in SOx, NOx, soot control in combustion. Symp. (Intl.) Combust. 19 (1), 1223–1242.

Liati, A., Scheriber, D., Alpert, P.A., Liao, Y., Brem, B.T., Corral, A., et al., 2019. Aircraft soot from conventional fuels and biofuels during ground idle and climb-out conditions: electron microscopy and X-ray micro-spectroscopy. Environ. Pollut. 247, 658–667.

Lin, Y., Zhang, C., Xu, Q., Sung, C.J., Liu, G., 2012. Evaluation of combustion performance of a coal-derived synthetic jet fuel. In: Proceedings of ASME Turbo Expo. Copenhagen GT2012–68604.

Link, D.D., Gormley, R.J., Baltrus, J.P., Anderson, R.R., Zandhuis, P.H., 2008. Potential additives to promote seal swell in synthetic fuels and their effect on thermal stability. Energy Fuels 22 (2), 1115–1120.

Llamas, A., García-Martínez, M.-J., Al-Lal, A.-M., Canoira, L., Lapuerta, M., 2012. Biokerosene from coconut and palm kernel oils: production and properties of their blends with fossil kerosene. Fuel 102, 483–490.

Lobo, P., Hagen, D.E., Whitefield, P.D., 2011. Comparison of PM Emissions from a commercial jet engine burning conventional, biomass, and Fischer-Tropsch fuels. Environ. Sci. Technol. 45 (24), 10744–10749.

Lobo, P., Rye, L., Williams, R., Christie, S., Uryga-Bugajska, I., Wilson, C.W., et al., 2012. Impact of alternative fuels on emissions characteristics of a gas turbine engine – Part 1: gaseous and particulate matter emissions. Energy Fuels 46, 10805–10811.

Lobo, P., Christie, S., Khandelwal, B., Blakey, S.G., Raper, D.W., 2015a. Evaluation of non-volatile particulate matter emission characteristics of an aircraft auxiliary power unit with varying alternative jet fuel blend ratios. Energy Fuels 29, 7705–7711.
Lobo, P., Hagen, D.E., Whitefield, P.D., Rapper, D., 2015b. PM emissions measurements of in-service commercial aircraft engines during the Delta-Atlanta Hartsfield Study. Atmos. Environ. 104, 237–245.
Marketwatch, 2011. Biofuel-powered Jet Flies across the Atlantic. https://www.marketwatch.com/story/biofuel-powered-jet-flies-across-the-atlantic-2011-06-18.
Mawid, M.A., 2007. Development of a detailed chemical kinetic mechanism for JP-8 & Fisher-Tropsch-Derived synthetic jet fuels. In: 43rd AIAA/ASME/SAE/ASEE Joint Propulsion Conference and Exhibit, Cincinnati, USA, AIAA 2007-5668.
Min, K., Valco, D., Oldani, A., Lee, T., 2015. An ignition delay study of category a and c aviation fuel. In: Proceedings of the ASME 2015 International Mechanical Engineering Congress and Exposition, Houston, Texas, 2015, IMECE2015-51713.
Mosbach, T., Gebel, G.C., Clercq, P.L., Sadr, R., Kannaiyan, K., Al-Sharshani, A., 2011. Investigation of GTL-like jet fuel composition on GT engine altitude ignition and combustion performance: Part II—detailed diagnostics. In: Proceedings of ASME Turbo Expo. Vancouver, GT2011–45510.
Munzar, J.D., Zia, A., Versailles, P., Jiménez, R., Bergthorson, J.M., Akih-Kumgeh, B., 2014. Comparison of laminar flame speeds, extinction stretch rates and vapor pressures of jet a-1/HRJ biojet fuel blends. In: Proceedings of ASME Turbo Expo 2014: Turbine Technical Conference and Exposition, Düsseldorf, Germany, GT2014–25951, pp. 1–9.
Mze-Ahmed, A., Dagaut, P., Hadj-Ali, K., Dayma, G., Kick, T., Herbst, J., et al., 2012. Oxidation of a coal-to-liquid synthetic jet fuel: experimental and chemical kinetic modeling study. Energy Fuels 26, 6070–6079.
Naik, C.V., Puduppakkam, K.V., Modak, A., Meeks, E., Wang, Y.L., Feng, Q., Tsotsis, T.T., 2011. Detailed chemical kinetic mechanism for surrogates of alternative jet fuels. Combust. Flame 158, 434–445.
Newatlas, 2012. World's First 100 Percent Biofuel-Powered Flight of Civil Aircraft. https://newatlas.com/nrc-biofuel-flight/24896/.
Prak, D.J.L., Romanczyk, M., Wehde, K.E., Ye, S., Mclaughlin, M., Luning Prak, P.J., et al., 2017. Analysis of catalytic hydrothermal conversion jet fuel and surrogate mixture formulation: components, properties, and combustion. Energy Fuels 31, 13802–13814.
Pucher, G., Allan, W., LaViolette, M., Poitras, P., 2011a. Emissions from a gas turbine sector rig operated with synthetic aviation and biodiesel fuel. J. Eng. Gas Turbines Power 133, 111502–1.
Pucher, G., Poitras, P., Allan, W., LaViolette, M., 2011b. Emissions from a gas turbine sector rig operated with synthetic aviation and biodiesel fuel. J. Eng. Gas Turbines Power 133, 111502.
Purcher, G., Allan, W., Poitras, P., 2013. Characteristics of deposits in gas turbine combustion chambers using synthetic and conventional jet fuels. J. Eng. Gas Turbines Power 135, 071502–1–8.
Riebl, S., Braun-Unkhoff, M., Riedel, U., 2017. A study on the emissions of alternative aviation fuels. J. Eng. Gas Turbines Power 139, 081503–1–11.
Rock, N., Chterev, I., Emerson, B., Seitzman, J., Lieuwen, T., 2017. Blowout sensitivities in a liquid fueled combustor: fuel composition and preheat temperature effects. In: Proceedings of ASME Turbo Expo 2017: Turbomachinery Technical Conference and Exposition, 2017, GT2017–63305. Charlotte, NC, USA.
Rye, L., Wilson, C.W., 2012. The influence of alternative fuel composition on gas turbine ignition performance. Fuel 96, 277–283.

Saffaripour, M., Thomson, K.A., Smallwood, G.J., Lobo, P., 2019. A review on the morphological properties of non-volatile particulate matter emissions from aircraft turbine engines. J. Aerosol Sci. https://doi.org/10.1016/j.jaerosci.2019.105467.

Sasol, 2010. Sasol Takes to the Skies with the World's First Fully Synthetic Jet Fuel. https://www.sasol.com/media-centre/media-releases/sasol-takes-skies-world-s-first-fully-synthetic-jet-fuel.

Saffaripour, M., Zabeti, P., Kholghy, M., Thomson, M.J., 2011. An experimental comparison of the sooting behavior of synthetic jet fuels. Energy Fuels 25, 5584–5593.

Schripp, T., Herrmann, F., Oßwald, P., Köhler, M., Zschocke, A., Weigelt, D., et al., 2019. Particle emissions of two unblended alternative jet fuels in a full scale jet engine. Fuel 256, 115903.

Sivakumar, D., Vankeswaram, S.K., Sakthikumar, R., BN, R., 2015. Analysis on the atomization characteristics of aviation biofuel discharging from simplex swirl atomizer. Int. J. Multiphas. Flow 72, 88–96.

Sivakumar, D., Vankeswaram, S.K., Sakthikumar, R., Raghunandan, B.N., Hu, J.T.C., Sinha, A.K., 2016. An experimental study on jatropha-derived alternative aviation fuel sprays from simplex swirl atomizer. Fuel 179, 36–44.

Slavinskaya, N.A., Zizin, A., Aigner, M., 2010. On model design of a surrogate fuel formulation. J. Eng. Gas Turbines Power 132, 111501–1–11.

Smith, P.M., Gaffney, M.J., Shi, W., Hoard, S., Armendariz, I.I., Mueller, D.W., 2017. Drivers and barriers to the adoption and diffusion of sustainable jet fuel (SJF) in the U.S. Pacific Northwest. J. Air Transport. Manag. 58, 113–124.

Sundararaj, R.H., Kumar, R.D., Raut, A.K., Sekar, T.C., Pandey, V., Kushari, A., Puri, S.K., 2019. Combustion and emission characteristics from biojet fuel blends in a gas turbine combustor. Energy 182, 689–705.

Vasu, S.S., Davidson, D.F., Hanson, R.K., 2008. Jet fuel ignition delay times: shock tube experiments over wide conditions and surrogate model predictions. Combust. Flame 152, 125–143.

Vukadinovic, V., Habisreuther, P., Zarzalis, N., 2012. Comparison of laminar flame speeds, extinction stretch rates and vapor pressures of Jet A-1/HRJ biojet fuel blends. J. Eng. Gas Turbines Power 134, 121504–1–9.

Wang, H., Oehlschlaeger, M.A., 2012. Autoignition studies of conventional and Fischer–Tropsch jet fuels. Fuel 98, 249–258.

Wang, Z., Bai, Z., Yelishala, S.C., Yu, G., Metghalchi, H., 2018. Effects of diluent on laminar burning speed and flame structure of gas to liquid fuel air mixtures at high temperatures and moderate pressures. Fuel 231, 204–214.

Wang, Z., Feser, J.S., Lei, T., Gupta, A.K., 2020. Performance and emissions of camelina oil derived jet fuel blends under distributed combustion condition. Fuel 271, 117685.

WIRED, 2011. KLM Completes First Scheduled Service Flight Using Biofuel. https://www.wired.com/2011/07/klm-completes-first-scheduled-service-flight-using-biofuel/.

Won, S.H., Veloo, P., Santner, J., Ju, Y., Dryer, F., 2013. Comparative evaluation of global combustion properties of alternative jet fuels. In: 51st AIAA Aerospace Sciences Meeting including the New Horizons Forum and Aerospace Exposition, Texas, USA, AIAA 2013-0156, pp. 1–8.

Won, S.H., Dooley, S., Veloo, P., Wang, H., Oehlschlaeger, M.A., Dryer, F., Ju, Y., 2014. The combustion properties of 2,6,10-trimethyl dodecane and a chemical functional group analysis. Combust. Flame 161, 826–834.

Won, S.H., Veloo, P., Dooley, S., Santner, J., Haas, F.M., Ju, Y., Dryer, F., 2016. Predicting the global combustion behaviors of petroleum-derived and alternative jet fuels by simple fuel property measurements. Fuel 168, 34–46.

Xu, Q., Liu, Z., Zhang, C., Hui, X., Lin, Y., 2017. Surrogate formulation for a coal-based jet fuel using a mixing model based on explicit equations and artificial neural network. Fuel 210, 262–271.
Xue, X., Hui, X., Singh, P., Sung, C.-J., 2017. Soot formation in non-premixed counterflow flames of conventional and alternative jet fuels. Fuel 210, 343–351.
Xue, H., Hui, X., Vannorsdall, P., Singh, P., Sung, C.-J., 2019. The blending effect on the sooting tendencies of alternative/conventional jet fuel blends in non-premixed flames. Fuel 237, 648–357.
Zel'dovich, J.B., 1946. Oxidation of nitrogen in combustion and explosion. Comptes Rendus (Doklady) de l'Academiedes Sciences de l'URSS 51 (3), 217–220.
Zhang, C., Hui, X., Lin, Y., Sung, C.J., 2016. Recent development in studies of alternative jet fuel combustion: Progress, challenges, and opportunities. Renew. Sustain. Energy Rev. 54, 120–138.
Zhao, J., Zhao, B., Wang, X., Yang, X., 2017. Atomization performance and TG analysis of Fischer-Tropsch fuel compared with RP-3 aviation fuel. Int. J. Hydrogen Energy 42, 18626–18632.
Zheng, L., Ling, C., Ubogu, E.A., Cronly, J., Ahmed, I., Zhang, Y., Khandelwal, B., 2018. Effects of alternative fuel properties on particulate matter produced in a gas turbine combustor. Energy Fuels 32, 9883–9897.
Zhu, Y., Li, S., Davidson, D.F., Hanson, R.K., 2015. Ignition delay times of conventional and alternative fuels behind reflected shock waves. Proc. Combust. Inst. 35, 241–248.

CHAPTER 5
Economics of biojet fuels

5.1 Introduction

The single largest hurdle in any product category entering the market as a mainstream product is in its economic feasibility. This is particularly true for alternative fuels, which already have a long-term incumbent in the form of petroleum. While the use of petroleum since the eras of ancient civilizations has been well documented, "modern"-day dominance of petroleum started in the midnineteenth century. This itself predated even the heavier-than-air human-made flight. By the point of the first flights around the turn of 20th century, the competing power sources were the petroleum-based gasoline, steam power, and also compressed gas for the key designs by the Wright brothers, Langley, and Herring, respectively. The Wright brothers' choice of a gasoline-powered internal combustion engine to rotate the propellers for thrust generation proved to be the right choice. With that, the dominance of petroleum has encroached even flight. Even the advent of jet engines only shifted the use of petroleum from the gasoline distillate to jet fuel fraction.

Since the mid-2000s, the dominance of fossil fuel for aviation applications was challenged by biojet fuels. This is made necessary by the need to combat climate change. Biojet fuels have all the right technical ingredients to supplant conventional jet fuel, as it can be produced sustainably, operate as a drop-in fuel for comparable flight performance, and help to decarbonize the sector. The only force slowing down its progress is market force as it is presently still not price competitive.

5.2 Biojet fuel prices

The price of petroleum is driven by and drives the world's economy, energy security concerns, and geopolitics. It is near impossible to disassociate petroleum price to the upstream, midstream and downstream products in the supply chain. In fact, it even dictates the price of

Biojet Fuel in Aviation Applications
ISBN 978-0-12-822854-8
https://doi.org/10.1016/B978-0-12-822854-8.00009-3

competing products. This is particularly true where crude oil price is directly correlated to aviation jet fuel price, which in turn affects biojet fuel price. Fig. 5.1 shows the conventional jet fuel price against crude oil price (IATA, 2020). In general, jet fuel price commands a slight premium over crude oil price, although the COVID-19 pandemic has caused an aberration in March 2020 where the inverse occurred. This happened because the demand for jet fuel plummeted due to the travel restrictions and lockdowns in place for many countries globally. The price has since been near identical although conventional jet fuel price is expected to regain its premium once air travel returns to normalcy.

Policies, legislation, and standards also play a part in determining or influencing prices. For biojet fuel, the price tie-in to conventional jet fuel is greater as all commonly accepted standards require "drop-in" for blending. This well-intended ruling to avoid changes to the operation of existing jet engines and infrastructures also meant that the biojet fuel in the blend with conventional jet fuel cannot be differentiated. Thus, both biojet fuel and conventional jet fuel must have very similar properties and perform similarly. This makes the use of biojet fuel and conventional jet fuel a zero-sum game in the financial sense as the market will always choose the cheaper option if products are functionally identical.

5.2.1 Sustainable aviation fuel price assessment

Unlike conventional jet fuel for the aviation sector, the biojet fuel market previously lacked a benchmark price. This changed amid the global

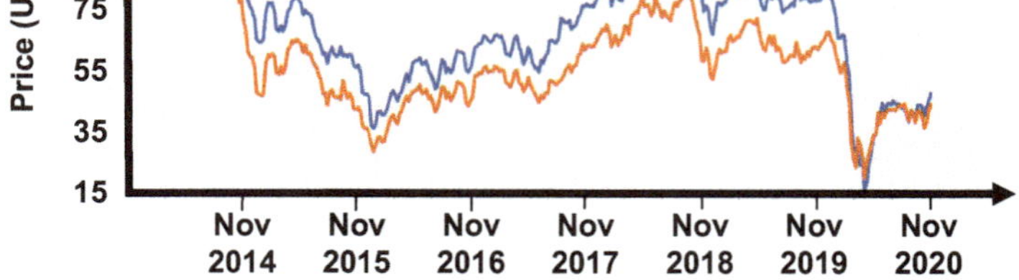

Figure 5.1 Jet fuel price versus crude oil price (Brent) from November 2013 to November 2020. *(Adapted from IATA, 2020. Jet Fuel Price Monitor. https://www.iata.org/en/publications/economics/fuel-monitor/. (Accessed 24 November 2020).)*

pandemic when S&P Global Platts ("Platts") launched the first-to-market Sustainable Aviation Fuel (SAF) price assessment in Europe (S&P Global, 2020c). It is called the "Sustainable Aviation Fuel Ex Works Northwest Europe." The daily price assessment started in August 17, 2020, and bears the symbolic point of the aviation industry resetting with the green agenda in mind. This brings biojet fuel from a niche product to be closer as a commodity and energy-like product in the open market. Biojet fuel price can now be compared against that of conventional jet fuel and crude oil price for price discovery, bringing about greater transparency as airlines join in the process of decarbonization of the industry. This is particularly important as spot market for biojet fuel is still in its infancy stages. In this respect, spot market refers to the trading of biojet fuel for immediate delivery.

The price assessment follows a cost-based system and mirrors the cost of SAF derived from used cooking oil on an ex-refinery basis in Northwest Europe. It also factors in hydrogen cost and fixed refinery costs, while also deducting for the by-product credits of propane, naphtha and diesel. In other words, it portrays the production cost of SAFs for blending into the petroleum-based jet fuel. The publicly published price assessment will immediately be of use to the seven airports in Europe that accepts batch deliveries from pilot SAF production plants. Presently, they are concentrated in the Northwestern region of Europe, or more precisely within the subregion of Scandinavia.

Platts followed up with the launch of two US West Coast SAF price assessments in September 21, 2020 (S&P Global, 2020b). They are the "Sustainable Aviation Fuel with credits US West Coast" and "Sustainable Aviation Fuel without credits US West Coast." Like their European counterpart, the price assessment will also be cost-based. However, the US assessments differ in having a variation where environmental credits are factored in and also reflecting SAF produced from tallow.

A comparison of the price assessments for biojet fuels is summarized in Table 5.1 (S&P Global, 2020b). Presently, the price assessments are only available for two geographical divisions. This is in contrast with conventional jet fuels, which has assessment from five major global trading, supply, and demand centers. This allows a global index, the "Jet Index Global" to be developed based on the weighted average of the regions. The weighting of each region of North America (38.61%), Europe and CIS (28.47%), Asia and Oceania (21.74%), Middle East and Africa (7.10%), and Latin America

Table 5.1 Sustainable aviation fuel price assessments.

Price assessment	Sustainable aviation fuel ex works northwest Europe	Sustainable aviation fuel with credits US West Coast	Sustainable aviation fuel without credits US West Coast
Geographical division	Europe, the Middle East, and Africa (EMEA)—covering 116 countries. Focal point is in northwest Europe	United States of America. Focal point is in California	United States of America. Focal point is in California
Basis of assessment	Cost-based price assessment (ex-refinery price)	Cost-based price assessment (ex-refinery price)	Cost-based price assessment (ex-refinery price)
SAF input	Cost of used cooking oil and hydrogen added to fixed renewable refinery costs, then deducting the by-product credits of propane, naphtha, and diesel	Cost of packer grade beef tallow and hydrogen (without carbon capture and storage) added to fixed renewable aviation fuel refinery costs, then deducting the by-products of gasoline, propane, and diesel	Cost of packer grade beef tallow and hydrogen (without carbon capture and storage) added to fixed renewable aviation fuel refinery costs, then deducting the by-products of gasoline, propane, and diesel
Environmental credits factored	No	Yes, credits from renewable identification numbers (under US RFS), CARB's low carbon fuel standard, and US biodiesel tax credit (where applicable)	No
Units	USD per metric ton	US cents per gallon USD per metric ton USD per barrel	US cents per gallon USD per metric ton USD per barrel
Frequency	Daily	Daily	Daily

and Caribbean (4.08%) is allocated based on uplift data and trading volume. As biojet fuel grows in usage, a future global index for biojet fuel would be useful for price monitoring.

Currently, biojet fuels on their own are not price competitive against conventional jet fuel. Since the advent of the European-based price assessment for biojet fuel to allow fair comparisons, it is traded at around 4.3× to 5.4× of conventional jet fuel as shown in Fig. 5.2 (S&P Global, 2020a). However, the gap can be narrowed through various incentives such as tradable carbon credits and increase in conventional jet fuel spot prices. Tradable credits might even outright make biojet fuel profitable, as the credits are often stackable and may even exceed the cost of production (see Fig. 5.3) (S&P Global, 2020a). A study of the US aviation industry modeled

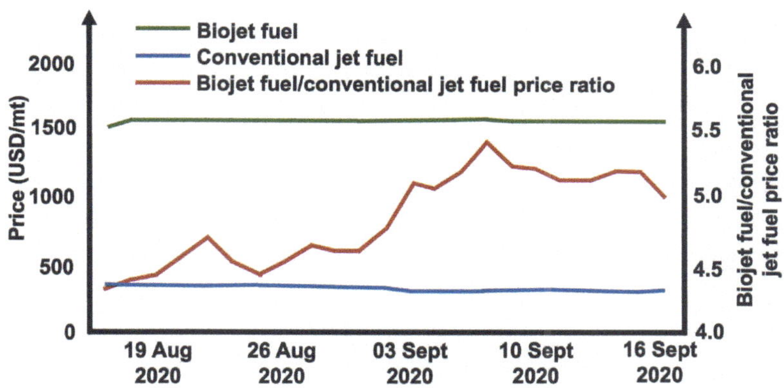

Figure 5.2 Price comparison of sustainable aviation fuel against conventional jet fuel (S&P Global, 2020a).

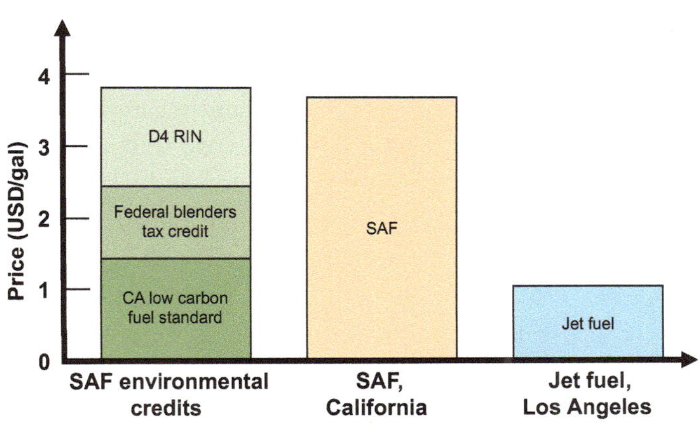

Figure 5.3 Price of sustainable aviation fuel, conventional jet fuel and stackable environmental credits (S&P Global, 2020a).

using soybean oil as feedstock showed that an implicit subsidy of USD 0.71 per liter of biojet fuel for producers is required if the Federal Aviation Administration's aims of using 1 billion gallons of renewable jet fuel annually is to realize in 2020 (Deane et al., 2017). The implicit subsidy could be reduced to just USD 0.09 per liter of biojet fuel as feedstock is changed to oilseed rotation crops.

The underlying reasons for the higher biojet fuel price can be pinpointed to the typical batchwise production method as opposed to the more economically feasible continuous production method. The technology to convert from batchwise to continuous is not the barrier; the lack of demand makes it unjustifiable to operate using continuous production methods (Deane et al., 2017). Additionally, the logistics, infrastructure, technology maturity, and economies of scale are not yet at the levels of conventional jet fuel. All these widens the gap in cost between the fuels.

The recency of biojet fuels meant that actual reports of the economic impacts related to the introduction of biojet fuels are scarce (Cremonez et al., 2015). The availability of price assessment for biojet fuels will allow its economic impacts on the aviation fuel chain to be assessed in the future. This will allow economic figures to be from the real world rather than projected numbers extrapolated from research and testing programs.

5.2.2 Economic viability

The economic viability of biojet fuel is primarily tied to the profitably of the producer. For biojet fuel and conventional aviation fuel, the key number is the minimum selling price (MSP). This can also be contextualized as the minimum jet fuel selling Price (MJSP). The fuel with the lower MJSP between biojet fuel and conventional jet fuel will have the cost advantage.

Diederichs et al. (2016) conducted a technoeconomic comparison of biojet fuel from lignocellulose, vegetable oil, and sugarcane juice. Lignocellulose represented second-generation biomass and was evaluated through three conversion pathways, namely the gasification and FT synthesis (GFT-J), biochemical conversion to ethanol with upgrading (L-ETH-J) and gasification, and syngas fermentation to ethanol with upgrading (SYN-FER-J) processes. First-generation feedstocks comprised of vegetable oil and sugarcane juice were evaluated through the hydroprocessing of vegetable oil (HEFA) and sugarcane juice to ethanol by sucrose fermentation with upgrading (S-ETH-J), respectively. A summary of the economic evaluation is shown in Table 5.2 (Diederichs et al., 2016).

Table 5.2 Economic evaluation and minimum jet fuel selling price (MJSP) for first- and second-generation feedstocks (Diederichs et al., 2016).

Parameter	Process				
	L-ETH-J	SYN-FER-J	GTF-J	HEFA	S-ETH-J
Generation	Second	Second	Second	First	First
Feedstock	Lignocellulose	Lignocellulose	Lignocellulose	Vegetable oil	Sugarcane juice
Raw material and waste disposal (million USD/annum)	120.23	69.75	62.51	112.60	85.36
By-product credits (million USD/annum)	24.77	13.03	38.16	25.44	38.28
Fixed operating costs (million USD/annum)	24.78	22.09	27.85	10.52	18.92
Total indirect costs (million USD)	274.2	232.8	321.3	91.7	184.1
Fixed capital investment (million USD)	482.6	409.7	565.5	161.4	324.0
Fixed capital investment/Annual jet fuel kilogram (USD/kg)	7.90	6.54	9.05	2.87	5.30
Total capital investment (million USD)	532.7	452.5	623.9	179.4	358.3
MJSP (USD per kg jet fuel)	3.431	2.495	2.444	2.223	2.541
(1) Main feedstock	0.961	0.938	0.940	1.992	1.175
(2) Trash	—	—	—	—	0.384
(3) Enzymes	0.722	—	—	—	—
(4) Catalysts	0.177	0.176	0.136	0.018	0.191
(5) Other raw materials	0.215	0.110	0.023	0.001	0.046
(6) Waste disposal	0.027	0.029	0.026	0.001	0.044
(7) Grid electricity	−0.197	0.0	−0.126	0.0	−0.542

Continued

Table 5.2 Economic evaluation and minimum jet fuel selling price (MJSP) for first- and second-generation feedstocks (Diederichs et al., 2016).—cont'd

Parameter	Process				
	L-ETH-J	**SYN-FER-J**	**GTF-J**	**HEFA**	**S-ETH-J**
(8) Fuel by-products	−0.208	−0.208	−0.484	−0.452	−0.208
(9) Fixed costs	0.406	0.353	0.446	0.187	0.371
(10) Capital depreciation	0.120	0.118	0.118	0.143	0.144
(11) Average income tax	0.173	0.143	0.191	0.062	0.142
(12) Average return on investment	1.035	0.836	1.174	0.271	0.794
Lowest MJSP from sensitivity analysis (USD per kg jet fuel)	3.02	2.09	1.99	1.37	2.22
Highest MJSP from sensitivity analysis (USD per kg jet fuel)	4.17	3.17	3.24	3.36	3.10
Largest influence	Enzyme cost	Feedstock cost	Fixed capital investment	Feedstock cost	Fixed capital investment

Vegetable oil–derived biojet fuel from the HEFA pathway was found to have the lowest baseline MJSP of USD 2.223 per kg jet fuel, while the L-ETH-J pathway with lignocellulose has the highest MJSP of USD 3.431 per kg jet fuel. From the sensitivity analysis, the MJSP for the L-ETH-J pathway could further rise to USD 4.17 per kg jet fuel if the unfavorable enzyme cost scenario of USD 1385 per MT broth were to materialize. Conversely, the MJSP of the HEFA pathway can be reduced to USD 1.37 per kg jet fuel if the feedstock price is reduced to USD 546 per MT oil. Even the most optimistic projection for the HEFA pathway is still greater than the maximum fossil-based jet fuel MJSP range of USD 0.42–1.28 per kg jet fuel. It should also be noted that since the main feedstock price of the HEFA pathway can be up to 89%, this represents a cause of uncertainty, which increases the risk for producers. It essentially makes the profitability of biojet fuel producers to be dependent on commodities price.

Hypothetically, a simultaneous reduction in feedstock cost to USD 546 per MT oil and increase in by-product price of naphtha to USD 1.52 per liter could feasibly bring the MJSP to USD 1.22 per kg jet fuel. This will bring the MJSP of biojet fuel to within the higher end jet fuel prices. The further reduction in feedstock cost is plausible as palm oil futures traded at USD 535.02 per MT as late as December 2018, although the high naphtha price scenario is unlikely with the highest price in the past decade being merely ~ USD 0.75 per liter in March 2012. The price of naphtha has since fallen to USD 0.26 per liter in November 2020.

If first-generation biojet fuel cannot yet compete with conventional jet fuel in cost, then second-generation biojet fuel might be at a further disadvantage due to the substantially greater fixed capital investment required. The fixed capital investment required for second-generation biojet fuel can be up to 3.5 times greater than that of the first-generation counterparts. With the main feedstock cost for second-generation biojet fuels contributing to 28.0%–38.4% of the MJSP, a favorable reduction in lignocellulose price through technological breakthrough will reduce the gap.

Martinez-Hernandez et al. (2019) performed an uncertainty analysis using Monte Carlo simulation to predict the MSP required for biojet fuel to cope with the uncertainties with feedstock cost, product price, and capital investment. To lower the risk of failure due to uncertainty, a liter of biojet fuel needs to have an MSP of USD 1.35. This allows an internal rate of return (IRR) projection of 10% to provide buffer for the uncertainty.

Fig. 5.4 shows the IRR and MSP variation against biojet fuel production capacity. Both IRR and MSP are sensitive to crude oil price. A low

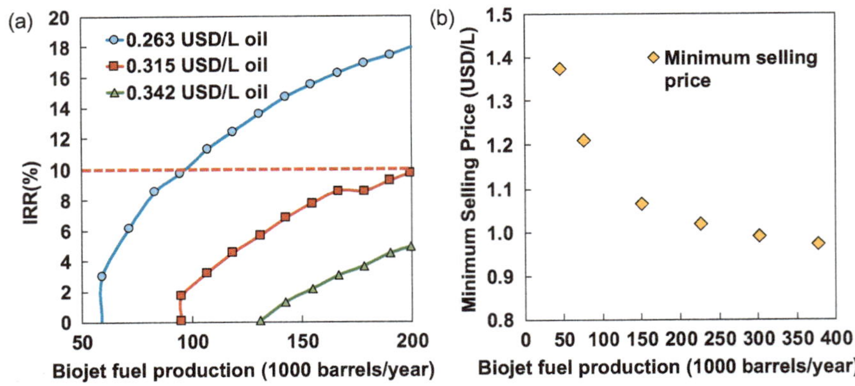

Figure 5.4 Internal rate of return (IRR) and minimum selling price against biojet fuel production capacity (Martinez-Hernandez et al., 2019).

crude oil price of USD 0.263 per liter (or USD 41.81 per barrel) will allow a 95,000 barrel per year biojet fuel plant to have a project IRR of 10%, although at the oil price no plants with a capacity under 58,000 barrels per year are expected to be profitable. With just an increase of oil price to USD 0.342 per liter (or USD 54.37 per barrel), a plant annual output of 200,000 barrels is needed for an expected IRR of 5%. Thus, it is more favorable for larger plants to absorb the price uncertainty of crude oil from sheer economies of scale. However, there is a limitation to how large a plant should be, as a 200,000 barrel per year plant requires around 459,000 barrels of vegetable oil. For scale, this represents around 61% of the national vegetable oil production capacity of Mexico.

The improvement to MSP shows a diminishing return with biojet fuel production capacity above 150,000 barrel per year. At 150,000 barrels per year, the predicted MSP of around USD 1.05 per liter is still more than double the typical fossil jet fuel price of USD 0.50 per liter. As such, only a blending mandate coupled with subsidies could make biojet fuel an attractive proposition with high IRR at the current state of technology and market penetration.

The study also predicted an annual cost breakdown of USD 22.7 million per year for a 75,000 barrel per year biojet fuel plant, with oil feedstock for the HEFA process taking up 66% of the cost. The other costs in descending order are annual capital cost (13%), waste treatment (8%), hydrogen (7%), labor (5%), and utilities (1%). Such a plant if located in Mexico will have an 85% chance of garnering an IRR >> 10%. The chances would improve with larger plant size and lower feedstock cost.

Ranganathan and Savithri (2016) conducted a discounted cash flow analysis of the third-generation wastewater-based microalgae as a feedstock for biofuel production. The analysis of a conceptual plant factored in the wastewater treatment, hydrothermal liquefaction, hydroprocessing, and hydrogen generation processes. Although the product is a mixture of gasoline, diesel, and biojet fuels, the bulk of the product with 69.6% is biojet fuel. From the sensitivity analysis, an increase in algae yield per nitrogen used holds the greatest process-based effects of reducing MSP, where a 10% increase in algae yield will favorably reduce MSP by 6%. A drop of 10% in biooil yield will adversely increase MSP by 13%. It is also shown that income tax rate from the government could also greatly swing the MSP required at −6.18% and +7.2% for income tax rate change of −10% and +10%, respectively. Unlike first generation–based biojet fuel, feedstock cost is a nonfactor.

For such an integrated facility, the likeliest MSP of the biofuel is USD 4.3 per gasoline gallon equivalent (GGE) or USD 1.01 per liter of kerosene. Such a conceptual plant may yet to be realistic as of present, but it shows that the use of third-generation feedstock could be comparable biojet fuel produced from first-generation feedstocks price–wise. Projections of biofuel MSP can differ greatly, where the MSP from 10 similar work ranging from USD 2.07–7.11 per GGE. This brings the MSP range to USD 0.49–1.67 per liter.

While there are many factors determining the MJSP, feedstock and pathway combinations would be the predominant factors. Table 5.3 tabulates the economic characteristics of the different biojet fuel production pathways from various feedstocks (Wei et al., 2019). The pathways of HEFA and CH with first-generation edible oil as feedstock allow for the lowest MJSPs as the conversion pathways are relatively mature and the feedstocks are also among the cheapest. Using the average jet fuel price of USD 2.07 per gallon in 2018, the use of camelina oil under the HEFA pathway could achieve price advantage over conventional jet fuel. On the other hand, the use of second- and third-generation feedstocks still commands a premium, notably with microalgae and HEFA having an MJSP of USD 31.98 per gallon.

The HEFA pathway is the most cost-effective pathway, but its feasibility is still dependent on feedstock price. In order of descending feedstock cost using the HEFA pathway are third-generation > second-generation nonedible oil > waste oils and animal fats > first-generation edible oil. The catalytic hydrothermolysis (CH) and hydroprocessed depolymerized

Table 5.3 Minimum jet fuel selling price of the different biojet fuel production pathways from various feedstocks (Wei et al., 2019).

Pathway	Type	Feedstock	MJSP (USD per gallon)
Hydrogenated esters and fatty acids	First-generation edible oil	Camelina oil[a]	1.63–4.62
		Soybean oil	3.82–4.39
	Second-generation nonedible oil	Jatropha	5.42–5.74
	Third generation	Microalgae	31.98
	Waste oils and animal fats	Yellow grease	3.33–4.01
		Tallow	3.98–4.73
Catalytic hydrothermolysis or hydrothermal liquefaction	First-generation edible oil	Camelina oil[a]	2.48–3.23
	Second generation	Lignocellulose	3.66–5.06
Hydroprocessed depolymerized cellulosic jet	Second generation	Lignocellulose	5.23–7.15
Fischertropsch process	Second generation	Lignocellulose	6.23–7.57
Alcohol-to-jet	First generation edible crop	Sugarcane	3.65–8.08
		Corn grain	3.84–6.63
	Second generation	Lignocellulose (biochemistry)	4.32–10.91
		Lignocellulose (thermochemistry)	7.30–7.82
Direct sugar to hydrocarbons	First-generation edible crop	Sugarcane	7.17
	Second generation	Lignocellulose—forest residue	18.55–20.61
		Lignocellulose—wheat straw	24.74–26.80
Aqueous phase reforming	Second generation	Lignocellulose	4.66–4.75

[a]Camelina can be classified as either first- or second-generation feedstock.

cellulosic jet (HDCJ) pathways have comparable MJSP to HEFA due to the relatively high yields and also low equipment costs.

Comparing like-to-like, Fischer–Tropsch (FT) pathway using second-generation feedstock has higher MJSP as compared with HEFA's equivalent, but the FT process typically produces a significant amount valuable by-product such as gasoline and LPG. Alcohol-to-jet (ATJ) presently has greater MJSP as compared with HEFA due to the cost of enzymes. Its large range is contributed by feedstock and capital costs. Among the various ATJ methods, the biochemistry pathway has the greatest potential for the lowest cost.

The direct sugar to hydrocarbon (DSCH) pathway is an interesting proposition as it has not only the greatest cost but also the highest value intermediates. One of the intermediates isoprenoids, namely farnesene that has applications in the lucrative pharmaceutical industry, is priced at USD 22.53 per gallon. The intermediates could serve to recoup a portion of the cost if the process could be tweaked to have higher selectivity for biojet fuel and farnesene. Industrious producers might even prioritize the production of farnesene and treat biojet fuel as the by-product to offset the costs. However, the process shares the same risk as the transesterification process to produce biodiesel, where the previously high-valued by-product of glycerine became close to worthless. This happened as there was a glut of glycerine flooding the pharmaceutical market due to the success of the biodiesel industry.

Pereira, MacLean, and Saville (Pereira et al., 2017) conducted a Monte Carlo analysis utilizing a discounted cash flow approach to evaluate the financial viability of different biojet fuel pathways. The analysis factored in internal uncertainty such as scaling up and external concerns such as crude oil price. Table 5.4 shows the financial analysis scenarios for the biojet fuel pathways (Pereira et al., 2017).

All of the evaluated pathways have positive IRRs with the exception of ATJ. The HEFA pathway leads with IRR range of 28.2%−29.0%. For the purpose of biojet fuel production, the use of camelina oil as feedstock has an advantage over soybean oil due to its higher biojet fuel yield of 16.77%. The CH pathway using first-generation edible oil has also shown high IRR of 17.9%−18.9%. The use of nonoil edible crop feedstock for the pyrolysis-to-jet and FT pathways produces less desirable IRR (8.7%−12.4%). ATJ pathways show negative IRR of up to 8.0%. Presuming that the minimum attractive rate of return (MARR) is fixed at 15%, then only HEFA and CH pathways would be attractive to investors.

Based on scenario projections of high−low oil price and optimistic−pessimistic technology development, there is a 100% probability (90% confidence interval) that the HEFA pathway will meet the MARR for high oil price scenarios. The CH pathway is also promising as a low risk option

Table 5.4 Financial analysis scenarios for the biojet fuel pathways (Pereira et al., 2017).

Pathway	Feedstock	Capital investment (USD million)	Annual operating cost (USD million)	Return on investment (USD million)	Annual revenues (USD million)	Biojet fuel (kg/metric ton of biomass)	Net present value (USD million)	IRR (%)
HEFA	Camelina	375	303	44	468	167.7	637	28.2
	Soybean	385	605	45	792	91.0	691	29.0
CH	Camelina	486	303	57	417	78.4	292	17.9
	Soybean	497	605	58	741	42.6	345	18.9
Pyrolysis-to-jet	Corn stover	472	148	55	204	178.5	8	10.5
	Sugarcane bagasse	472	136	55	204	178.5	76	12.4
FT	Corn stover	470	100	55	144	115.5	−42	8.7
	Sugarcane bagasse	470	88	55	144	115.5	22	10.9
ATJ	Corn	311	140	36	142	163.6	−214	−8.0
	Sugarcane	428	93	50	95	38.5	−270	−5.0

where a high oil price economic climate will lead to a 99% and 85% chance of meeting the MARR for optimistic and pessimistic technology scenarios, respectively. However, the CH pathway is predicted to only meet MARR 50% of the time if oil price tumbles.

This analysis also highlights the fact that biojet fuel is often a minor product in the production process and the calculation of MJSP is sensitive to the yields and prices of coproducts. Even at its peak, biojet fuel as a product is only 17.85% of total biomass used as feedstock. In fact, biojet fuel only contributes to roughly 16% and 14% of revenue for HEFA and CH plants, respectively. Both pathways would rely on revenues obtained from protein meal produced to the tune of 72% for HEFA (soybean) and 60% for CH (camelina). This is expected as even for oil crops such as soybean and camelina, protein meal outweighs oil by mass.

5.2.3 Process cost and investment cost

While it is clear that the MJSP of biojet fuel will determine the economic viability of the renewable energy, the estimation of the current biojet fuel production costs remained unclear. Production cost estimates have a large degree of uncertainty as most values are only obtained from economic models, pilot studies, or lab-scale extrapolations, instead of actual commercial-scale values (Chiaramonti et al., 2014). Thus, this section attempts to use process cost from available literature to construct a more generalized process cost breakdown.

In the broadest sense, production costs can be attributed to capital expenditure (CAPEX), operational expenditure (OPEX), and biomass feedstock cost. By-product credits can be used to offset the production costs. The biomass feedstock cost is deliberately separated from OPEX as it could be the single most costly item in some pathways. Table 5.5 shows the process costs for various pathway—feedstock combinations by major categories. The values are calculated through a normalization process from (Pereira et al., 2017; Atsonios et al., 2015). It should be noted that the values are skewed toward a producer entering the industry in its formation years. Such an approach was adopted as the largest barrier to enter a potentially profitable industry or an industry supported through subsidies is the CAPEX requirement. While the size of the plant is important, it would be impossible to project for a variety of plant sizes. Also, working capitals are not considered as part of the analysis as they infer financial liquidity instead of the cost associated to the process. Governmental interventions such as subsidies and grants are not considered to provide an estimation resembling the free market.

Table 5.5 Biojet fuel process costs by major categories for various pathway–feedstock combinations.

Conversion pathway	Biomass feedstock	Biomass feedstock costs (%)	Capital expenditure (%)	Operational expenditure (%)	By-product credits (%)	Normalized MJSP (–)	Jet fuel yield (kg/metric ton of biomass)
Alcohol-to-jet (ATJ)	Sugarcane	13.9%	78.1%	8.0%	−19.1%	1.88	38.5
ATJ	Corn	22.9%	65.1%	12.0%	−32.8%	1.36	163.6
ATJ (with modFT)	Wood chip	25.0%	45.9%	29.1%	−13.4%	2.24	112
ATJ (with modMeOH)	Wood chip	24.5%	44.8%	30.8%	−10.5%	1.93	138
Direct fermentation-to-jet (DFJ)	Sugarcane	12.2%	80.1%	7.8%	−14.6%	2.26	36.8
DFJ	Corn	18.8%	69.8%	11.4%	−23.7%	1.88	156.2
Hydroprocessed ester and fatty acid (HEFA)	Soybean	56.2%	34.8%	9.0%	−78.7%	1.00[a]	91
HEFA	Camelina	38.2%	49.3%	12.6%	−67.6%	1.05	167.7
Catalytic hydrothermolysis (CH)	Soybean	51.5%	41.2%	7.4%	−67.5%	1.67	42.6
CH	Camelina	33.7%	56.4%	9.8%	−53.3%	1.71	78.4
Pyrolysis-to-jet (PTJ)	Sugarcane bagasse	7.1%	72.5%	20.4%	−34.5%	1.81	178.5
PTJ	Corn stover	9.0%	71.4%	19.6%	−33.9%	1.86	178.5
Fischer–Tropsch (FT)	Wood chip	28.7%	43.1%	28.2%	−28.7%	1.87	97
Gasification with Fischer–Tropsch (GFT)	Sugarcane bagasse	7.8%	79.2%	13.0%	−26.7%	1.85	115.5
GFT	Corn stover	9.9%	77.9%	12.2%	−26.3%	1.89	115.5

[a]MJSP is normalized against biojet fuel produced from soybean through the HEFA pathway.

In general, the use of first-generation feedstocks and lipid-based pathways such as HEFA and CH will lead to biomass feedstock cost being dominant, even up to more than half of the cost. However, a switch to second-generation feedstock like camelina will substantially lead to capital expenditure being the major costs. In both cases, the lipid-based pathways will lead to great amount of saleable meal which can offset costs. In fact, meals are expected to provide nearly 5 times and 10 times the revenue of biojet fuels for HEFA and CH, respectively. The HEFA—edible oil combination will require the lowest MJSP among all with the CH methods generally be ~60% greater.

The nonlipid pathways of DFJ, PTJ, ATJ, and FT are CAPEX intensive with expenditure in the category exceeding 60% of total costs in all of them. None of the methods are expected to rival the lipid-based pathways in the near term. The methods of DFJ, PTJ, ATJ, and FT are greater than the HEFA—edible oil combinations by at least 88%, 81%, 36%, and 85%, respectively. The use of corn for ATJ is promising with the gap against HEFA being only 36% due to the lower installed equipment cost required, the high jet fuel yield, and higher offset due to by-product credits.

If we were to consider the best combination of cost and sustainability, the use of second-generation camelina oil as feedstock under the HEFA pathway is among the best. Its relative MJSP is only about 5% greater than the best case but would not bear the baggage of first-generation feedstocks. A detailed technoeconomical study by Li et al. looks at the production of biojet fuel from camelina at commercial scale using the HEFA process (Li et al., 2018). The economic analysis is broken down into total capital costs, operating costs, and costs related to scale and profitability.

The total capital costs are associated to the investment amount, which covers all upstream and downstream sections of the process. The total plant direct cost (TPDC), total plant indirect cost (TPIC), direct fixed capital (DFC), and total investment (TI) are calculated through Eqs. (5.1)—(5.4). The coefficients associated to the equations are summarized in Table 5.6. Eq. (5.1) calculates TPDC using

$$\begin{aligned} TPDC = {} & Total\ equipment\ purchase\ cost + Process\ ping + Instrumentation \\ & + Insulation + Electrical + Buildings + Yard\ improvement \\ & + Auxiliary\ facilities + Installation \end{aligned} \tag{5.1}$$

Table 5.6 Coefficients use for the HEFA plant economic analysis calculations.

Cost category	Coefficients
Total equipment purchase cost (PC)	1.20 × listed equipment purchase cost[a]
Process ping	0.20 × *PC*
Instrumentation	0.35 × *PC*
Insulation	0.40 × *PC*
Electrical	0.10 × *PC*
Buildings	0.45 × *PC*
Yard improvement	0.15 × *PC*
Auxiliary facilities	0.40 × *PC*
Installation	0.50 × *PC*
Engineering	0.25 × *TPDC*
Construction	0.35 × *TPDC*
Contractor's fee	0.05 × (*TPDC* + *TPIC*)
Contingency	0.10 × (*TPDC* + *TPIC*)
Start-up and validation cost	0.05 × *DFC*
Labor basic rate	USD 30 per hour
Benefits factor	0.40
Supplies factor	0.10
Supervision factor	0.15
Administration factor	0.60
Laboratory and quality control cost	0.15 × *LC*

[a]PC factors in unlisted equipment purchase cost which is assumed to be 20% of listed equipment purchase cost.
Adapted from Li, X., Mupondwa, E., Tabil, L., 2018. Technoeconomic analysis of biojet fuel production from camelina at commercial scale: case of Canadian Prairies. Bioresour. Technol. 249, 196–205.

Eq. (5.2) calculates TPIC through

$$TPIC = Engineering + Construction \tag{5.2}$$

Eq. (5.3) calculates DFC from

$$DFC = TPDC + Engineering + Construction + Contractor's\ Fee + Contingency \tag{5.3}$$

Eq. (5.4) calculates TI using:

$$TI = DFC + Start\text{-}up\ and\ validation\ cost \tag{5.4}$$

The plant operating cost factors in labor, material, utility and co-products cost used in the process. Eq. (5.5) calculated the labor cost (LC) where

$$LC = [\textit{Labour basic rate} \times (1 + \textit{Benefits} + \textit{Supplies} + \textit{Supervision} + \textit{Administration}) \times (\textit{Labour} / \textit{hours})] \times (1 + \textit{Laboratory and quality control cost}) \quad (5.5)$$

The key material costs consist of camelina oil (USD 0.80/L), hydrogen (USD 2.90/kg), catalyst (USD 330/kg), and hexane (USD 2.00/kg). Costs of water required are for cooling tower water, chilled water, cooling water, steam, and high-pressure steam at USD 0.04/MT, USD 0.40/MT, USD 0.05/MT, USD 12.00/MT, and USD 20.00/MT, respectively. Water has to be disposed at a price of USD 0.85 per m^3, while power can be bought at USD 0.08/kWh. Cost can be recouped through the sales of coproducts. The key saleable coproducts are propane (USD 0.19–0.73/L), LPG (USD 0.19–0.76/L), naphtha (USD 0.22–1.08/L), and neat biodiesel (USD 0.80–1.24/L).

Economies of scale refer to the cost advantage obtained through increasing output level. However, there is a limit to how large an operation should scale before it reached the point of diminishing returns. For HEFA-based plants, a minimum scale of 340–570 million L per year is recommended by the Transportation Research Board. For each plant, the optimum point is arrived at when the marginal cost (MC) and average cost (AC) for the biojet fuel production plant intersect. MC refers to the incremental costs of the last unit of biojet fuel produced, while AC is the total cost divided by the quantity of biojet fuel produced. Thus, their intersection will imply the minimum of the average cost curve.

The profitability of the biojet fuel production plant can be determined using the common net present value (NPV), which is calculated as the difference between present values of cash inflows from sales of biojet fuel and cash outflows due to production of biojet fuel. Underpinning this relationship is the MJSP required as a financial breakeven point. The model again reiterated that competitiveness of biojet fuel is heavily dependent on feedstock costs and prevailing conventional het fuel price.

Neuling and Kaltschmitt (2018) modeled the investment costs required for four different conversion pathways, each with two feedstocks. The pathways modeled include HEFA, biomass-to-liquid (BtL), biogas-to-liquid (BioGtL), and ATJ. Fig. 5.5 shows the annualized costs for four different biojet fuel conversion pathways. The plants are assumed to have a life span of 20 years.

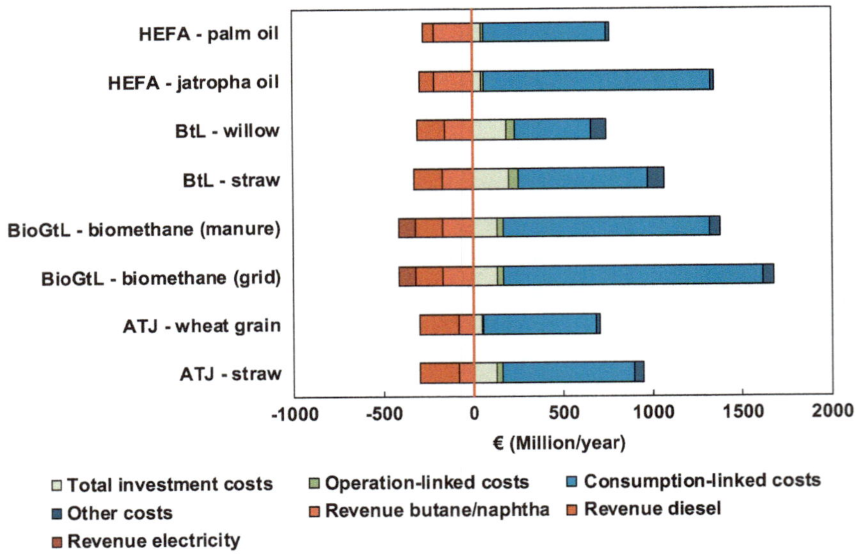

Figure 5.5 Annualized costs for four different biojet fuel conversion pathways. *(Adapted from Neuling, U., Kaltschmitt, M., 2018. Techno-economic and environmental analysis of aviation biofuels. Fuel Process. Technol. 171, 54–69.)*

For pathways at deployment phase, the HEFA–palm oil or HEFA–edible oil combination again proved to be very competitive pricewise as investment costs are low, bringing down the nett cost. The dominant cost associated to HEFA is the consumption-linked costs. As the second-generation feedstock of jatropha oil cost is still high as compared with first-generation feedstocks, the HEFA–jatropha oil combination is not attractive. On the other hand, the usually uncompetitive ATJ pathway has the lowest nett cost among all methods for the ATJ–wheat grain combination. This can be single-handedly pinpointed to the extremely low investment cost, which is in the same range as HEFA's investment costs. Putting things into perspective, the production costs for HEFA–palm oil, HEFA–jatropha oil, and ATJ–wheat grain are 890 €/t, 2000 €/t, and 827 €/t, respectively. The cost competitiveness of ATJ against HEFA is surprising as ATJ has a technology readiness level of 8 and HEFA is at 9.

For the qualification phase pathways, biogas-to-liquid methods are generally expensive due to the use of biomethane as feedstock. Despite the added revenue from electricity generation leading to the highest by-product credits to offset production costs, the BioGtL methods are the most expensive. The production costs for biomethane from manure and grid are 2178 and 2854 €/t, respectively. The BtL—willow combination is also promising due to the low feedstock cost. If not for the high total investment costs, this method that has a production cost of 1054 €/t could have been competitive with ATJ—wheat grain and HEFA—palm oil.

Moving away from the more mature technologies, Yang et al. (2018) analyzed the capital costs for biojet fuel production through the microwave-assisted catalytic pyrolysis pathway. The 47,882.74 gallon biojet fuel/day modeled plant is integrated with mild hydrogenation that investigated four scenarios varying for solvent selection, availability of heat integration, and selling of coproducts for extra credits. The capital cost comparison is tabulated in Table 5.7. Despite having the second highest capital investment of USD 285.51 million and total operating cost of USD 52.5 million, scenario 2 has the lowest production cost. This can be attributed primarily to the sales of extra H_2 and secondarily to the superior solvent/product separation from the use of hexane. The risk to such operation is the uncertainty of H_2 prices in the open market, where the higher capital investment for H_2 production requires a high H_2 price.

5.2.4 Impacts of subsidies and taxes

Renewable energies such as wind, solar, biodiesel, and bioethanol are now success stories in their own rights, but all of them started with suitable subsidies. The subsidies were useful as consumers understood the ecological benefits of using renewables but were unwilling to be the early adopters that will pay for the premium associated with new renewables. In a scenario where consumers are unwilling or unable to pay for the premium, then public policy in the form of subsidies can be pushed forward for the greater good.

Reimer and Zheng (2017) modeled the effects of subsidies on aviation bioenergy supply chain from 21 sectors based on oilseed camelina for the Pacific Northwest region of the United States. In the study, subsidy for biojet fuel and taxation on conventional jet fuel were considered.

Table 5.8 shows the price gap between biojet fuel price and conventional jet fuel price for various subsidy and tax scenarios. From the projected scenarios, a 17.2% subsidy for biojet fuel on one end or 22.6% taxation on conventional jet fuel would allow biojet fuel to be worked without needing consumers to pay for a huge premium. The prices will be brought to close to parity. The subsidy scenario will bring down both fuel prices, while the tax

Table 5.7 Capital cost (in USD million) comparisons considering the total purchased equipment cost (TPEC) and total installed equipment costs (TIC) for microwave-assisted catalytic pyrolysis for biojet fuel production (Yang et al., 2018).

	Scenario 1: Hexane as solvent, with heat integration and selling of biochar and extra H_2		Scenario 2: Hexane as solvent, with heat integration and selling of biochar and extra syngas		Scenario 3: Hexane as solvent, no heat integration and selling of biochar		Scenario 4: Heptane as solvent, with heat integration and selling of biochar and extra H_2	
Production process	**TPEC**	**TIC**	**TPEC**	**TIC**	**TPEC**	**TIC**	**TPEC**	**TIC**
Biomass preheating	0.11	0.22	0.11	0.22	0.11	0.22	0.11	0.22
Microwave-assisted catalytic pyrolysis	5.48	13.16	4.93	13.76	4.92	13.73	4.93	13.79
Biooil collection	4.88	5.98	4.86	5.92	4.63	5.89	4.90	6.00
Hydrogen production via steam reforming	2.55	4.51	2.60	5.35	2.32	4.51	2.54	4.62
Biojet fuel production via hydroprocessing	1.85	3.95	1.81	3.97	1.62	3.31	2.13	4.69
Separation process in distillation columns	35.52	69.15	14.61	28.55	14.50	28.21	36.65	70.20
Cooling water system	0.95	1.90	0.95	1.90	0.95	1.90	0.96	1.91
Auxiliary utilities	0.66	1.37	0.66	1.37	0.66	1.37	0.66	1.37
Wastewater treatment and recycling	3.32	7.97	3.32	7.97	3.32	7.97	3.32	7.97
Total cost	55.32	108.21	33.85	69.01	33.03	67.11	56.19	110.77
Total capital investment	285.51		183.07		174.93		291.50	

Table 5.8 Price gap between biojet fuel price and conventional jet fuel price for various subsidy and tax scenarios.

Scenario	Demand change	Biojet fuel subsidy (%)	Conventional jet fuel tax (%)	Biojet fuel price (USD per gallon)	Conventional jet fuel (USD per gallon)	Price gap (USD per gallon)
Baseline	No	–	–	3.69	3.06	0.63
1	Yes	Reference	Reference	3.71	3.05	0.66
2	Yes	17.2	0	3.06	3.05	0.01
3	Yes	0	19.5	3.79	3.66	0.13
4	Yes	0	22.6	3.80	3.75	0.05
5	Yes	9.0	9.0	3.36	3.33	0.03

Adapted from Reimer, J.J., Zheng, X., 2017. Economic analysis of an aviation bioenergy supply chain. Renew. Sustain. Energy Rev. 77, 945–954.

conditions will keep the prices for fuel high. The former will artificially support the biojet fuel industry while reducing revenue to a country, while the latter might unfairly punish the oil and gas (O&G) industry.

A combined 9% biojet fuel subsidy and 9% conventional jet fuel tax would be a more favorable scenario as this mitigates the gap without overly distorting either industries. This method will not reduce motivation from the biojet fuel industry to seek free market profitability, nor will it be seen to overly penalize the existing O&G industry. While short-term subsidies and taxation can kickstart a new industry, governments need to avoid making it a long-term answer as it will result in competitive distortions. Both the subsidies and taxation could also be made indirect using market-based measures (MBMs) such as carbon offsetting mechanism.

5.2.5 Impacts of biojet fuel on travel costs

The Energy Information Administration (EIA) projected the conventional jet fuel price in 2020 to be 0.54 €/L, while the IEA predicted the dominant biojet fuels produced to be in the range of 0.96−1.45 €/L (Deane et al., 2017). This meant a price delta of 0.42−0.91 €/L that translates to a 1.20−4.30 €/passenger cost increase for a 1000-km flight. They are modeled under the presumption that costs are spread across all domestic and intra-EU-28 flights in 2020, with biojet fuel contributing to 4% of jet fuel volume demand.

In 2020, the world was hit by a global pandemic which all but decimated air travel and skewed all projections. From a recalculation by the authors using mid-September 2020 prices, the expected cost increase for a 1000-km flight has risen to 7.02 €/passenger as compared with conventionally fueled flights. This rise is due to the plunge in global crude oil price and the lower demands of both biojet fuels. Unlike typical supply and demand scenario, the fall in demand for biojet fuel does not reduce price as cost rises due to smaller batch being produced. However, if adjusted for the previously predicted values, the gap reduces to 5.82 €/passenger cost increase for a 1000-km flight.

The increase in gap does not augur well for the biojet fuel industry from a free market sense, as the lower cost increase would implicitly mean that biojet fuels has achieved greater economies of scale. It is not entirely clear if the increase in relative cost for biojet fuel is due to the renewable fuel not making inroads into the market or the pandemic setting back the industry. The recovery of the biojet fuel industry is dependent on the recovery trajectory that the aviation sector follows. A V-shape recovery will be preferable to a U-shape recovery. Regardless, stronger legislations in the form of blending mandates and emissions limits; and greater incentives in the form of subsidies, pioneering statuses, and environmental credits might be required in the near term for the industry to regain its footing.

5.3 Potential feedstock

Biojet fuel can be derived from various biomass such as edible vegetable oil, animal fat, waste cooking oil (WCO), cellulose, and algae. Unlike the more established biofuels of bioethanol and biodiesel, the methods available to produce biojet fuel are more diverse. From a technical point of view, the choice of feedstock is primarily determined by the conversion pathway. However, conversion pathway is itself secondary when more localized circumstances such as availability of feedstock, supply chain maturity, local cost of production, political decisions, trade restrictions, and availability of crude oil will be the more practical determinant. In other words, the most optimum method will not be the same for all producers.

Potential feedstock for biojet fuel can be classified by their main groups as shown in Table 5.9. The first-generation feedstock is taken from food source, and usage for biojet fuel will cause competition to feed mouths. This sparks the "food versus fuel" debate. These feedstocks are in the position where the accompanying technological pathways such as HEFA are relatively mature. Also, the boom-and-bust cyclical nature of these feedstocks meant that most of them have at some point in time been touted as economically viable options for biojet fuel production. For first-generation feedstocks, the availability, land usage concerns, and moral dilemma are the factors plaguing their use as biojet fuel feedstocks.

5.3.1 First-generation feedstock

First-generation feedstocks are dominated by edible oil, animal fats, and sugar/starch crops. The global quantity of key oils crops processed as published by the Food and Agriculture Organization of the United Nations (FAO) is shown in Fig. 5.6. From it, palm dominates with one-third of global oil crop processed, followed by soybean (slightly above a quarter), rapeseed (15%), and sunflower (9%) to round off the top four. The big four feedstocks contribute to an extremely high 83.6%, which is further increased to 87.4% if palm kernel is factored in due to palm oil and palm kernel oil coming from the same source. This is out of a total of roughly

Table 5.9 Potential biojet fuel feedstocks.

Generation	Common characteristics	Category	Key feedstock
First generation	First-generation feedstocks typically come from edible sources from plants and animals. Unless surplus of feedstock, the use of these feedstocks for biojet fuel will compete with human diet. It might also exacerbate deforestation if virgin forest is used for cultivation.	Edible oil crop	Castor[a], coconut, palm, peanut, rapeseed, soybean, sunflower
		Edible animal fat	Lard, poultry fat, tallow
		Sugar and starch crop	Cassava, cereals, corn, sugarcane
Second generation	Second-generation biojet fuels could be drawn upon from a larger pool of feedstock which are generally inedible and hence will not be in competition with food. These feedstocks do not require fertile arable land which can be reserved for food-based agriculture. The feedstocks are likely to meet sustainable criteria.	Nonedible oil	Babassu, jatropha, karanja, camelina[b]
		Cellulosic material (may be dedicated energy crops)	Alfalfa, grasses (bamboo, switchgrass, reed canary) plant residues, straw, sugarcane bagasse, woody energy crops (eastern cottonwood, green ash, poplar, silver maple, sycamore), wood by-products
		Wastes	Agricultural waste, municipal solid waste (biomass fraction), waste cooking oil, yellow grease sewage, tyre, flue gas
Third generation	Third-generation feedstock typically has very high oil content and fast growth rate. They can be cultivated using marginal lands, or in some cases even be nonterrestrial. The feedstocks will likely meet sustainability criteria. However, extensive downstream processes are often accompanied with these feedstocks.	Oil for energy	Algae (microalgae), bacteria, insect, seaweed (macroalgae), yeast

[a]Castor is nonedible but classified as an oil crop.
[b]Camelina can be classified as both first- and second-generation feedstocks.

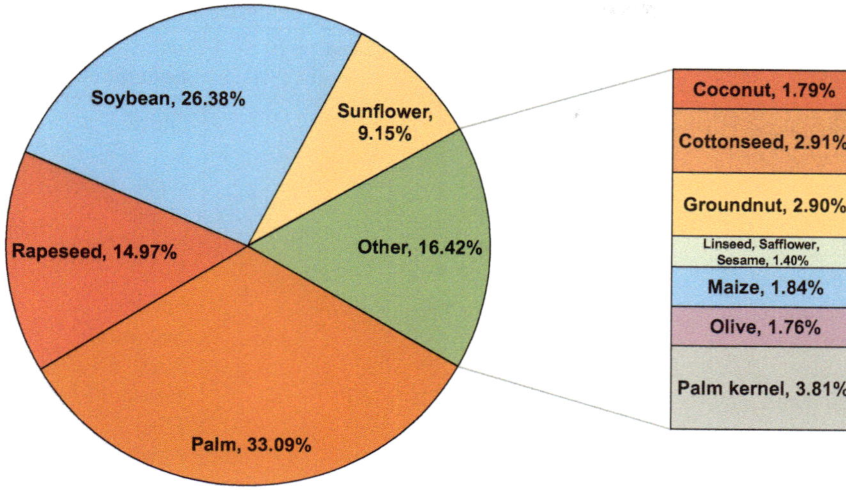

Figure 5.6 Global quantity of key oils crops processed (FAO, 2020b).

173.3 million tons of oil crops processed globally. There are two schools of thoughts, which is to either use a dominant feedstock as surplus is a possibility or use a minor oil crop so that it can be designated as the feedstock of choice for biojet fuel production. For this, two oil-based first-generation feedstocks namely, palm and coconut, stand out as high-potential feedstock for biojet fuel production.

Oil palm for the production of palm oil as feedstock also represents a potential opportunity as it has the highest oil yield exceeding 8000 kg per hectare annually, although actual average yields are closer to 3300 kg oil per hectare annually (Woittiez et al., 2017). In fact, the maximum theoretical yield is calculated to be 18,500 kg oil per hectare annually. The actual yield depends on palm age, pollination, harvesting, water availability, pests, diseases, manpower, planting material, planting density, canopy management, machineries, and available technologies. Regardless, even the lower estimates of 3300 kg oil/ha/yr are greater than the typical yield of other widely regarded high yielding edible oil crops such as coconut and jatropha as shown in Table 5.10.

There is a perception that the rapid expansion of oil palm plantations is the main cause of deforestation in the Southeast Asia and Latin America regions. However, key palm oil—producing nation such as Malaysia stated that the palm oil industry has grown responsibly and sustainably. Malaysia was focused on improving productivity and yield of oil palm plantations,

Table 5.10 Typical oil yield from various oil-producing biomasses.

Crop	Oil yield (L/ha)	Seed oil content (%)	Generation
Microalgae (30%–70% oil by weight)	58,700 –136,900	–	Third
Oil palm	3300 –18,500	40	First
Macauba	6000	–	Second
Coconut	2338–2806	–	First
Karanja	1800–3600	30–40	Second
Jatropha	1892–3000	–	Second
Castor	1413	–	First
Rapeseed	1029–1216	35	First
Canola	1190	–	First
Peanut	1059	–	First
Sunflower	702–982	32–49	First
Camelina	915	–	First/Second
Chinese tallow	907	–	First
Safflower	779	–	First
Soybean	374–514	21	First
Cotton	327–421	14–20	First
Corn	168–187	–	First
Rubber seed	80–120	–	First

Adapted from Scarlat and Dallemand (2019), Woittiez et al. (2017), Chuck et al. (2016), Hari et al. (2015), Cruz et al. (2020).

rather than expanding land usage with cultivation area capped at 6.55 million hectares by 2023. The European Palm Oil Alliance (EPOA) recognized the problem of deforestation but argued that much of oil palm plantation expansion occurred on land previously used for coffee or rubber, and not all deforested land was primary forest before clearing. The World Wildlife Fund (WWF) stands by their position to support sustainable palm oil. While the issues of deforestation attributed to palm oil are not in doubt, but there is potentially a scope for exaggeration. An EU report in 2018 showed that soybean (5.4%), maize (3.2%), pasture (24%), beef/meat/leather (24%), and animal feed (8%) all contributed to a higher global share of deforestation as compared with oil palm (2.3%) (European Comission, 2018). As such, oil palm plantations cannot be pinpointed as the main driver for deforestation. The actual 2.3% global share of deforestation is a huge difference to the 40% value attributed to palm oil as heavily reported in media or already planted in the psyche of the general public.

Coconut oil has unique properties as feedstock for biojet fuel production. The oil is primarily saturated, with the shortest average carbon chain length among all major oil crops. As the dominant fatty acids for coconut oil are caprylic acid (C8:0), capric acid (C10:0), lauric acid (C12:0), myristic acid (C14:0), and palmitic acid (C16:0) which are in the C8–C16 range, they also overlap well with the typical biojet fuel carbon chain length of C8–C16 range. This meant that coconut oil–derived biojet fuel will potentially have higher yield and maybe even skip the cracking step of the HEFA process as required for longer carbon chain length feedstocks. Coconut has high oil yield but without the perception that it is unsustainable as a crop. The only drawbacks are the present low production quantity at less than 2% of global vegetable oil and relative high price compared with other major vegetable oils. The prices of various oils are shown in Table 5.11. Coconut oil can be a candidate to be cultivated for the purpose of biojet fuel production, just like how jatropha was planted in some regions for the sole purpose of producing biodiesel. When produced in larger quantities, price will invariably fall due to economies of scale. This will benefit coconut planters as it will be converted from an expensive niche product to a more moderately priced large-volume commodity.

Animal fats are typically produced at one-sixth to one-seventh of the fats and oils produced. A substantial proportion of the animal fats produced are not fit for human consumption, which severely limits their market appeal as compared with vegetable oil. Thus, they are usually utilized in the

Table 5.11 Vegetable oil prices.

Vegetable oil	Global rank in quantity	Proportion of oil crop processed globally (%)	Quantity of oil crop processed globally (million tons)	Price (USD/MT) in 2018
Palm oil	1	33.1	57.329	535–709
Sunflower oil	4	9.2	15.848	703–806
Soybean oil	2	26.4	45.705	728–871
Rapeseed oil	3	15.0	25.945	793–854
Coconut oil	9	1.8	3.106	787–871
Palm kernel oil	5	3.8	6.602	708–1265
Groundnut oil	7	2.9	5.031	1433–1477

animal feed, pet food, and soap-making industries. As feedstock for these industries, the price that animal fats command is low. Animal fats may contain contaminants such as phospholipids (gums) and polyethylene. Animal fats also have high sulfur contents, which is undesirable as the combustion of biojet fuels containing sulfur will form SO_2. Proper filtering of the contaminants and vacuum distillation to remove sulfur will be required to put it on par with vegetable oil as a feedstock from a technical standpoint. The greatest obstacle to animal fat as a major feedstock for biojet fuel production is the fact that it is merely the by-product of the meat supply chain. It will not be feasible to raise animals for the production of animal fats, so supply cannot meet potential spike in demand.

While the use of edible oil and animal fat as feedstock can be controversial, the conscientious use of these feedstock will be acceptable. The world has a hunger problem with 10.8% of the population considered to be undernourished and 9.8% facing severe food insecurity (FAO, 2017). However, it is not from the lack of food production. Instead the problem comes from the poor distribution of food and urban food wastage. Globally, the average dietary energy supply adequacy has increased every decade with 115%, 116%, and 122% in 1997, 2007, and 2017, respectively. The growth of vegetable oil production of 1.45% from 2019 to 2028 is expected to outpace growth in its requirement as food on a weight per capita basis of 0.81% for the same 10-year period (OECD-FAO, 2019).

For sugar and starch crops, sugarcane is front-runner as a biojet fuel feedstock. The direct use of sugar or the derived ethanol from sugarcane can serve as a feedstock for biojet fuel production. It is photosynthetically efficient and is one of the most grown crops globally. It is already widely used for bioethanol, with three quarters being used in the food manufacturing sector and the rest for biofuels production. As a sugar crop, sugarcane has the upper hand as compared with sorghum and sugar beet in terms of total production and yield as shown in Table 5.12 (FAO, 2019). Both the Americas and Asia produced more sugarcane than the rest of the world combined for sorghum and sugar beet. In particular, Brazil, India, China, and Thailand are powerhouses for sugarcane production. These countries are also located in the correct climate and hence have the largest yield in the range of 731,441—734,768 hg/ha. The major drawback of using sugarcane as a feedstock is the relative immaturity of the conversion pathways as opposed to the more advanced stage of development for oil-to-jet pathways. The success of biojet fuel will see the gradual shift of

Table 5.12 Area harvested, production, and yield for sugar crops (FAO, 2019).

Sugar crop	Region	Area harvested (ha)	Production (tonnes)	Yield (hg/ha)
Sugarcane	Africa	1,547,616	94,925,364	613,365
	Americas	13,919,856	1,022,785,798	734,768
	Asia	10,279,738	751,902,468	731,441
	Europe	37,739	2,280,152	604,188
	Oceania	484,869	35,130,948	724,545
	World	26,269,819	1,907,024,730	725,938
Sorghum	Africa	29,710,774	29,782,406	10,024
	Americas	5,300,555	19,244,420	36,306
	Asia	6,432,088	7,973,788	12,397
	Europe	236,174	1,079,149	45,693
	Oceania	463,554	1,262,340	27,232
	World	42,143,146	59,342,103	14,081
Sugar beet	Africa	275,038	15,021,745	546,169
	Americas	475,234	33,007,156	694,545
	Asia	743,392	41,743,674	561,530
	Europe	3,315,825	185,113,732	558,274
	Oceania	—	—	—
	World	4,809,490	274,886,306	571,550

sugarcane from a source of sucrose to a source of energy. This will pose a unique problem of sugarcane being in a three-way usage competition for the sugar, bioethanol, and biojet fuel industries.

5.3.2 Second-generation feedstock

Second-generation feedstocks will not compete with the food market and can theoretically be cheaper as a feedstock in the long term. However, prices are still high due to the inability to grow nonedible oil on marginal land and collection of wastes in large quantities. Most operations are still at small scale as compared with the vast collective operations of the oil crop and O&G industries. Second-generation feedstock also has the advantage of democratizing the availability of feedstocks as cellulosic materials and wastes are found in significant quantities for nearly all countries. The use of these feedstocks will be dominant once cost parity with first-generation feedstock is achieved.

Among second-generation feedstocks, jatropha oil, camelina, and WCO are the current oil-based feedstock candidates. Jatropha gained its reputation as a wonder crop for biofuel during the biodiesel industry boom due to its high oil content (>30%), high yield (~3000 kg oil/ha.yr), and ability to be grown in marginal land without irrigation. The plant is also potentially drought and pest resistant (Hari et al., 2015). These factors allowed jatropha to enter the biofuel industry as a viable feedstock without being embroiled in the food versus fuel debate. Fast forwarding a couple of decades, the results of jatropha as feedstock for biofuels is mixed as efforts to plant jatropha in marginal lands regularly failed in areas without sufficient land moisture, like in Ethiopia (Wendimu, 2016). The Indian efforts on jatropha as feedstock for biofuels, inclusive of biojet fuels, are more successful. This is reflected in the area of jatropha plantation from 2006 to 2012 where Asia, Africa, and Latin America contributed to 85%, 13%, and 2%, respectively (van Eijck et al., 2014).

Camelina, which is traditionally used as edible oil and lamp oil, has second generation—like characteristics of being able to be cultivated at marginal land. This ability will reduce precious arable land usage, freeing up more arable land for food production. While not unique to camelina, the oil crop can serve as a rotational crop with wheat for soil preservation.

WCO has the base properties of edible cooking oil used as feedstock in the HEFA process but does not incur additional direct land change usage. The use of WCO solves the trifecta problems of diverting edible oil for fuel production, reducing oil waste, and minimizing the incentive to reuse WCO for food. The last of which is the "gutter oil" practice where restaurants reuse WCO collected from restaurants, grease traps, and slaughterhouse wastes. WCO has high variability in quality and composition due to the locality, collection method, and period of the year when different vegetable oil is consumed (Chiaramonti et al., 2014). The biggest issue facing WCO is the contaminant. This represents the largest technical challenges to be dealt with before the catalytic hydrotreatment process.

There is a large quantity of WCO produced annually as shown for the selected countries in Table 5.13. The selected countries alone have the

Table 5.13 Quantity of waste edible oil for selected countries (Kalam et al., 2011).

Country	Quantity of waste edible oil (million tons/year)
United States	10.0
China	4.5
Europe	0.7—1.0
Malaysia	0.5
Japan	0.45—0.57
Canada	0.12

capacity to provide 16.5 million tons of WCO per annum, which easily places WCO among the top four feedstocks if it were an edible vegetable oil. Technical concerns aside, WCO is more likely to be curtailed by logistic concerns. The collection of WCO will represent a challenge as the supply system, collection methods, and processing capacity require augmentation into the existing waste management supply chain.

For dedicated energy crops, lignocellulose-type feedstock can be cultivated for biojet fuel production. Table 5.14 tabulates some key energy crops and the suitable geographic locations. Using the lower bounds for productivity, it is apparent that productivity of energy crop is greater as the suitable climate transits from temperature to subtropical or tropical. It makes sense to suggest that countries with temperate climate should not prioritize lignocellulosic energy crops as feedstocks for biojet fuel unless the conditions are favorable for crops such as miscanthus, switchgrass, and poplar to reach the upper bounds.

Agricultural wastes are great sources of lignocellulosic materials as a result of various agricultural operations. They are usually the leftover biomass after the edible portions of the plants are harvested. It is inaccurate to classify the wastes as unwanted as they are often returned to the land in

Table 5.14 Productivity and suitable climate for energy crop (Chuck et al., 2016).

Crop	Productivity (tonne/ha/year)	Suitable geographic location
Energy cane (*Saccharum* sp.)	33–400	Subtropical, tropical
Elephant grass (*Pennisetum purpureum*)	22–31	Temperate, subtropical, tropical
Sorghum (fiber)	16–43	Warm climates worldwide
Giant reed (*Arundo donax*)	10–30	Temperate, subtropical, tropical
Eucalyptus	10–21	Temperate, subtropical, tropical
Miscanthus (*Miscanthus* sp.)	5–43	Temperate
Switchgrass (*Panicum virgatum*)	5–35	Temperate
Willow	5–11	Temperate
Poplar	2–34	Temperate
Canary grass (*Phalaris arundinacea*)	2–10	Temperate
Alfalfa	1–17	Temperate

various waste-to-wealth initiatives. Depending on the types of wastes, they can be used to improve fertility of soil, as animal feed and direct use for low-grade fuel. Despite the utility, they are often not fully consumed and also represent a loss in opportunity costs as they could be converted into the high value biojet fuel. The amount of agricultural wastes globally is estimated to be in the range of 1.47−3.84 billion tons per year as shown in Table 5.15.

Staple foods such as rice and wheat, which already provide food for more than 90% of the global population, could further contribute to the large proportion of the agricultural residues. Rice residues are dominant in the tropic regions of Asia, while wheat residues from Europe are prominent in temperature regions due to its domestication as a winter crop. Next down the list are C4 plants such as maize and sugarcane. C4 plants, which use the C4 or Hatch-Slack pathway, thrive in tropical climates; hence, the residuals from the aforementioned plants are mostly from Americas and Asia. On the other hand, barley, oats, oil palm, rye, and sunflower have high residue to crop ratio, which renders the collection of agricultural waste to be potentially worthwhile. All nine crops mentioned here are good candidates as feedstock as the cost will be low with relatively high quantities for collection.

5.3.3 Third-generation feedstock

Third-generation feedstocks can be classified as any feedstock with the capability to accumulate lipids. They are usually fast growing and require little additional natural resources to cultivate. The leading lights of this generation are from microorganisms such as microalgae and yeast. Microalgae can be harvested all-year long, located at marginal lands, and use saline or wastewater streams (Noh et al., 2016). Microalgae has high lipid contents as shown in Table 5.16 (Giwa et al., 2018). Presently, the microalgal market is led by *Chlorella*, *Spirulina*, and *Dunaliella*. There are other microalgae with theoretically higher oil content, but optimum growth is dependent on many factors.

Although research at lab scale has shown promising results such as having lipid to land usage ratio that dwarves even that of oil palm, this has not yet translated to any large-scale microalgae or yeast farms. At the upper bounds of oil yields, microalgae's oil yield of 136,900 L/ha/yr is 7.4 times greater than that of oil palm yield of 18,833 L/ha/yr. Also, the production of biojet fuel from microalgae grown in the Netherlands is currently

Table 5.15 Agricultural waste estimations for various crops (Chuck et al., 2016).

Crop	Global production of edible portion (billion tons)	Residue to crop ratio	Estimate of annual residue production in 2010 (billion tons)						Calorific value (GJ/ton)
			Africa	Americas	Asia	Europe	Oceania	World	
Barley	144.8	2.25	10	24	30	111	11	186	16.9
Cassava	276.7	0.088	—	—	—	—	—	—	18.4
Coconut	62.0	0.419	—	—	—	—	—	—	18.1–18.6
Cotton	44.5	1.5	—	—	—	—	—	—	12.4–16.3
Maize	—	2.0	0–64	149–445	34–246	29–85	0.2–0.5	203–840	12.4–16.8
Millet	1016.7	—	23	0.4	24	0.5	0.06	48	—
Oats	29.9	2.5	0.2	5	1	12	1	20	17.4
Oil palm	55.8	2.6	—	—	—	—	—	—	9.8–17.9
Potatoes	368.1	0.25	5	10	38	27	0.4	8	17.4
Rapeseed	72.5	—	—	—	—	—	—	—	16.5
Rice	745.7	1.5	21–35	37–56	668–949	3.9–6.5	1.7	731–1045	12.6–16.3
Rye	16.7	2.25	0.1	4	3	35	0.8	39	17.4
Sesame	4.8	1.5	—	—	—	—	—	—	15.9
Sorghum	61.4	1.07	32	34	15	1	2	84	13.9
Soybean	276.4	0.85	—	—	—	—	—	—	16.3
Sugar beet	250.2	—	3	8	9	38	0	58	—
Sugarcane bagasse	1877.1	0.3	11–22	88–241	75–156	0	6–8	181–428	15.8–18.1
Sunflower	44.8	5.0	—	—	—	—	—	—	18.6
Wheat	713.2	1.5	5.3–33	63–169	145–439	133–306	8–34	354–981	15.9
Total	6602	—	—	—	—	—	—	1470–3836	—

Table 5.16 Lipid contents of microalgae (Giwa et al., 2018).

Microalgae species	Oil content (% dry wt)
Botryococcus braunii	25–75
Chlorella sp.	28–32
Crypthecodinium cohnii	20
Cylindrotheca sp.	16–37
Dunaliella primolecta	23
Isochrysis sp.	25–33
Monallanthus salina	>20
Nannochloris sp.	20–35
Nannochloropsis sp.	31–68
Neochloris oleoabundans	35–54
Nitzschia sp.	45–47
Phaeodactylum tricornutum	20–30
Schizochytrium sp.	50–77
Tetraselmis suecica	15–23

approximated to be 60 times greater than that of conventional jet fuel (Deane et al., 2017). Without scaling up, the cost of using third-generation feedstock remains high, and such projects will have a longer-term horizon rather than immediate industrial application. While it makes sense to accelerate the progress of third-generation feedstock for the supernormal gains, it is more prudent to focus on improving the more immediately achievable price parity of second-generation feedstocks.

While microalgae comes to the public's mind for third-generation feedstocks, macroalgae (seaweed) also have huge potential as feedstock for biojet fuel production. Seaweed has the resource advantage over terrestrial crops and microalgae, as it does not require fertiliser and freshwater, while also forming stable ecosystems if cultivated responsibly (Chuck et al., 2016). As compared with terrestrial crops, seaweed has at least four times the solar efficiency, exceeding 8%. Presently, only 16 million tons per year of seaweed biomass is cultivated globally. This represents a large unrealized potential as conservative estimates show that offshore farms, coastal farms, and open sea colonies have the potential to generate 110 EJ, 35 EJ, and 6000 EJ, respectively. Among the macroalgae classes, the green variant has the best potential for the HEFA conversion pathway. The key properties of macroalgae are shown in Table 5.17. The high ash contents for all macroalgal classes meant that thermal conversion and fermentation pathways will be less efficient.

Table 5.17 Key properties of macroalgae (Chuck et al., 2016).

Description	Macroalgal class		
	Green	Red	Brown
Key species	*Ulva lactuca*, *Ulva pertusa*	*Gelidium amansii*	*Laminaria japonica*, *Sargassum fulvellum*
Carbohydrate content (%)	25–65	30–85	40–60
Carbohydrate type	Starch, cellulose	Floridian starch, galactans, agar, carrageenan, cellulose	Laminarin, mannitol, alginate, fucoidan, cellulose
Lipids content (%)	<6	<1.1	<2
Protein (%)	7–20	8–14	12–19
Ash (%)	18–25	3–9	Up to 46%
Sugar released on hydrolysis	Up to 60%	Up to 57%	Up to 37%
Main sugar type	Glucose	Glucose, galactose	Glucose, mannitol

Yeast, more particularly oleaginous yeast can be a possible rival to microalgae. It has almost all of the advantages of microalgae such as short life cycle, high growth, and high oil yield but without the need for light source. In fact, yeasts even have higher cell densities while having shorter doubling time. The fermentation of yeasts may also produce other saleable coproducts, which can be used to offset production costs. For the purpose of biojet fuel production, the *Saccharomyces cerevisiae* yeast is currently favored although it is nowhere near to being sufficient for industrial usage. The oleaginous yeast *Yarrowia lipolytica* is widely researched for biofuels as it can produce a high amount of lipids. Like *S. cerevisiae*, the *Y. lipolytica* yeast would need major optimization before it can make the step-up for economically viable usage in the industry. Primarily, improvements have to be made on mass of lipids produced per unit of sugar consumed, which is the most important parameter for biojet fuels derived from single cell oils.

The term "fourth-generation feedstock" has also been coined, with the general idea of them being genetically engineered feedstocks. Feedstock metabolism could be altered to improve yield, while composition could be optimized for biojet fuel production. The ability to capture and store carbon could also be enhanced in which more carbon could be captured at cultivation stage and less carbon released postcombustion. For example, strains of microalgae, which have the most desirable existing properties, could be a candidate for gene alteration to improve cellular metabolism. This will further accelerate growth and increase oil yield. Conversely, the microalgae's composition could also be tweaked to prioritize either sugar, starch, or oil production depending on the desired conversion pathways to be used. Other example includes the altering of sugarcane (which is not usually associated as an oil crop) to produce triglycerides using metabolic engineering and plant genetics techniques. Oil content could increase by 200 times to oil crop levels for the bioengineered sugarcanes as compared with their wild-type plant counterparts.

Pennycress (*Thlaspi arvense* L.) from the Brassicaceae family is another highly regarded feedstock for biofuels produced from oil-based conversion pathways (Claver et al., 2016). It has no specific water requirement, requires no pesticide, does not need fertilizer, and can be planted in rotation (pennycress-soybean rotation). Pennycress could be planted in nonagricultural lands, hence not competing with food crops for land usage. It also has the added benefit of being a winter annual, making it the equivalent of the tropic region–centric jatropha for the temperate climates of North America and Europe. The use of biotechnology techniques to control the gene expressions of TaFAE1, TaSGAT1, and TaWRI1 will maximize the production of fatty acids for biojet fuel production. The relatively high oil content will make it a possible candidate for sustainable biojet fuel production.

The term "fourth generation" has not gained general consensus as some researchers consider it as an extension of third-generation feedstocks. Regardless, this method holds great promises as the potential upside is huge, justifying the risk-reward evaluation to support the high initial R&D costs.

Interestingly enough, the success of biojet fuel is hinged upon the trajectory of biodiesel and bioethanol industry as they all share the biomass-based feedstock. The increased use of first-generation feedstock will serve to raise the price of feedstock for all three industries as currently the dominant conversion pathways are still pivoted toward edible oil, sugar, and starch crops. This leads to a four-way competition between the three industries and the food market. On the other hand, any gravitation toward second-generation feedstock will improve the feedstock cultivation and collection supply chain, which will diversify the availability of feedstock, leading to lower feedstock costs for both first- and second-generation feedstocks.

It should be noted that the categorization differs as some feedstock arguably matched multiple categories, and the classification of generation is also a point of contention. As such, the term "advanced biofuels" is gaining wider acceptance and generally refers to biofuels with superior carbon footprint.

5.3.4 Feedstock cost implications

While there is not yet widespread production of biojet fuel in the scale of bioethanol and biodiesel, the effects can best be predicted from the biodiesel industry. This is because the leading biojet fuel production pathway of HEFA bears similarity to the biodiesel transesterification process, in which they both rely on lipids as feedstocks. Fig. 5.7 shows the international prices for oil crop products and price index from FAO as of October 2020.

From 2002 to 2008, crude oil price went on a bull run due to the 2003 Iraq war (~USD30/barrel), 2005 Hurricane Katrina (~USD50/barrel), 2006 North Korean missile launch incident (~USD79/barrel), 2007 Turkish tension (~USD90/barrel), 2008 Nigerian militant attack (~USD117/barrel), and reducing US dollar strength. These catalysts brought crude oil price to an impossibly high peak price of USD 145.85 per barrel on July 3, 2008. This made them expensive to produce, while

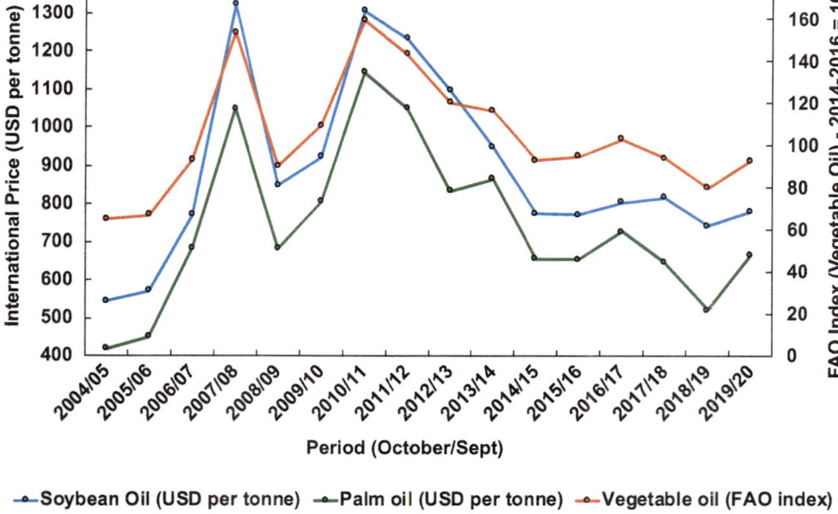

Figure 5.7 International prices for oil crop products and FAO vegetable oil price index (FAO, 2020a).

making biodiesel to be very profitable. Many biodiesel projects were declared to capitalize on the high crude oil price. As biodiesel projects relied only on edible vegetable oil at that time, price spiked due to demand as biodiesel feedstock. It is more profitable to use the feedstock for fuel than to feed mouths. This is on top of the natural mirroring of vegetable oil and crude oil prices since 2003. Year 2008 saw the highest food inflation since the early 1990s.

Thus, it is not inconceivable to believe that if biojet fuels become profitable due to extremely high petroleum price, the price of vegetable oil will face a sudden spike before it renormalizes. Two opposing economic forces in the supply—demand play might skew the outcome. First, a greater spike in vegetable oil price is expected as both biodiesel and biojet fuel will use the same feedstocks, increasing demand in vegetable oil. Second, demand will not rise as much as expected as the dominant biojet fuel conversion pathway might gravitate away from HEFA and use other feedstocks instead. Irrespective of the spike, market will renormalize by either increasing supply-side vegetable oil or having lower demand as higher feedstock cost will make biojet fuel production less desirable. Fears of hyper food inflation should not happen in the more efficient modern market era. The post-COVID-19 period is still an unknown quantity, but early signs are showing that vegetable oil price and crude oil price might be decoupling.

5.4 Global biojet fuel production

The US Department of Energy (US DOE) stated that the global demand for aviation jet fuel is 106 billion gallons in 2019 and expected to grow to 230 billion gallons in 2050 (US DOE, 2020). This translated to an actual 2019 and predicted 2030 global demand of 401 billion liters and 871 billion liters, respectively. Aviation industry is reliant on crude oil as the primary power source, as it can neither rely on electricity unlike road transport nor can it rely on solar and wind which favor stationary usage. When it comes to power generation for propulsion, liquid fuel is still the only choice that provides sufficient energy density volumetrically. This has led to the use of 6.5% of all oil refined worldwide for jet fuels. For biojet fuel to supplant or complement conventional jet fuel to meet the broad-based 2020 and 2050 aviation industry goals, large quantities of feedstock will be required. The present target for biojet fuel production is around 2—3 million tons by 2020, with longer-term targets rising exponentially to at least 58 million tons by 2050.

The International Air Transport Association (IATA) shows that 215,000 commercial flights from 40 airlines had use biojet fuels, and there were 6 billion liters (or 4.8 million tons) in airline forward purchase agreements as of December 2019 (Soone, 2020). These numbers are encouraging, but the actual market penetration remained low. The International Energy Agency (IEA), European Aviation Safety Agency (EASA), and European Commission (EC) have estimated that biojet fuel accounted for less than 0.1% (2018), at 0.004% (2017), and at 0.05% (2017) of total aviation fuel consumption, respectively. The global production volume of biojet fuels is difficult to be projected accurately at the moment as legislations, mandates, and policies regarding its use are still in a nascent stage. Reporting is also typically on a voluntary basis and often delayed by a couple of years. This is further made complicated where there is a gap between targets, actual usage, and production data.

As such, the International Civil Aviation Organization (ICAO) SAF Stocktaking process, which "take stock" of global progress of the development and deployment of biojet fuels might provide the best estimates (ICAO, 2020). Fig. 5.8 shows the actual global biojet fuel production

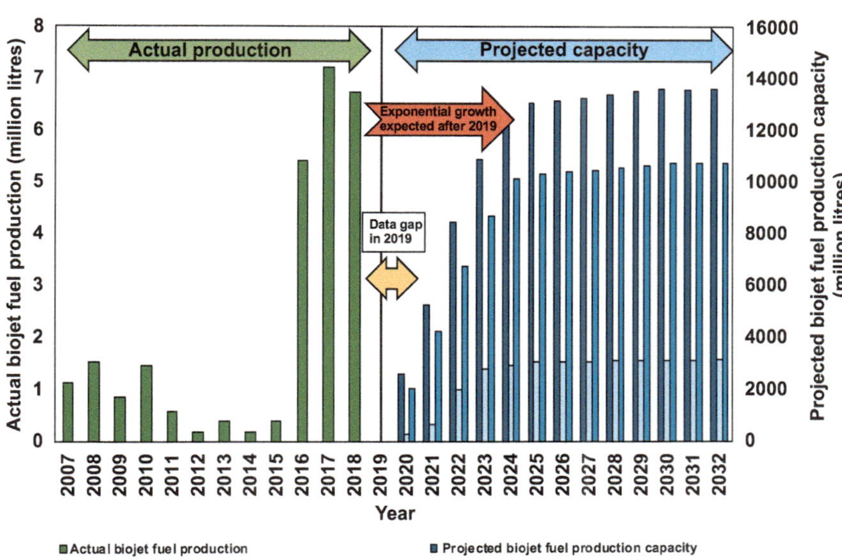

Figure 5.8 Actual global biojet fuel production (2007–18) and the projected global biojet fuel production capacity (2020–32). *(Adapted from ICAO, 2020. SAF Stocktaking - What is it about? https://www.icao.int/environmental-protection/Pages/SAF_Stocktaking.aspx.)*

(2007–18) and the projected global biojet fuel production capacity (2020–32). The stocktaking result as of July 2020 shows that the 3-year average 6.45 million liters per year for the 2016–18 period. This represented 22.2 times increase from the 2013–15 timeframe, which were the nadir for the biojet fuel industry. The industry is also projected to have an extended period of exponential growth until 2024, followed by a plateauing period where global biojet fuel production capacity is expected to be 13.6 billion liters.

The projected global biojet fuel production capacity for 2020 is about 2.6 billion liters, which could potentially contribute to 0.65% of total aviation fuel demand. However, having the capacity and actual production is not the same. It is still unclear on how the postpandemic biojet fuel industry will respond. It is also equally uncertain if the doubling of production capacity would achieve the 5.23 billion liters projection for 2021. Assuming a low ratio of 10% capacity being utilized, biojet fuel is expected to only fulfill 0.64 billion liters or roughly 0.07% of global jet fuel demand. An optimistic high ratio of 80% will lead to about 1.05% of global need fulfilled by biojet fuel.

A summary of the planned biojet fuel production capacity is shown in Table 5.18. Major projects include those from Marathon, World Energy, Diamond Green, Lanzatech, Hollyfrontier, Gevo, and Fulcrum in the United States; Total, Neste, Preem, and UPM in Europe; Neste in Asia; and ECB in Latin America.

5.5 Barriers to commercialization

The sustainability credentials of biojet fuel are undisputed; likewise is the business opportunity offered by biojet fuel. Scientific and technical solutions are sprouting to move biojet fuel production toward commercialization and mass-scale adoption. Nonetheless, there are still barriers to commercialization. Table 5.19 summarizes the main challenges of the biojet fuel market as compiled from experts globally, concentrated in "Europe and North America" and in Brazil.

5.5.1 Economic barriers

The aviation industry is a very price-sensitive industry without abilities of airlines to absorb any increase in fuel price (Hudson et al., 2016). As fuel costs can account for 30%–50% of operating costs, the concern of high cost or lack of price parity between biojet fuel and conventional jet fuel is of

Table 5.18 Planned biojet fuel production projects around the world (ICAO, 2020).

Producer	Location	Production year	Full production capacity (Mgal/year)
REG	Geismar, United States	In operation	75
Neste	Porvoo, Finland	2019	31
RedRock	Oregon, United States	2020	15.1
Total	La Mède, France	2020	165
Lanzatech	Port Talbot, Wales	2021 onwards	26.4
Marathon	Dickinson, United States	2021	184
World Energy	Los Angeles, United States	2021 (already in operation with 38 MGPY)	306
ARA licensee 1	–	2022	40.62
ARA licensee 2	–	2022	46
ARA licensee 3	–	2022	81.25
Diamond Green	Louisiana, United States	2022 (already in operation with 275 MGPY stated production capacity)	675
ECB	Paraguay	2022	252.95
Neste	Singapore	2022	317
Lanzatech	Georgia, United States	2022 (existing 10 MGPY)	100
ST1 Oy	Gothemburg, Sweden	2022	66
Hollyfrontier	United States	2022	90
Hollyfrontier	New Mexico, United States	2022 (full capacity in 2023)	125
Neste	Rotterdam, netherlands	2023	145.3
Refuel YYZ	Toronto, Canada	2024 (starting in 2023 with 4.76 MGPY)	9.51
Caphenia	Germany	2024	0.08
PREEM	Gothemburg, Sweden	2024	264

Continued

Table 5.18 Planned biojet fuel production projects around the world (ICAO, 2020).—cont'd

Producer	Location	Production year	Full production capacity (Mgal/year)
UPM	Kotka, Finland	2024	165
Flexjet project	Baden-Württemberg, Germany	2025	5.3
SAF plus consortium	Montreal, Canada	2025	7.9
Velocys	Immingham, United Kingdom	2025	13.2
Gevo	Minnesota and Texas, United States	2029 (starting with 50 kGPY in 2020)	100
Fulcrum	Nevada, United States	– (starting with 10.5 MGPY, 2020)	300
Velocys	Mississippi, United States	–	20
LTU Greenfuels	Piteå, Sweden	–	–

utmost importance. In short, biojet fuel must be price-competitive over the long term, although it could rely on carbon offsets in the short term. Despite the numerous economic and financial projections showing the potential profitability of biojet fuels, the stark reality remains where the feedstock cost of biojet fuel often eclipses the price of conventional jet fuel.

The discrepancy is primarily due to projections done at full scale, while the present phase of the biojet fuel industry is the "ramping-up" phase. Before scaling up is completed, such price disparity may exist. Many projections are also done with favorable macroeconomics conditions in place, including scenarios where demands of biojet fuel are high, leading to globalized economies of scale effects. Such optimism is placing a burden on early adopters to be profitable. For the industry to flourish, the early adopters must be profitable. However, for the early adopters to be profitable, the industry must flourish to form a strong supply chain.

Presently, there are two groups of biojet fuel producers, the large corporations attempting to diversify from their core businesses and the smaller start-ups entering the market. Any increase in biojet fuel producers

Table 5.19 Biojet fuel industry challenges to commercialization.

Challenges	Description of challenges	Global (Hari et al., 2015)	Europe and North America (Gegg et al., 2014)	Brazil (de Souza et al., 2018)
Environmental challenges	Social and environmental impacts inclusive of deforestation, afforestation, biodiversity, soil destruction, water issues, and land usage	X	X	X
Environmental controls	Overly strict environmental hurdles and tendency to overlook slightly less environmentally beneficial technologies in favor of "holy grail"–type technologies		X	
Production issues	Cost-effectiveness of the process, feedstock flexibility, suitability of catalyst, consistency of production process, and immature technology of biojet fuel production routes	X		X

Continued

Table 5.19 Biojet fuel industry challenges to commercialization.—cont'd

Challenges	Description of challenges	Global (Hari et al., 2015)	Europe and North America (Gegg et al., 2014)	Brazil (de Souza et al., 2018)
Distribution problems	Fuel infrastructure, by-product marketing regulations, and coordination of investors and biomass suppliers	X		X
Investment	Investment for aviation biofuel technologies and government investment, difficulty in obtaining credit due to global economic downturn		X	X
High costs	Uncompetitive cost of biojet fuel leading to large gap to price parity with conventional jet fuel	X	X	X
Feedstock availability and sustainability	Feedstock yield, water intake, fertilizer requirements, food-to-fuel issues, availability of biomass feedstock, supply feedstock supply chain, feedstock research, and unclear sustainability criteria	X	X	X

Table 5.19 Biojet fuel industry challenges to commercialization.—cont'd

Challenges	Description of challenges	Global (Hari et al., 2015)	Europe and North America (Gegg et al., 2014)	Brazil (de Souza et al., 2018)
Compatibility with conventional fuel	Fuel properties (such as sulfur and aromatics contents, freezing point, autoignition temperature, thermal stability and impurities contents), performance, safety, and storage stability	X	X	
Certification	Complications in certification, quality testing, and certification done abroad		X	X
Legislation	Legislation, imbalance playing field between road and aviation biofuels, mandates, tax breaks, subsidies, and poor knowledge flow between legislators and aviation biofuel community		X	

that has a sustainable business model will help the industry to flourish. The issue for entry to the market faced by larger and smaller producers alike is the reduced availability of cheap credits. Industrial experts in Europe, North America, and Brazil have identified the lack of eager investors as a barrier to commercialization. The lack of enthusiasm stemmed from the present sentiments of being risk averse, requiring quick returns and the lack of conviction from producers to overcome the payback hurdles. Investors want regulators to derisk their investments. Improving the coordination between investors and would-be producers is imperative to grow the biojet fuel industry.

Feedstock cost is the main operating culprit as it is the costliest component in biojet fuel production. This is especially true presently where the oil-based HEFA pathway is the dominant commercially available pathway for biojet fuel production. The soybean oil, palm oil, rapeseed oil, sunflower oil, and coconut oil monthly prices in November 2020 are USD 973.88, USD 917.81, USD 1047.78, USD 890.00, and USD 1368.95 per metric ton, respectively. The average crude oil spot price of Brent, Dubai, and West Texas Intermediate for the same period is USD 42.30 per barrel. Comparing like-to-like, the prices of soybean oil, palm oil, rapeseed oil, sunflower oil, and coconut oil per barrel are approximately USD 141.98, USD 134.89, USD 152.42, USD 130.00, and USD 196.53, respectively. This shows that the selected oil-based feedstock costs for biojet fuels are 3.07−4.64 times the feedstock cost for conventional jet fuel.

While high feedstock cost is a problem of its own, the uncertainties associated with fluctuating feedstock price are an equally large concern as producers will find it difficult to tame the bullwhip effect. Producers and feedstock sellers will have to participate in the practice of hedging, rationing, and gaming, which causes distortion to the market. There is also the issue of demand forecasting as the nascent biojet fuel industry does not yet have a centralized body where members could update demand forecasts to predict end-customer demand.

From a macroeconomic point of view, shift in trends due to the emergence of biojet fuel industry will affect the supply and demand equilibrium. As biojet fuel production is still at low quantity, there is no direct example of this happening at large scale. A parallel example relates to the biofuel development in China to meet the Medium and Long-Term Development Plan for Renewable Energy in 2007 (Chang et al., 2012). This economic supply of feedstock barrier shows that since biofuel feedstocks of cassava follow market price, then production cost will fluctuate based on the current market and future demand of the crop. Fig. 5.9 shows the import scenario of cassava and cassava starch for China for years 2000−08.

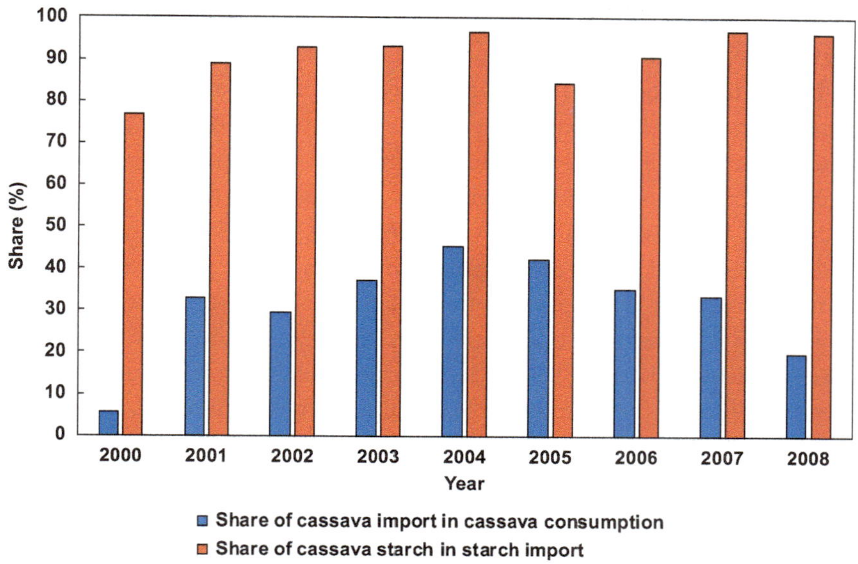

Figure 5.9 Import scenario of cassava and cassava starch. *(Adapted from Chang, S., Zhao, L., Timilsina, G. R., Zhang, X., 2012. Biofuels development in China: technology options and policies needed to meet the 2020 target. Energy Pol. 51, 64–79.)*

As China was already the largest cassava importer in the world since 2001, any demand side increase will add pressure on the supply side as share of imports would have to be increased, sparking an even higher proportion of cassava import in cassava consumption. This will drive feedstock price up and place a market driven barrier to commercialization of the industry. The higher import dependence also increases amount of foreign exchange outflow that could have benefitted the farmers and might discourage them to plant cassava domestically. The higher dependency on foreign imports will put producers at the mercy of price fluctuations, which lead to a second economic barrier to commercialization.

5.5.2 Sustainability barriers

The concerns on the environmental challenges posed by the biojet fuel industry are unanimous. Concerns on deforestation, afforestation, and water shortage are appropriate as it will be counterproductive to concentrate on decarbonization of the aviation industry, only to swap one set of issues to another. Present-day sustainability debates are focused on the food-to-fuel issue and the discrepancy in indirect land usage calculations to determine sustainability of a feedstock.

The food-to-fuel debate has cooled down in recent years due to the vegetable oil surplus and growing unused stockpile, for example, palm oil in Malaysia and Indonesia. It is increasingly seen as a food distribution and affordability issue than outright availability problem. Instead, the sustainability issues have come to the forefront with respect to feedstocks. Scientists could not attain consensus on how sustainability parameters are calculated. Meanwhile advocates and lobbyists from various regions are pushing for the adoption of calculation methods which portrays their own region's dominant feedstock in a better light. As such, some feedstocks are given poor sustainability ratings due to deforestation concerns despite the science being disputed. This has led to the blanket ban of palm oil in the European Union, be it sustainable or unsustainable palm oil. While it is understandable to ban unsustainable palm oil plantation practices, but the boycott calls that affect also sustainable palm oil will deprive the biojet fuel industry of a possibly sustainable feedstock.

In addition to deforestation, other sustainability concerns include soil pollution, decreased land service expectancy, habitat interruptions, eutrophication, overstretched water usage, waterway damage, and uncertainty in carbon footprint generated. The sustainability criteria, in particular the assigned coefficients and formula used, need to be improved and evidence based. It has to be purely viewed from a scientific viewpoint without prejudice or vested economic interests. Protectionism would have to take a step back so that potentially sustainable feedstock for biojet fuel production could flourish. Another solution is the identification of a purposely grown oil crops as feedstock for biojet fuel production, which meets all sustainability criteria.

If not for the global pandemic, the wildfires of western United States, Australia, and Siberia would have dominated news headlines. While the availability of biomass in forests on its own does not contribute to forest fires, the use of such biomass will reduce the burning of organic matter from forest floors. A robust supply chain could be developed to utilize forest biomass, which helps to improve sustainability and also abate the exacerbating wildfire problems. The barrier to this is the lack of supportive forest regulations explicitly prioritizing biomass usage for as biojet fuel feedstock.

5.5.3 Operational barriers

One of the main concerns from regulators and the general public in a transition occurring within the transport sector is safety. This is extremely crucial for the aviation industry as planes cannot risk stalling at high altitude as compared with cars stalling on road. For biojet fuel to be accepted, it has to be proved that it will not jeopardize the safety of flights. None of the 215,000 commercial flights involving biojet fuels ended in tragedy. However, some stakeholders in the aviation fuel supply chain remained concerned with biojet fuel safety (Smith et al., 2017). More laboratory experiments and actual flight tests have to be conducted to assuage the safety concerns, especially if neat biojet fuels are used and their properties be allowed by regulators to deviate from the existing standards.

The quality perception facing biojet fuel is unwarranted. Many considers biojet fuel to be different from conventional jet fuel. This is true for their origins but is no longer distinguishable at use point. Biojet fuel must meet the prevailing standards such as the ASTM D7566's Jet A or Jet A-1 standards prior to blending. The properties of biojet fuels conforming to the standards are less of a concern than the lack of certification agencies in some countries to certify the fuel. Some countries do not have the resources to regulate and monitor the development of biojet fuels to ensure that aircraft and engine manufacturer standards are met.

Logistics is a real concern to commercialization. It is often stated that biojet fuel could directly use the infrastructure of conventional jet fuel due to their compatibility and interchangeability inside jet engines. However, this is not entirely true as logistics goes beyond the end user. Conventional jet fuels are often delivered to airports through pipelines, and producers are located strategically. Biojet fuel producers do not have the means or scale to set up pipelines and often required to deliver biojet fuel via ships, rail, and road transports. The distance traveled is often far as biojet fuel producers are often located closer to feedstock source than airports. Logistically, this is undesirable. One mitigation would be to tap into the existing pipelines for fuel delivery. However, there are whole other issues of biojet fuel being viewed as a contaminant, which may invalidate warranties. Furthermore, the vested interests between traditional oil companies and biojet fuel producers may not yet be aligned.

5.5.4 Societal barriers

If the promises of third-generation feedstocks like microalgae or yeast can be fulfilled, the quantum leap in gains will be phenomenal. At present, larger-scale algal farms have not been able to replicate the microalgal proliferation rates obtained in laboratories. Algae were often found to be vulnerable to infections and predators. Efforts need to be concentrated to reidentify algal strains, which are resistant to local infectors. Another method is to culture the microalgae to their original habitat with sufficient inoculum (Saad et al., 2019). Genetic engineering of microalgae may enhance its survivability, growth rate, and compatibility for use as biojet fuel feedstock. However, the increase use of a single strain, be it genetically modified or not, increases the risk of a new infection destroying an entire microalgae farm. Furthermore, societal barrier exists in the form of public rejection of genetically modified organisms. Safety concerns about genetically engineered microalgae are part of the public psyche.

The positive findings on biojet fuel may not yet come into public mindset. The general public view renewable energy favorably but often have poor understanding of biofuels and may be fixated on renewables such as solar power and wind energy. The fact that biojet fuel is premixed with conventional jet fuel also made it nonvisible as compared with solar panels on the roof and large wind turbines dotting the rural landscapes. Some also question the need for biojet fuels in the aviation sector as opposed to electrifying airplanes like in road transport. The general public need to be receiving information that electrifying airplanes poses a greater challenge than land vehicles as aircrafts are more sensitive in coping with extra weight from the electrical propulsion system and electricity storage. Due to the requirement of airborne transport to be lighter-than-air, energy use in aircrafts is directly proportional to its weight unlike the less stringent power-to-weight ratio required for land-based vehicles. Outreach efforts could be increased to better inform the public on the benefits of biojet fuel, and the public may reciprocate in kind to support it in combating climate change.

5.6 Summary

The biojet fuel industry is still currently in a nascent stage, lagging behind the more illustrious road transport sector biofuels such as biodiesel and bioethanol. Unlike bioethanol and biodiesel for road transports, which are facing strong competition from electric vehicles, biojet fuels do not face any direct renewable energy competitors due to the need for long haul travel and high density fuels. This means that the main hurdles for the industry are purely to make it profitable through the use of cheaper feedstocks, develop mature technologies, and increase output for improved economies of scale. While the ability for nations to produce biojet fuel is more democratized as compared with other forms of renewable energies due to availability of natural feedstocks, there is still a resource-technology conundrum. Countries with cheaper feedstocks often lag behind in the development of production technologies and also availability of skilled engineers for setting up the facilities. The current trajectory points toward a subsidy-supported industry to allow eventual organic growth for the industry. Such organic growth will eventually bring down cost of production and make biojet fuel profitable globally. In addition to economic barrier, the industry also faces challenges in the form of sustainability and operational and societal barriers. Biojet fuel as a green energy is no longer just a "corporate social responsibility." It has the capability to transform the economies of countries while also decarbonizing the aviation sector and improving energy securities. Although the climate change planetary crisis sparks the need for biojet fuels, it will ultimately be economic factors that will decide if the biojet fuel industry is a long-standing success.

References

Atsonios, K., Kougioumtzis, M.-A., Panopoulos, K.D., Kakarasa, E., 2015. Alternative thermochemical routes for aviation biofuels via alcohols synthesis: process modeling, techno-economic assessment and comparison. Appl. Energy 138, 346–366.

Chang, S., Zhao, L., Timilsina, G.R., Zhang, X., 2012. Biofuels development in China: technology options and policies needed to meet the 2020 target. Energy Pol. 51, 64–79.

Chiaramonti, D., Prussi, M., Buffi, M., Tacconi, D., 2014. Sustainable bio kerosene: process routes and industrial demonstration activities in aviation biofuels. Appl. Energy 136, 767–774.

Chuck, C.J., McManus, M., Allen, M.J., Singh, S., 2016. Feedstocks for Aviation Biofuels, in Biofuels for Aviation, pp. 17–34.

Claver, A., Rey, R., Lopez, M.V., Picorela, R., Alfonso, M., 2016. Identification of target genes and processes involved in erucic acid accumulation during seed development in the biodiesel feedstock Pennycress (*Thlaspi arvense* L.). J. Plant Psychol. 208, 7–16.

Cremonez, P.A., Feroldi, M., Oliveira, C.D.J., Teleken, J.G., Alves, H.J., Sampaio, S.C., 2015. Environmental, economic and social impact of aviation biofuel production in Brazil. N. Biotech. 32 (2), 263–271.

Cruz, G., Silva, A.V.S., Da Silva, J.B.S., Caldeiras, R.d.N., de Souza, M.E.P., 2020. Biofuels from oilseed fruits using different thermochemical processes: opportunities and challenges. Biofuels Bioprod. Biorefining 14 (3), 696–719.

de Souza, L.M., Mendes, P.A.S., Aranda, D.A.G., 2018. Assessing the current scenario of the Brazilian biojet market. Renew. Sustain. Energy Rev. 98, 426–438.

Deane, P., Gallachóir, B.Ó., Shea, R.O., 2017. Biofuels for aviation. In: Europe's Energy Transition - Insights for Policy Making, pp. 79–88.

Diederichs, G.W., Mandegari, M.A., Farzad, S., Görgens, J.F., 2016. Techno-economic comparison of biojet fuel production from lignocellulose, vegetable oil and sugar cane juice. Bioresour. Technol. 216, 331–339.

European Comission, 2018. Feasibility Study on Options to Step up EU Action against Deforestation. https://ec.europa.eu/environment/forests/pdf/KH0418199ENN2.pdf.

FAO, 2017. Food and Agriculture Data. http://faostat.fao.org/static/syb/syb_5000.pdf.

FAO, 2019. Foof and Agriculture Organisation of the United Nations. FAOSTAT. http://www.fao.org/faostat/en/#data/QC.

FAO, 2020a. Food and Agriculture Organization of the United Nations. Food Outlook. http://www.fao.org/3/cb1993en/cb1993en_oilcrops.pdf.

FAO, 2020b. Food and Agriculture Organization. Crops Processed. http://www.fao.org/faostat/en/#data/QD.

Gegg, P., Budd, L., Ison, S., 2014. The market development of aviation biofuel: drivers and constraints. J. Air Transport. Manag. 39, 34–40.

Giwa, A., Adeyemi, I., Dindi, A., García-Baños Lopez, C.G.-B., Lopresto, C.G., Curcio, S., Chakraborty, S., 2018. Techno-economic assessment of the sustainability of an integrated biorefinery from microalgae and Jatropha: a review and case study. Renew. Sustain. Energy Rev. 88, 239–257.

Hari, K., Yaakob, T.Z., Binitha, N.N., 2015. Aviation biofuel from renewable resources: routes, opportunities and challenges. Renew. Sustain. Energy Rev. 42, 1234–1244.

Hudson, L., Jefferson, A., Bauen, A., Nattrass, L., 2016. Sustainable Aviation: A uk Roadmap for Sustainable Aviation Fuels, in Biofuels for Aviation, pp. 315–337.

IATA, 2020. Jet Fuel Price Monitor. https://www.iata.org/en/publications/economics/fuel-monitor/.

ICAO, 2020. SAF Stocktaking - What is it about? https://www.icao.int/environmental-protection/Pages/SAF_Stocktaking.aspx.

Kalam, M.A., Masjuki, H.H., Jayed, M.H., Liaquat, A.M., 2011. Emission and performance characteristics of an indirect ignition diesel engine fuelled with waste cooking oil. Energy 36 (1), 397–402.

Li, X., Mupondwa, E., Tabil, L., 2018. Technoeconomic analysis of biojet fuel production from camelina at commercial scale: case of Canadian Prairies. Bioresour. Technol. 249, 196–205.

Martinez-Hernandez, E., Ramírez-Verduzco, L.F., Amezcua-Allieri, M.A., Aburto, J., 2019. Process simulation and techno-economic analysis of bio-jet fuel and green diesel production — minimum selling prices. Chem. Eng. Res. Des. 146, 60–70.

Neuling, U., Kaltschmitt, M., 2018. Techno-economic and environmental analysis of aviation biofuels. Fuel Process. Technol. 171, 54–69.

Noh, H.M., Benito, A., Alonso, G., 2016. Study of the current incentive rules and mechanisms to promote biofuel use in the EU and their possible application to the civil aviation sector. Transport. Res. Transport Environ. 46, 298–316.

OECD-FAO, 2019. OECD-FAO Agricultural Outlook, Oilseeds and Oilseed Products. http://www.agri-outlook.org/commodities/Oilseeds.pdf.

Pereira, L.G., MacLean, H.L., Saville, B.A., 2017. Financial analyses of potential biojet fuel production technologies. Biofuels Bioprod. Biorefining 11 (4), 665–681.

Ranganathan, P., Savithri, S., 2016. Techno-economic analysis of microalgae-based liquid fuels production from wastewater via hydrothermal liquefaction and hydroprocessing. Bioresour. Technol. 284, 256–265.

Reimer, J.J., Zheng, X., 2017. Economic analysis of an aviation bioenergy supply chain. Renew. Sustain. Energy Rev. 77, 945–954.

S&P Global, 2020a. As Jet Fuel Market Craters, Sustainable Aviation Fuel Prepares for Takeoff. https://www.spglobal.com/platts/en/market-insights/blogs/agriculture/092320-as-jet-fuel-market-craters-sustainable-aviation-fuel-prepares-for-takeoff.

S&P Global, 2020b. Platts Sustainable Aviation Fuel (SAF) and HVO Assessments. https://www.spglobal.com/platts/PlattsContent/_assets/_files/en/our-methodology/methodology-specifications/platts-saf-and-hvo-faq.pdf.

S&P Global, 2020c. S&P Global Platts Launches First Sustainable Aviation Fuel Price Assessments in Europe. https://www.spglobal.com/platts/en/about-platts/media-center/press-releases/2020/081720-platts-launches-first-sustainable-aviation-fuel-price-assessments-europe.

Saad, M.G., Dosoky, N.S., Zoromba, M.S., Shafik, H.M., 2019. Algal biofuels: current status and key challenges. Energies 12 (10), 1920.

Scarlat, N., Dallemand, J.-F., 2019. Future role of bioenergy. In: The Role of Bioenergy in the Bioeconomy, pp. 435–547.

Smith, P.M., Gaffney, M.J., Shi, W., Hoard, S., Armendariz, I.I., Mueller, D.W., 2017. Drivers and barriers to the adoption and diffusion of sustainable jet fuel (SJF) in the U.S. Pacific northwest. J. Air Transport. Manag. 58, 113–124.

Soone, J., 2020. European Parliamentary Research Service, Sustainable Aviation Fuels. https://www.europarl.europa.eu/RegData/etudes/BRIE/2020/659361/EPRS_BRI(2020)659361_EN.pdf.

US DOE, 2020. Sustainable Aviation Fuel-Review of Technical Pathways. https://www.energy.gov/sites/prod/files/2020/09/f78/beto-sust-aviation-fuel-sep-2020.pdf.

van Eijck, J., Romijn, H., Balkema, A., Faaij, A., 2014. Global experience with jatropha cultivation for bioenergy: an assessment of socio-economic and environmental aspects. Renew. Sustain. Energy Rev. 32, 869–889.

Wei, H., Liu, W., Chen, X., Yang, Q., Li, J., Chen, H., 2019. Renewable bio-jet fuel production for aviation: a review. Fuel 254, 115599.

Wendimu, M.A., 2016. Jatropha potential on marginal land in Ethiopia: reality or myth? Energy Sustain. Dev. 30, 14–20.

Woittiez, L.S., van Wijk, M.T., Slingerland, M., van Noordwijk, M., Giller, K.E., 2017. Yield gaps in oil palm: a quantitative review of contributing factors. Eur. J. Agron. 83, 57–77.

Yang, Z., Qian, K., Zhang, X., Lei, H., Xin, C., Zhang, Y., et al., 2018. Process design and economics for the conversion of lignocellulosic biomass into jet fuel range cycloalkanes. Energy 154, 289–297.

CHAPTER 6

Sustainability of aviation biofuels

6.1 Introduction

From the perspective of greenhouse gas (GHG) emissions, the aviation industry contributes to a mere 2% of the global anthropogenic carbon dioxide (CO_2) emissions, but the rapid expansion in the airline industry is expected to lead to an increase of emissions. Unlike ground transportation, the difficulty for aircraft to follow the path of electrification means that the reliance on hydrocarbon fuels will continue in the foreseeable future. It has been projected that the jet fuel consumption within the transportation sector would increase from 11% to 14% in the next 20 years, with international aviation accounting for approximately 65% of global aviation fuel consumption. The growth in the aviation sector is expected to contribute 1.1−1.5 billion tons of CO_2 by 2035 (ICAO, 2016). Under the United Nations Framework Convention on Climate Change (UNFCCC), the tabled Kyoto Protocol specifies that strong cooperation between countries is needed to limit and reduce the GHG emissions. Although GHG emissions from international aviation are not included in national GHG inventories or targets, the Kyoto Protocol specifies that industrialized countries shall pursue limitations or reductions of GHG emissions from aviation through the International Civil Aviation Organization (ICAO) (UNFCCC, 1998).

The ICAO is a United Nation agency that spearheads the role of addressing the issue of GHG emissions for the aviation sector by working with 191 member states and industry groups to develop policies, standards, and recommended practices for the civil aviation sector. To facilitate carbon-neutral growth for international civil aviation after 2020, the ICAO introduced the Carbon Offsetting and Reduction Scheme for International Aviation (CORSIA), which is a global market-based measure system to offset international aviation emissions growth, along with other proposed GHG mitigation strategies such as improving operational and technology

Biojet Fuel in Aviation Applications
ISBN 978-0-12-822854-8
https://doi.org/10.1016/B978-0-12-822854-8.00005-6

efficiency measures, and adopting alternative aviation fuels to meet the carbon-neutral emission goal.

To meet the aspiration of carbon negative growth by 2050, a large bulk of carbon reduction is expected to be achieved from the use of low carbon alternative jet fuels derived from nonpetroleum sources such as biomass, wastes, vegetable oils, etc. Certified fit for purpose synthetic jet fuels that meet the standard specifications can be used as "drop-in" fuels to replace conventional jet fuel with no modification to the engine or fuel distribution system. There are currently seven types of alternative jet fuel approved under the ASTM D7566 standard; each of the synthetic jet fuel is produced from distinctively different pathways and specific feedstocks. Details of the production method and properties of the alternative jet fuels are described in Chapters 2 and 3, respectively. To evaluate the potential of a specific alternative jet fuel, there are several aspects that need to be considered apart from the technical performance in engine, including the economics and environmental sustainability, such as the environmental impacts, effects on the water consumption, local air quality, land usage and energy resources, technical feasibility, and economics related to the fuel production pathways, to ensure the long-term viability of alternative jet fuels. This chapter aims to provide an overview on some of the methods used to assess the environmental sustainability of alternative jet fuel, using methodology such as the life cycle assessment (LCA) of GHG emissions, energy balance analysis, and sustainability indices.

6.2 Life cycle assessment of aviation jet fuel

The goal set by ICAO to keep the global net carbon emissions of aviation industry at neutral growth rate from year 2020 is to be partially met with the usage of alternative jet fuels. However, the potential emission reductions from alternative jet fuel need to be quantified from the life cycle emissions perspective. An Alternative Fuels Task Force was created within the ICAO technical body to develop a methodology to acquire a full understanding on the potential benefits of aviation biofuels and its associated impacts to the environment. The methodology used to estimate the carbon intensity of a given alternative jet fuel is LCA. LCA is defined as a compilation and evaluation of inputs, outputs, and the potential environmental impact of a product throughout its life cycle. The LCA approach for an industrial product is typically based on the "cradle-to-grave" concept, starting with the natural resources that are used as raw materials in the processing and ending with the residues generated by the consumption of the products that return to nature.

The International Organization for Standardization (ISO) specifies the LCA methodology and framework in ISO 14040:2006, which broadly consists of four phases (ISO 14040, 2006). The goal and scope definition phase defines the system boundary and level of detail. The system boundary defines the unit processes to be included in the system and the inputs and outputs through the boundaries. For aviation jet fuel, the system boundary typically covers the well-to-wake (WTWa) cycle, where the boundary defines the raw material acquisition stage up until the end-use stage. The functional unit is defined at this stage to provide a reference to which the inputs and outputs are related. All subsequent analyses are based on the functional unit to ensure comparability of the LCA results.

The inventory analysis phase that is related to the inventory analysis phase involves the evaluation of the data collected and calculation to quantify relevant inputs (e.g., resources) and outputs (e.g., emissions) of a product system, based on the defined boundary systems and the functional unit. The calculation procedures include the validation of the data of each unit process and reference flow of the functional unit. For an industrial process that involves multiple end products and recycling streams, a proper allocation procedure is needed to ensure a realistic distribution of the impact assessed.

The significance of the potential environmental impacts of the product is assessed via the impact assessment phase of LCA by using the results calculated from the life cycle inventory (LCI) analysis. The impacts of the life cycle can be assessed by associating the inventory data with specific environmental impact categories and indicators, such as global warming and water resources depletion, among others. The most assessed impact category for alternative jet fuel is perhaps the global warming potential (GWP), as it directly concerns with the GHG emissions, although the assessment of other impact categories related to environment and human health is also found in many studies. The inventory and impact assessment phases provide the information necessary for the interpretation phase, where recommendations and decisions are made in accordance with the goal and scope defined. The interpretation of the result is based on the relative approach, where they indicate potential environmental effects instead of predicting the actual impacts on category endpoints. The relationship between the phases is illustrated in Fig. 6.1. It should be noted that LCA studies differ from one another and that the limitations and recommendations of the

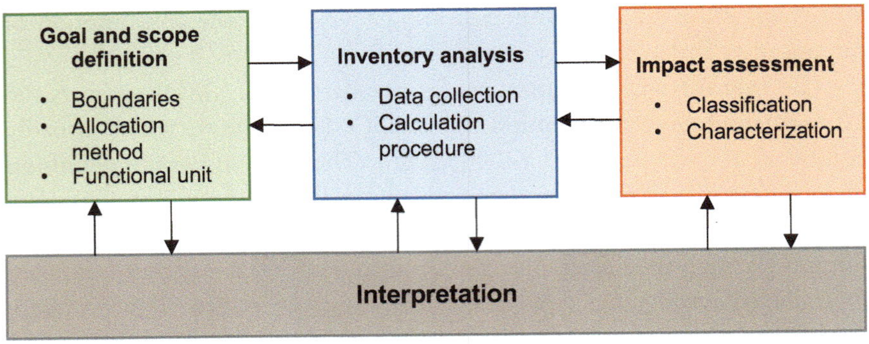

Figure 6.1 Life cycle assessment framework according to ISO 14040 (2006).

study should be clearly stated in the report. The LCA method is useful to identify opportunities to improve the environmental performance of products at various points in their life cycle.

The life cycle of jet fuel constitutes of multiple steps, covering feedstock extraction, production, and the final use in engine. At each of these steps, GHG emissions are likely to be produced; hence the LCA approach can be adopted to ascertain the total carbon footprint of the fuel emitted from each of these steps. The carbon intensity of a given fuel is estimated using LCA methodology and is typically expressed in gCO_2 equivalent per MJ of fuel (i.e., its carbon intensity). The life cycle of fossil-based jet fuels begins at the moment it is extracted from the well to the production facility, undergoing processing and refining into practical fuels, transportation and distribution of the fuels to the aircraft tank, and finally combusted in the engine. The net total impact of all of these life cycle phases is commonly referred to as the WTWa, which consists of the fuel production and distribution, "well-to-tank," and fuel combustion, "tank-to-wake." The life cycle of the fossil jet fuel is illustrated in Fig. 6.2.

For biomass-derived jet fuel, the life cycle differs from that of fossil jet fuel, as the emissions are now associated with the cultivation, harvesting, transport, and conversion of feedstock. Although the emissions of the

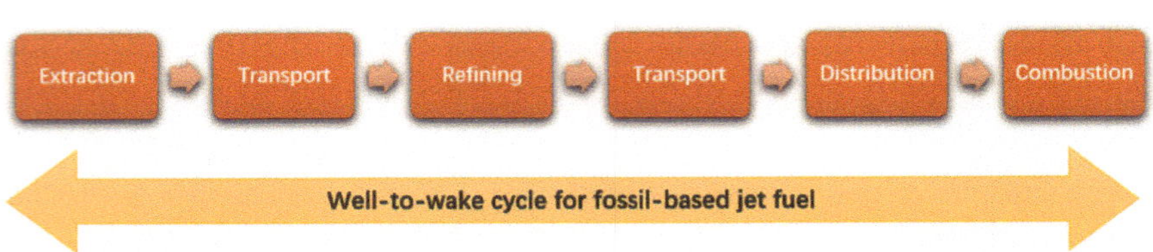

Figure 6.2 Life cycle of fossil-based jet fuel (Stratton et al., 2010).

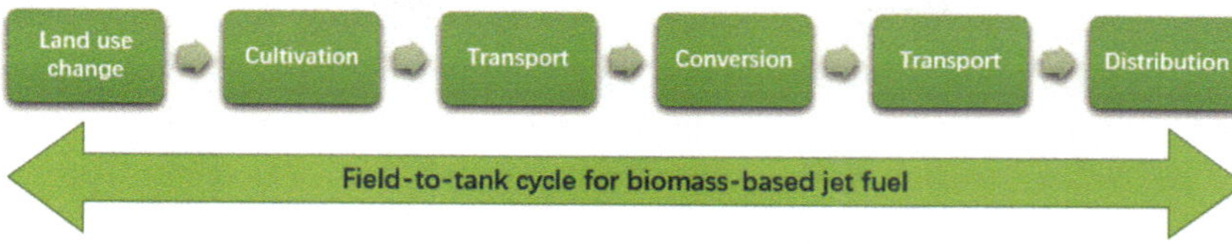

Figure 6.3 Life cycle of biomass-derived fuels (Stratton et al., 2010).

biofuels are produced as a result of combustion in engine, it is considered as "neutral," based on the assumption where the CO_2 emissions from the combustion are absorbed by the biomass during their growth phase (Hileman and Stratton, 2014). Such sequestration of the CO_2 by the biomass offsets the CO_2 in the life cycle assessment; thus, biofuels have "biomass credit." The biomass credit is the primary difference between biomass and fossil fuels in terms of their GHG emissions. Therefore, the life cycle of the biofuel produced from biomass is considered based on the "field-to-tank" approach, as illustrated in Fig. 6.3. Depending on the feedstock type and agricultural practices, the cultivation part can represent a significant portion of the GHG emissions. The GHG emissions reduction potential of the alternative jet fuel is ascertained by subtracting the carbon intensity of the alternative jet fuel from the baseline carbon intensity of petroleum-derived jet fuel. A lower GHG emission for the full life cycle of alternative jet fuel has positive impact on the environment and can be considered as a potential fuel to meet the ICAO's carbon reduction goal.

6.2.1 Product allocation

The production of biojet fuel typically results in the creation of coproducts in addition to the primary fuel product. Product allocation is implemented to partition the environmental impacts from the process based on product flows. For example, during the extraction and pretreatment processes, the biomass coproducts generated could be used as solid fuel or processed into fertilizer. Liquid fuel coproducts derived during the jet fuel production stage such as hydrocarbons with shorter or higher chain length can be used as fuels for different purposes. These coproducts have values, which can be quantified based on physical metrics, and displace some of the values of the intended products. This is important considering the net GHG reduction for the whole life cycle is needed to achieve the carbon reduction target as per required by local policy. For instance, under the EU Renewable Energy Directive 2018/2001/EU—Recast to 2030 (REDII), the alternative jet fuels produced must achieve a 50% reduction of GHG emissions for production facilities before October 2015, while a mandatory 50% reduction is required for production facilities after the date. For sustainable jet fuel produced in facilities that begin operations after January 2021, a 65% of carbon reduction must be achieved (EU, 2018). Therefore, proper accounting of the carbon intensity and energy input that takes into account coproduct allocation is essential to meet the regulatory set targets.

Nonetheless, allocation is not a straightforward division of environmental impacts between products and coproducts. A set of protocol set forth by ISO 14044:2006 provides the basis for the allocation (ISO 14044, 2006). There are four methods that can be used to assign the life cycle GHG emissions between the primary fuel products and coproducts, which are (1) mass allocation, (2) energy allocation, (3) market-value allocation, and (4) displacement (system expansion) methods. Each of the methods differs in allocation methodologies that can invariably lead to different results, in particular for biofuel pathways with significant quantities of coproducts. LCA practitioners can choose the method that best fits their scenarios to achieve the goal of the study. The absence of a fixed methodology is one of the reasons that makes comparison among LCA studies difficult. It is important to clarify the allocation approach adopted in any LCA work.

The ISO 14044:2006 (E) specifies that processes shared with other product systems shall be identified and dealt with by preferentially using process disaggregation, system expansion, allocation by an underlying physical relationship, and economic value, in this order. Inventories are based on material balances between input and output; therefore, allocation procedures should attempt to approximate such fundamental input/output relationships and characteristics (ISO 14044, 2006). The mass and energy allocation approaches distribute the life cycle GHG emissions based on respective mass or energy content of the coproducts of the fuel. The work by Stratton et al. (2010) utilized the energy allocation method to allocate the energy and emissions between products of Fischer–Tropsch (FT) process and hydroprocessed renewable jet fuels. The basis of this choice is due to the same functionality of coproduct (liquid fuel) compared with the primary jet fuel product.

Allocation of the coproducts based on the market value approach is subjected to the sensitivity of the market forces, and the values can change over time. Contrary to the mass or energy allocation approaches, which are fixed in values, the market value of coproducts fluctuate depending on their utilization. For example, the market value method will allocate more of the emissions to the main fuel produced if the coproducts displace some existing product. Such method has been used to allocate the emissions based on market valuation between coproducts of the hydroprocessed esters and fatty acids (HEFA) production pathways (Stratton et al., 2010). The displacement method can be deployed if the coproducts displace the production of a substitute product, e.g., counterpart of the fossil coproduct, so that the emissions credit from the displaced product is aggregated onto the primary product. Li and Mupondwa (2014) used this method in their LCA of GHG emissions of camelina oil in Canadian Prairies for biojet fuel production. The displacement method may give negative emission intensities if the emission credits exceed the total system emissions, contrary to energy allocation which strictly yields positive emissions intensities except when carbon capture and storage is adopted (de Jong et al., 2017). A significant difference has been reported in the GHG emission intensity between energy allocation and the hybrid method using the displacement method, especially when high amount of nonfuel coproducts was produced that effectively displace carbon-intensive products. This effect was evident for conversion pathways such as FT, jatropha-based HEFA, or corn-based ATJ. Nevertheless, the advantage of the displacement method is its reflection of the potential GHG emissions mitigation effects of producing coproducts, as stated by the ISO (de Jong et al., 2017). However, the challenge of this method is the requirement to determine the displacement ratio and life cycle of the GHG emissions of the displaced product; thus, the level of uncertainty is greater. For example, the allocation of the land use change (LUC) emissions to the biofuel complicates the application of the displacement method.

6.2.2 Effect of land use change on emissions

The use of biomass feedstocks for alternative jet fuel products can cause direct and indirect emissions due to LUC. The emissions and sequestration of CO_2 are closely related to changes in the biomass, soil, and organic waste contained within the land, which can significantly impact the GHG emissions. Direct LUC occurs when land is converted to grow feedstock for bioenergy. An example is the clearing of forest for feedstock growing, which causes a shift in that land's carbon stocks due to changes in land management practices. The indirect land use change (ILUC) refers to the indirect conversion of arable land to land catered for the demand of energy crops, most often driven by the indirect effect of biofuel policies. ILUC considers the change in emissions from land conversion that is used for the growth of energy crops and is usually quantified through LCA modelling. In such analysis, it is commonly assumed that the biomass supplied to the global food market is satisfied; hence, the subsequent LUC for energy crops is used to fulfill the demand of biofuels. Emissions from ILUC is due to existing cropland being used to farm energy crops to meet the biofuel demand, thereby resulting in the displacement of other agricultural products with high carbon stocks or other ecosystem services. The determining factors that are critical in the modeling of emissions caused by ILUC include the elasticity of food demand for price, elasticity of yield to price, choice of crops, utilization of coproducts, elasticity of area to price, and carbon stock of new land. These factors need to be taken into account to reflect the realistic scenario of ILUC emissions (Malins et al., 2014). Valin et al. (2015) utilized the Global Biosphere Management Model (GLOBIOM) to assess the biofuel-induced LUC and its impact on GHG emission. The model accounts for 12 sources of GHG emissions covering crop cultivation, livestock, LUC, and soil organic carbon based on advanced accounting framework. The study included various emissions sources and sinks related to biomass and soil carbon stocks, but the emissions related to feedstock cultivation and processing, biofuel production, transport, and distribution are not considered. Since the study was set in the EU context, the emissions values that are factored in LUC effect may differ significantly if the scenario is set in a different region. The type of land used for conversion into cropland was shown to have a significant effect on the emissions. For example, peatland is known to be efficient carbon sinks and long-term repositories of terrestrial organic carbon formed by dead organic matter and concentrated biomass in oxygen-low water environment that have been accumulated over centuries. Conversion of peat land into agricultural farm land can induce significant amount of CO_2 released into the atmosphere, with approximately 105 tons of CO_2 equivalent per hectare per year when averaged over 20 years (Page et al., 2011). The change in agricultural practice by rotating the plantation of woody biomass and perennial plants can induce negative LUC emissions, owing to the increase of carbon stock on the land (Valin et al., 2015). Contrary to the jet fuel produced from biomass, the effect of LUC for fossil fuel extraction and fuel processing is considered to be negligible due to the high throughput of fuel volume and mass created per unit land converted (Stratton et al., 2010).

6.3 Alternative jet fuel production pathway

The establishment of the system boundary for alternative jet fuel requires the unit processes to be defined, as well as the input and output flows into the system. Identification of the unit processes for each of the production pathway assists in the establishment of list of inventories. Fig. 6.4 depicts some of the typical feedstocks and their possible conversion pathways. Although details on the production pathways for biojet fuels have been discussed in depth in Chapter 2, the major production routes are briefly presented here to assist in defining the system boundaries for the LCA. In general, the production pathways for biojet fuel can broadly be categorized into three main groups, i.e., lipid conversion, thermochemical conversion, and biochemical conversion methods. The lipid conversion pathway converts the fatty acids and esters extracted from oil-bearing crops such as vegetable oils, animal fat, waste cooking oil (WCO), or algae into synthetic jet fuel via hydroprocessing, where the oxygen molecules in the oil are removed and the double bonds in the fuels are saturated via hydrogenation. The produced oil is isomerized and cracked into jet fuel and other hydrocarbon products. This production method is the most technologically matured pathway and is currently at commercial stage. Another method of lipid conversion is via the hydrothermolysis treatment, which involves hydrothermal conversion and hydrotreating operations to produce biojet fuel that contains aromatics.

The production of jet fuel via thermochemical conversion route can be performed via gasification and pyrolysis techniques. Biomass in solid form can be gasified at elevated temperature to produce synthesis gas (syngas), consisting of CO and H_2, which are then purified and used as precursor gases to synthesize jet fuel via the Fisher-Tropsch (FT) process. The FT hydroprocessing method is a certified production route for biojet fuel under the ASTM D7566 (ASTM D7566-19b, 2019). Although biooil derived from pyrolysis process can further be upgraded via hydrotreatment processes to derive jet fuel quality hydrocarbons, this method is still at development phase and has yet to be certified as fit for purpose jet fuel.

The biochemical conversion route for production of biojet fuel is via the fermentation of sugar or starch derived from organic resources such as lignocellulosic biomass. The alcohol produced from the fermentation is dehydrated, oligomerized, and hydroprocessed into jet fuel. The production of jet fuel via the direct sugar to hydrocarbon conversion (DSHC) method has been shown to be feasible via the use of modified yeast, producing farnesene (C15 hydrocarbon molecule) that can be further processed into jet fuel via hydroprocessing and fractionation. This processing method demonstrated by the pilot plant of Amyris (US) was shown to be able to achieve GHG emissions reduction up to 80% (US DOE, 2012). The ATJ and DSHC production methods are both

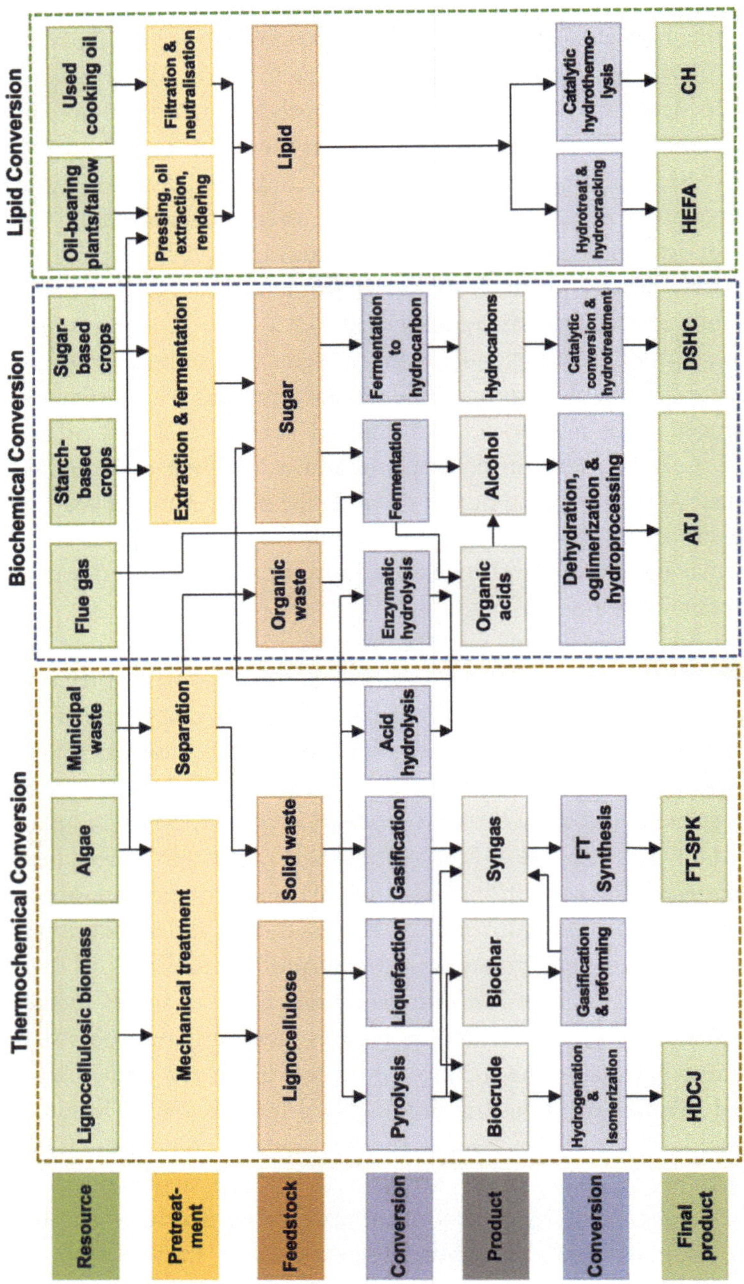

Figure 6.4 Production pathways for different feedstocks.

ASTM-certified production pathways (ASTM D7566-19b, 2019). Details of the production pathway are important in developing the life cycle inventory, but the pre- and postproduction processes, including the cultivation, transportation, and distribution, among others, are also needed to complete the system boundary in LCA.

6.4 Life cycle greenhouse gas emissions for different production pathways

The life cycle GHG emissions of alternative jet fuel depend on the production pathways. Each of the conversion routes is discussed, followed by a comparison of the GHG emissions across different feedstocks and production pathways. It should be noted that different methodologies were used in these studies, such as the different approaches used in the product allocation; hence, comparison of the LCA values across different studies should be done with care. It is important for the practitioners of LCA to clearly indicate the assumptions and allocation methods used in their studies. This also reflects the need for the harmonization of LCA to better represent the GHG impacts from the fuels (Takriti et al., 2017).

6.4.1 Biochemical conversion

Conversion of lignocellulosic biomass into biojet fuel via biochemical route has shown promising GHG emissions reduction potentials. Alternative jet fuel produced from sugarcane ethanol in Brazil can lead to GHG reductions of 70%−90%, or relevant GHG mitigation in comparison to fossil kerosene, depending on the methodological approach (Capaz et al., 2018). Lignocellulosic biomass used for the biojet fuel production generally refers to a wide range of materials such as woody biomass, eucalyptus, wheat straw, or sugarcane bagasse, depending on the local availability. The feedstock is usually pretreated to derive the sugars before undergoing enzymatic hydrolysis and saccharification to produce fermentable sugars. Subsequently, the sugars are converted into isoparaffinic kerosene using fermentation and oligomerization process. This process is commonly known as the alcohol-to-jet (ATJ) production route. Ganguly et al. (2018) assessed the environmental implications of alternative jet fuel produced from residual woody biomass by conducting an LCA study. It was shown that the potential GWP can be reduced by 78% compared with fossil-based jet fuel. The main portion of the GWP is attributable to the recovery and processing of woody biomass. For mallee eucalypt-based jet fuel, the carbon emissions were found to be 85%–93% lower than fossil jet fuel (Crossin, 2017). Transporting distance can occupy up to 35% of the carbon emissions, while the impact of sheep farming and supply of hydrogen can account for over 50% of the mallee jet fuel GHG impacts. Wheat straw was shown to have high GHG emissions reduction potential with only 30.4 gCO_2eq/MJ kerosene, but the energy efficiency is relatively poor due to high losses during the conversion process (Neuling and Kaltschmitt, 2018). On the other hand, wheatgrass showed relatively higher GHG emissions of 70.3 gCO_2eq/MJ kerosene with moderate energy efficiency.

Triisobutane produced from cellulosic isobutanol has been identified as a potential jet fuel blend component. It is an isoalkane with similar boiling specific range and molecular weight as Jet A-1, which can be produced via thermochemical upgradation process. The conversion process from isobutanol to triisobutane contributes to 6.60 gCO_2eq/MJ; thus the final net GHG emissions for the production of triisobutane from cellolusic biomass via the ATJ route are 64.54 gCO_2eq/MJ, which is ~28% lower compared with Jet A-1 fuel (89 gCO_2eq/MJ) (Vela-García et al., 2020). The processes that emit the most CO_2 are the cultivation, harvesting, pretreatment of raw material, and the conversion of cellulose into simple sugars via strong acids, ionic liquids, or enzymatic reaction. Development of catalysts is needed to increase the efficiency of the conversion process. Jet fuel produced from waste gas was shown to reduce GHG emissions by 67% owing to the utilization of carbon contained in the gas (Handler et al., 2016). The large carbon credit gained is attributable to net gas absorption during the ethanol production phase. The carbon-containing waste gas can be sourced from steel mill or gasified biomass. Unlike other biomass such as corn stover, switchgrass, and forest residue, there are no GHG emissions due to waste gas procurement. Budsberg et al. (2016) reported the GHG emissions of jet fuel produced from poplar biomass via the biochemical conversion route to be ranging between 32 and 73 gCO_2eq/MJ jet fuel, which is lower compared with those of conventional jet fuel (93 gCO_2eq/MJ). The values were derived under different scenarios, such as using natural gas steam reforming and lignin gasification to produce the hydrogen needed for hydroprocessing. The usage of hog fuel to provide heat and steam further lowers the GHG emissions.

6.4.2 Thermochemical conversion

Thermochemical conversion of biomass into biojet fuel can be performed via the FT processing route, where the biomass is gasified to produce synthesis gas and subsequently used as precursor gas for biojet fuel production. Negative GHG values have been reported for FT fuel produced from switchgrass owing to the carbon sequestration effect due to the changes in soil (Stratton et al., 2010). The potential for offsetting CO_2 emissions via soil carbon changes depends on the rate of carbon addition, the capacity of soil for carbon storage, and the stability of sequestered soil carbon over time. In a study that compares the LCA GHG emissions of different production routes, the FT production pathway was shown to produce lower GHG than the HEFA route for biojet fuel production using integrated biorefineries with ethanol distillers in Brazil (Klein et al., 2018). Two scenarios of FT were studied. The first was using lignocellulosic materials of sugarcane bagasse and straw, while the second scenario uses eucalyptus and sugarcane bagasse. The use of biomass does not incur additional impact to the agricultural phase, while eucalyptus crops require fewer inputs such as fertilizer than sugarcane. Both scenarios show the GWP to be 9.4 and 9.3 gCO_2eq/MJ jet fuel. The potential of FT route is also reflected in another LCA study conducted by Carter et al. (2011), in which switchgrass-based FT biomass-to-liquid (BTL) was found to emit 12−26.1 gCO_2eq/MJ jet fuel. The carbon footprint is even lower than camelina HEFA with 19.9−34.1 gCO_2eq/MJ, as compared with the baseline conventional jet fuel of 87.5 gCO_2eq/MJ. These studies indicate that FT process with switchgrass as feedstock has the potential to achieve lowest GHG emissions among all feedstocks and production methods.

Han et al. (2013) compared the FT and pyrolysis conversion processes of corn stover into biojet fuel. Both methods showed a potential GHG emissions reduction of 89% and 68%−76%, respectively, but the coproduct allocation selection can affect the result significantly. Although FT shows the most promising GHG reductions among all processing routes, the capital costs for biomass-to-liquid (BTL) facilities can be significantly higher compared with other production methods (Carter et al., 2011).

6.4.3 Lipid conversion

The technology for lipid conversion into biojet fuel through HEFA processing pathway is presently the most mature and commercial-ready among all methods. From the life cycle GHG emissions perspective, several studies have shown positive impacts toward to environment, albeit with a large variation depending on the type of feedstock used. Han et al. (2013)

showed that the life cycle GHG emissions can be reduced by 41%–63% for HEFA-SPK derived from different oil seeds including soybean, palm, rapeseed, jatropha, and camelina. Among all, palm shows the highest GHG emission reduction potential of 63%. Another study shows that the carbon emission intensities for HEFA derived from camelina, jatropha, and microalgae are 71%, 64%, and 57.5% lower compared with Jet A, respectively (Lokesh et al., 2015). The study was conducted using mass-based allocation method, and the emissions related to coproducts were accounted. The high biomass output of microalgae results in lower GHG emissions owing to higher allocation to the coproduct. Other microalgae-based jet fuels have also shown a wide variation of carbon intensities due to different cultivation conditions and assessment methodologies used (Takriti et al., 2017). Several studies have compared several feedstocks for the HEFA production. The GHG emissions intensity was found to be sensitive to the yield of the feedstock. Among the feedstock compared, palm-based HEFA has been reported to emit the lowest GHG emissions of 52.0 gCO_2eq/MJ fuel due its high yield while assuming no LUC effect. The matured HEFA technology and high yield of palm enable the production cost to be competitive, which is estimated to be 890 €/t compared with the nonedible jatropha-based HEFA of 2000 €/t (Neuling and Kaltschmitt, 2018). However, when compared with the production cost of conventional jet fuel (~340 €/t) (IATA, 2020), even the best performing HEFA crop case is still not market-competitive. The oil yield from the cultivation of energy crops simply cannot match the high throughput from oil drilling. Further, there are other costs associated to the cultivation stage that contributes to the carbon cost. The GHG emissions for energy crops are sensitive to the use of fertilizer and pesticides during the cultivation stage. For example, soy oil–based jet fuel was shown to produce three times higher GHG emissions than that of palm oil (Vasquez et al., 2019); thus the cost of biojet fuel production is also correspondingly increased.

In an LCA study of camelina biojet fuel using system expansion method (Li and Mupondwa, 2014), camelina oil from the Canadian Prairies region could significantly reduce GWP by 65%–97% compared with fossil-based jet fuel (88 gCO_2eq/MJ). The main contributors to GWP have been identified as yield of oil, natural gas, and electricity consumption. Among them, the GHG emissions are particularly sensitive to oil yield. Higher seed yield in camelina could reduce GHG emissions by 77% and 96% compared with low yield scenario. In another study that uses jatropha curcas as jet fuel feedstock, an efficient plantation management and displacement of high emission product such as recycling of jatropha oil cake were shown to be essential to ensure a positive impact to the environment (Liu and Qiu, 2019).

The HEFA production pathway requires hydrogen for the hydrotreatment process. A study by de Jong et al. (2017) assumed that the hydrogen used in their LCA study was supplied through steam methane reforming of natural gas, which is the main industrial practice for hydrogen production. The GHG emissions of the waste cooking oil (WCO)-based HEFA jet fuel were shown to be lower than conventional jet fuel by 68%. It is expected that the use of renewable hydrogen can further lower the GHG emissions. Study has shown that renewable hydrogen produced from palm biomass gasification and subsequent syngas reforming can reduce the carbon footprint of palm biojet fuel from 14.2 gCO_2eq/MJ to 4.8 gCO_2eq/MJ, whereas using hydrogen from water electrolysis can reduce the carbon footprint of soy biojet from 40.1 to 33.9 gCO_2eq/MJ compared with basedline jet fuel, which amounts to 95% and 63% emissions reduction, respectively (Vasquez et al., 2019). The electrolysis method requires high energy input to break the molecular bond in water to produce hydrogen, which in turn contributes a significant amount of GHG emissions. However, as the biomass utilized is considered as carbon neutral, the overall net GHG emissions is still considerably lower than conventional jet fuel, implying that renewable hydrogen is an attractive option to be included in the biojet fuel production process.

The conversion of fatty acid in WCO via the HEFA route is another promising feedstock for biojet fuel production. Although there is no cultivation of feedstock required, the collection of WCO from various locations in urban areas poses a logistical challenge, which could incur significant costs and GHG emissions. Nonetheless, WCO biojet fuel showed a potential reduction of GHG by 68% (de Jong et al., 2017). The production method can invariably affect the GHG emissions. A study was conducted to compare two different production methods of WCO biojet fuel. A technique known as the catalytic transfer of hydrogenation (CTH), which utilizes isopropanol as hydrogen donor, was compared with the HEFA pathway. The GWP for CTH method is lower than HEFA-WCO by 8%, under the scenario with no CO_2 sequestration through coproduct offsets. The higher GWP potential for HEFA-WCO is attributable to the higher heat input and electricity consumption compared with CTH. The dominant factor for GWP in CTH is the low-pressure flash, while the

HEFA's GWP is dominated by WCO transport emissions (Barbera et al., 2020). This shows that given the same feedstock, improvements to the lipid conversion process can lead to GHG emissions reduction.

Hydrothermal liquefaction (HTL) is an emerging lipid conversion method that uses subcritical water to convert biomass into a carbon-rich biocrude. It was reported that the GHG emissions for algal biojet fuel produced using HTL at a waste water plant are much lower than the jet fuel produced in a refinery, as the use of waste water algae can reduce the GHG emissions by 55.4%–76% compared with conventional jet fuel. The main contributing factors to the life cycle GHG emissions of algae jet fuel are the transportation of biomass and waste nutrients (Fortier et al., 2014).

6.5 Life cycle emissions values for CORSIA eligible fuel

In 2019, the ICAO has presented a report that details the methodology and calculation of the core life cycle GHG emissions for different sustainable aviation fuel (SAF) pathways (CORSIA, 2019). The purpose of the core LCA values is to enable the implementation of the CO_2 offsetting obligation agreed upon under CORSIA by aircraft operators. Details on the implementation of CORSIA scheme and calculation of the CO_2 offsetting requirements can be found in the recent environmental report published by the ICAO (2019a). The alternative jet fuels that are certified under ASTM D7566 and meet the sustainability criteria are known as CORSIA Eligible Fuels (CEF), which include SAFs and low-carbon aviation fuels (LCFs). At the time when the report was published, the GHG values of various feedstock produced from five ASTM-certified production pathways have been included, including the FT, HEFA, SIP, FT-SKA, ATJ-SPK (Annex 1–5 of ASTM D7566), and coprocessing route (update in ASTM D1655). With the emergence of new production pathways, the list of CORSIA default life cycle emissions values (core LCA and ILUC) is expected to be updated, provided the criteria are met. The production pathway must be a certified conversion process and validated for commercial production. Furthermore, there must be sufficient data available (including feedstock, ILUC) for LCA modeling. The core life cycle GHG values of the SAF are calculated based on the LCA attributional approach or "process-based" approach, which accounts for the mass and energy flows for the whole value chain. This means that no displacement effects are accounted for the coproducts. The effect of ILUC is accounted based on a consequential approach, in which the total life cycle GHG can be derived by adding the core LCA values. Table 6.1 shows part of the CORSIA default life cycle emissions values for CORSIA eligible fuels (ICAO, 2019b).

Table 6.1 CORSIA life cycle emissions values for CORSIA eligible fuels (ICAO, 2019b).

Fuel production pathway	Fuel feedstock	Core LCA value	ILUC LCA value	$LS_f(gCO_2e/MJ)$[a]
Fischer–Tropsch	Agricultural residues	7.7	0	7.7
	Miscanthus (herbaceous energy crops)	10.4	−22.0	−11.6
	Switchgrass (herbaceous energy crops)	10.4	−3.8	6.6
Hydroprocessed esters and fatty acids	Used cooking oil	13.9	0	13.9
	Tallow	22.5	0	22.5
	Soybean oil	40.4	24.5	64.9
	Rapeseed oil	47.4	24.1	71.5
	Palm oil—closed pond	37.4	39.1	76.5
Alcohol (isobutanol)-to-jet	Agricultural residues	29.3	0	29.3
	Switchgrass (herbaceous energy crops)	43.4	−14.5	28.9
Alcohol (ethanol)-to-jet	Sugarcane	24.1	8.7	32.8
	Corn grain	65.7	25.1	90.8
Synthesized isoparaffins	Sugarcane	32.8	11.3	44.1

[a]Total life cycle GHG value (LS_f) = core LCA value + ILUC LCA value.

6.6 Comparison of greenhouse gas emission performance

Fig. 6.5 shows the compilation of the GHG emissions values taken from several literature sources, classified according to their feedstock-production technology combinations and compared with the LCA values of CORSIA eligible fuels. The GHG emissions for baseline jet fuel derived from the WTWa cycle is represented by the dashed lines with reported values of 80.7 and 109.3 gCO_2eq/MJ, proposed by the working group in MIT under the initiative of Partnership for Air Transportation Noise and Emissions Reduction (PARTNER). Four alternative jet fuels derived from either coal or natural gas via the FT routes are included for comparison, with two FT fuels derived under the carbon capture and sequestration (CCS) scenario. For biomass-based jet fuels, the GHG emission values reported are derived from the field-to-wake cycle without considering the effect of LUC.

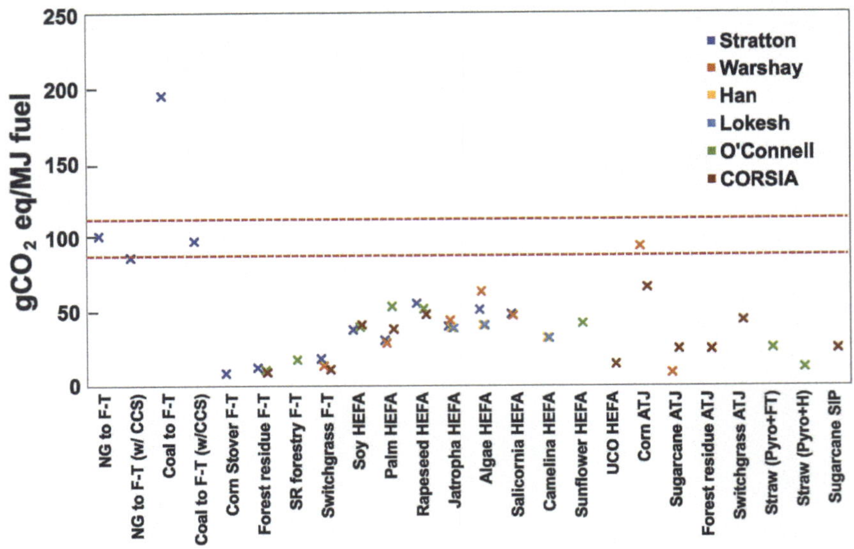

Figure 6.5 Compilation of the carbon intensities of alternative jet fuels without considering land use change effect. *ATJ*, alcohol-to-jet; *FT*, Fischer–Tropsch; *HEFA*, hydroprocessed esters and fatty acids; *Pyro*, pyrolysis; *SIP*, synthesized isoparaffins. *(Source data from Stratton, R.W., Wong, H.M., Hileman, J.I., 2010. Life Cycle Green House Gas Emissions from Alternative Jet Fuels. Partnership of Air Transportation Noise and Emissions Reduction. Institute of Technology, Cambridge, MA: Massachusetts. http://web.mit.edu/aeroastro/partner/reports/proj28/partner-proj28-2010-001.pdf), Warshay, B., Pan, J., Sgouridis, S., 2011. Aviation industry's quest for a sustainable fuel: considerations of scale and modal opportunity carbon benefit. Biofuels 2, 33–58. Han, J., Elgowainy, A., Cai, H., Wang, M.Q., 2013. Life-cycle analysis of bio-based aviation fuels. Bioresour. Technol. 150, 447–456, Lokesh, K., Sethi, V., Nikolaidis, T., Goodger, E., Nalianda, D., 2015. Life cycle greenhouse gas analysis of biojet fuels with a technical investigation into their impact on jet engine performance. Biomass Bioenergy 77, 26–44, O'Connell, A., Kousoulidou, M., Lonza, L., Weindorf, W., 2019. Considerations on GHG emissions and energy balances of promising aviation biofuel pathways. Renew. Sustain. Energy Rev. 101, 504–515, ICAO, 2019b. Annex 16 - Environmental Protection, CORSIA Default Life Cycle Emissions Values for CORSIA Eligible Fuels. https://www.icao.int/environmental-protection/CORSIA/Pages/default.aspx. The* red dotted line *represents the CO_2 emissions range for baseline jet fuel Stratton, R.W., Wong, H.M., and Hileman, J.I., 2010. Life Cycle Green House Gas Emissions from Alternative Jet Fuels. Partnership of Air Transportation Noise and Emissions Reduction. Institute of Technology, Cambridge, MA: Massachusetts. http://web.mit.edu/aeroastro/partner/reports/proj28/partner-proj28-2010-001.pdf.)*

Synthetic jet fuels derived from fossil-based feedstocks generally produce higher GHG emission intensities. The coal-to-liquid route without CCS produces 123% more CO_2 than baseline jet fuel (87.5 gCO_2eq/MJ), while the incorporation of CCS technology is able to reduce the carbon footprint significantly, but the net carbon emission intensity is still 11.1% higher than baseline. Using natural gas as feedstock to produce FT fuel via the GTL route results in 15.4% more CO_2 emissions than baseline jet fuel, which is lower than Coal-to-liquid (CTL) route owing to the absence of feedstock gasification process. The incorporation of CCS technology with natural gas FT process would reduce the CO_2 emissions to the level comparable with baseline jet fuel, with no positive impact to the environment. However, gasifying lignocellulosic biomass such as corn stover, forest residue, and switchgrass could subsequently lead to jet fuel production via the FT pathway, which resulted in significant GHG emissions savings. The FT pathway has lower emission than the lipid conversion route via hydrotreament as the former does not require hydrogen input. The heat required for gasification can be obtained from waste heat while the coproduction of electricity can further reduce GHG emissions. The pyrolysis pathway is another emerging technology that can be utilised to produce biojet fuel from lignocellulosic materials with promising carbon reduction potential. Pyrolyzing straw, for example, can produce synthesis gas, which can be processed into jet fuel via the FT route. The combined pyrolysis and hydroprocessing pathway used to process straw into jet fuel is considered to produce low GHG emissions comparable with those of ATJ pathways owing to the nature of self-sufficiency (O'Connell et al., 2019). The need for hydrogen during hydrotreatment can be resolved by producing hydrogen during the pyrolysis process, while the heat and electricity needed for the process are generated within the plant. In addition, the production of coproducts such as biochar and biogas enables product allocation, which further lowers the net GHG emissions.

The conversion of lignocellulosic biomass such as forest residues and switchgrass into jet fuel via the ATJ route produces lower emission intensities compared with conventional jet fuel, even if LUC effect is taken into account (ICAO, 2019b). Direct conversion of sugar into hydrocarbon via the synthesized isoparaffins (SIP) route is less carbon intensive, especially when sugarcane is utilized as feedstock for the biochemical conversion process (ICAO, 2019b). Converting the sugar from sugarcane and lignocellulosic feedstock, i.e., forest residues and switchgrass, via the ATJ route, shows promising low carbon intensities of <50 gCO_2eq/MJ fuel, but the use of starch from corn grain tends to yield higher carbon footprint. The CEF states that ATJ conversion with corn grain produces 65.7 gCO_2eg/MJ fuel without considering LUC. Warshay et al. (2011) reported the GHG value of corn-based ATJ is comparable with the baseline jet fuel.

A significant variation of GHG emission intensity can be observed for the hydroprocessing of lipid into jet fuel pathway. For oil-bearing crops such as camelina, jatropha, palm, rapeseed, sunflower, and soybean, the life cycle GHG emissions are less carbon intensive compared with conventional jet fuel when LUC effect is not considered. Sunflower HEFA has a carbon footprint of about 41.4 gCO_2eq/MJ, which is 52.6% lower compared with baseline jet fuel GHG emissions (O'Connell et al., 2019). Other oil crops such as palm, soybean, rapeseed, and camelina generally show lower carbon intensities than fossil jet fuel in the case with no LUC, although the values might differ depending on the LCA conducted. Palm can produce 23–41 gCO_2eq/MJ of GHG emissions under the scenario with no LUC. If LUC is factored in, the GHG emission can increase quite significantly, depending on the type of land conversion. Clearing of tropical rainforest or peatland can release significant GHG emissions into the atmosphere. It has been estimated that about 600 gCO_2eq/MJ of GHGs will be emitted from clearing the forest for palm cultivation (Stratton et al., 2010), whereas the cultivation of palm oil on mineral soils without LUC yields meaningful carbon reduction effect, with only 50 and 30 gCO_2eq/MJ of emissions if methane is captured from the mill effluent pond (O'Connell et al., 2019). Nevertheless, if the demand for biojet fuel continues to soar, it can be expected that LUC will occur to give way to energy crop cultivation to meet the production capacity. This will inadvertantly result in negative implication to the environment with significant rise in GHG emissions.

The degree of saturation of the fatty acids in the crop oil was found to have an effect on the yield and hydrogen input requirements. Highly unsaturated fatty acids require more hydrogen during the hydrotreatment process; thus the carbon emission intensities for rapeseed, sunflower, and soybean oil (7.8–8.3 gCO_2eq/MJ) are higher than those of oil palm (6 gCO_2eq/MJ). The use of urban waste such as WCO can potentially reduce the carbon intensity by 73.8% compared with baseline jet fuel. Lipid conversion from microalgae via the HEFA route shows a large variability in GHG emissions. Although no LUC effect is considered, the requirement of nutrient inputs for the cultivation of algae could increase the GHG emission level (Capaz et al., 2018).

Fig. 6.6 compares the breakdown of GHG emissions for several production route of biojet fuels (O'Connell et al., 2019). For HEFA process, the cultivation stage for oil seed crops produces the largest part of GHG, whereas the conversion stage of hydroprocessing accounts for 15% of the total GHG emissions. As compared with the FT and pyrolysis process using forest residue and straws, the advantage is that no cultivation is involved; hence, the GHG

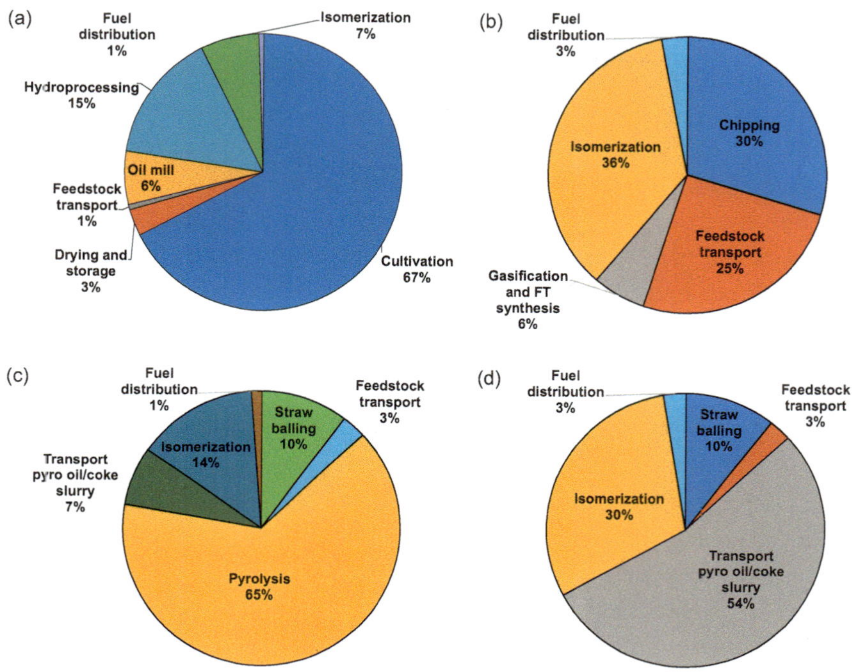

Figure 6.6 Breakdown of the GHG emissions for production of (A) rapeseed HEFA, (B) forest residue—FT fuel, (C) straw pyrolysis + FT, (D) hydrotreated pyrolysis oil from straw. *FT*, Fischer—Tropsch; *GHG*, greenhouse gas; *HEFA*, hydroprocessed esters and fatty acids. *(Adapted from O'Connell, A., Kousoulidou, M., Lonza, L., Weindorf, W., 2019. Considerations on GHG emissions and energy balances of promising aviation biofuel pathways. Renew. Sustain. Energy Rev. 101, 504—515.)*

emissions are mainly attributed to the mechanical processing and thermochemical conversion process. For the FT production route by gasifying forest residue, about 55% of the GHG emissions are attributed to chipping and transportation. Although gasification and the FT process emits only 15% of the GHG, the subsequent isomerization process used to produce the biojet fuel is a carbon-intensive process that takes up 36% of the total GHG emissions. The process of isomerization is to convert the small fuel molecules into jet fuel-like molecules, i.e., similar chain length and physicochemical properties as Jet A-1. This process tends to emit a significant part of GHG for the pyrolysis and the hydrotreated pyrolysis oil routes. The heating required for pyrolyzing straw results in 65% of the GHG emissions, whereas the transporting of pyro oil or coke slurry dominates with 54% of the total GHG emissions. The distribution of GHG emissions can vary quite significantly depending on the process involved.

6.7 Energy balance analysis

Analysis of the mass and energy flows of several aviation biojet fuels production pathway models has been conducted by Neuling and Kaltschmitt (2018). Fig. 6.7 shows the Sankey diagram with all energy flows for the process of GTL, i.e., jet fuel production from biomethane sources. The heat and power streams used in the production stage is illustrated, while the energy of the mass flow was calculated based on the lower heating value. The produced hydrocarbon products contain kerosene, naphtha, and diesel, with jet fuel kerosene constituting the largest part. Comparison of the mass flow analysis shows that HEFA option produced the highest kerosene mass fraction (~72 t/h), followed by ATJ (~65 t/h) and FT process via the BTL and Bio-BTL route (~58–60 t/h). The HEFA process with palm oil and jatropha oil as feedstock shows the highest overall energy efficiency of 90% with kerosene conversion efficiency of 58%–60%, while the processing of lignocellulosic biomass and FT routes shows lower overall energy efficiency (30%–38%) and kerosene fraction (18%–21%). The process simulation provides the basis to calculate the economics of production, with HEFA refineries and ATJ process with wheat showing the approximate

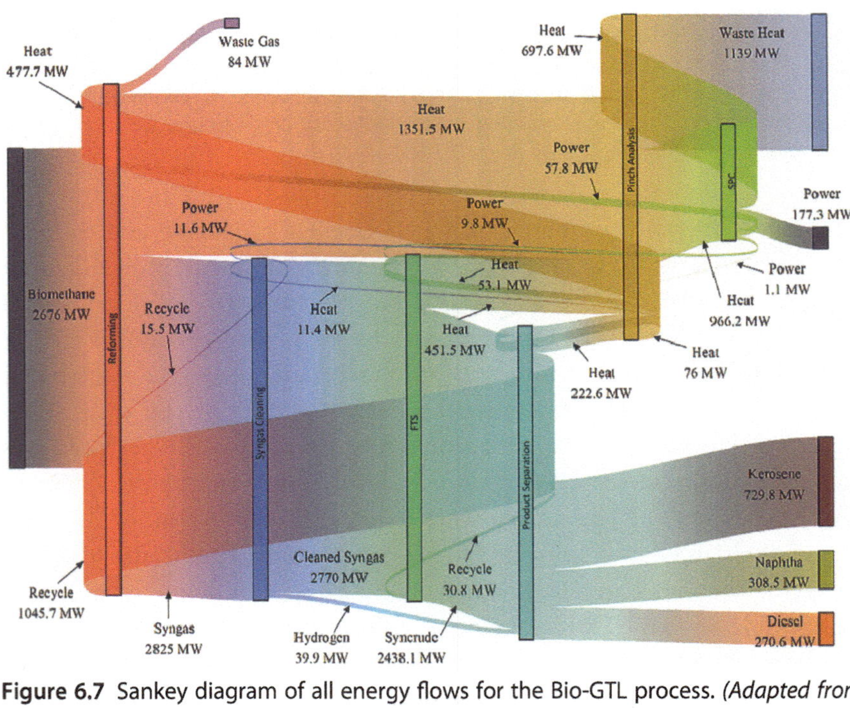

Figure 6.7 Sankey diagram of all energy flows for the Bio-GTL process. *(Adapted from Neuling, U., Kaltschmitt, M., 2018. Techno-economic and environmental analysis of aviation biofuels. Fuel Process. Technol. 171, 54–69.)*

costs of 670 and 655 M€, respectively, as opposed to the BTL process of using willow and wheat straw as feedstock with thermochemical gasification and heat-induced pretreatment processes that result in high investment costs of approximately 2570 and 2800 M€, respectively.

Identification of the energy flow throughout the conversion process provides a clear view on the energy consumption in each process. Fig. 6.8 shows the energy consumption breakdown for each of the biojet fuel produced from different feedstock and technology (O'Connell et al., 2019). For lignocellulose materials, gasification and pyrolysis processes constitute the main portion of energy consumed, which is significantly higher compared with the main hydrotreatment process (hydroprocessing + isomerization) of oil-bearing crops. This is due to the energy required to transform the solid biomass into gaseous form before converting into liquid fuel via the FT process. The energy usage for HEFA production varies with the type of seed oil used and the process involved. Palm-based oil was shown to exhibit higher energy loss during the processing stage in the oil mill. Processing the palm oil seeds constitutes more than 70% of the energy loss in all the scenarios considered, i.e., types of palm oil mill effluent ponds (open or closed) and the usage of land (mineral soil or peat land). Compared with

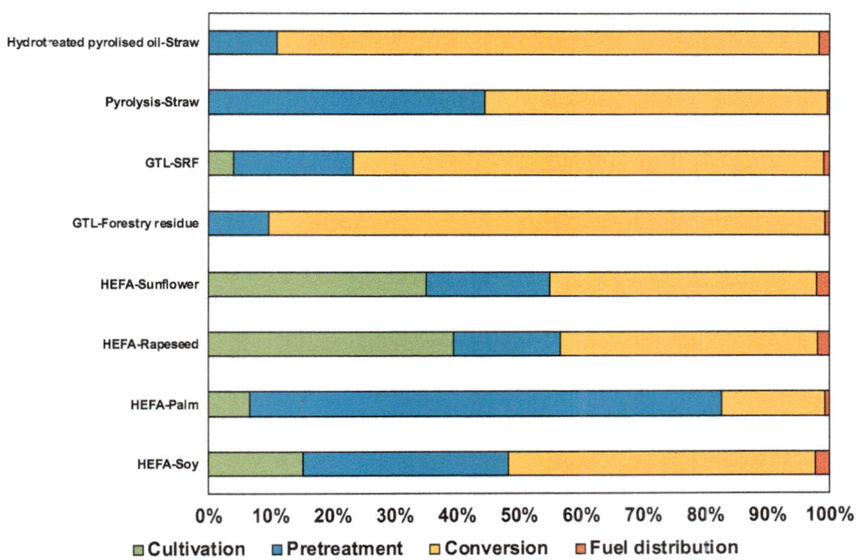

Figure 6.8 Percentage of energy loss (MJ/MJ) due to different processes in alternative jet fuel production. *(Adapted from O'Connell, A., Kousoulidou, M., Lonza, L., Weindorf, W., 2019. Considerations on GHG emissions and energy balances of promising aviation biofuel pathways. Renew. Sustain. Energy Rev. 101, 504–515.)*

other oil seeds such as soy or rapeseed, alternative jet fuel production from palm oil is comparatively more energy-intensive in spite of showing the best performance from GHG emissions perspective among the oil crops, even in the sustainably practiced case where no LUC is involved. The combined methods of pyrolysis and gasification for the production of alternative jet fuel from biooil present the highest energy usage compared with HEFA or ATJ routes, but hydrotreating the pyrolysis oil produced from straw presents a viable energy consumption profile comparable with nonpalm oil crops. Even though energy crops such as rapeseed, soy, sunflower, and straw demand relatively low energy input for processing, their low yields per hectare incur stresses on the required arable land and thus could potentially lead to LUC and other environmental impacts. The low energy intensity for the process is advantageous from the production cost, but further steps such as improvement on the processing technique, recuperation of waste heat, and improvement in biomass collection system can be taken to further reduce the energy utilization.

6.8 Energy—water—food nexus

The global population and economic expansion have led to growth in consumption of three key resources, namely energy, water, and food. Concerns over the consumption are warranted, as supply of these resources is finite. Traditionally, the resources of energy, water, and food were managed independently with little interactions with each other. This meant that in the pursuit of development to uplift living standards, trade-offs were often made in the exploitation of a resource at the expense of another. The complex and dynamic interlinkages between the three resources were often underemphasized. For example, the food versus fuel debate showed how the well-intentioned search for sustainable biofuels for energy caused a food crisis in 2008. Agricultural practices need water, and sometimes, water resources have to be stretched to obtain higher crop yield. The production of water is an energy-intensive process. As such, a nexus approach linking up the three resources will enhance coordination and integration among the sectors. For this, a paradigm shift is required where energy, water, and food are looked upon as a system instead of three siloed resources. Such an approach is called the "energy—water—food" nexus, or the EWF nexus.

EWF nexus is a core component in the 2030 Agenda for Sustainable Development, which contributes directly to three of the 17 Sustainable Development Goals. The holistic framework of the EWF nexus (as shown in Fig. 6.9) places greater emphasis on the efficiency of the overall system

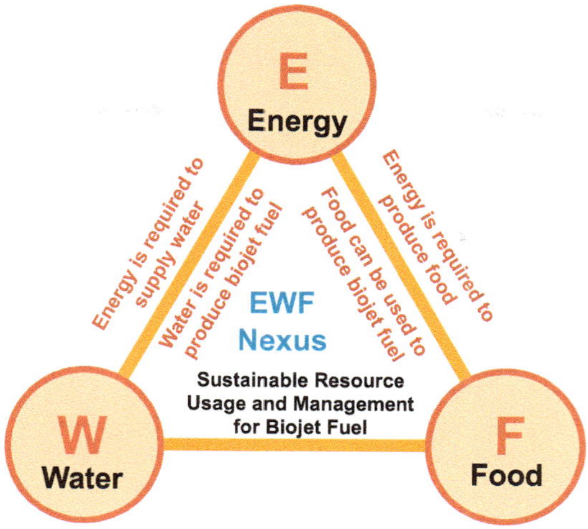

Figure 6.9 EWF nexus holistic framework. *EWF*, energy–water–food.

rather than any individual component. Trade-offs are part and parcel of the nexus but will lead to an improved allocation of resources, reduced adverse environmental impacts, and improved economic efficiency, while still developing in a sustainable manner.

EWF nexus as a concept has gained traction but policy design mostly remained sectoral-based, with mandates focused on individual sectors of energy, water, and agriculture (Chiaramonti and Goumas, 2019). Cross-sectoral effects are accounted in the more recent policies although primarily in nonformalized statements of intent. The most common nexus considered is between the agricultural and water policies. Cross-sectoral thinking has also led to radical but promising ideas such as introducing a diet change among EU citizens by reducing meat intake. For example, by halving meat intake from 200 g/day per person to 100 g/day per person, resources will be made available to multiply biofuel production by a 7.7-fold. This will have the social impacts of healthier citizens and triple environmental benefits of reducing agricultural GHG by 24%, decreasing GHG from EU transport by 14% and displacing 16.1% of EU fossil fuel for the transport sector.

6.8.1 Energy–water nexus in biojet fuel production

Water footprint of a crop can be evaluated by dividing further into green, blue, and gray water footprints. Green water footprint denotes rainwater consumed, blue water footprint refers to the surface and groundwater consumed, and gray water footprint quantifies the volume of freshwater

Table 6.2 Global average water footprint of primary crop products and derived crop products for potential biojet fuel feedstocks (Mekonnen and Hoekstra, 2011).

Category	Primary crop products and derived crop products	Global average water footprint (m^3 per ton)			
		Green	Blue	Gray	Total
Edible oil	Castor oil	21,058	2,938	744	24,740
	Coconut oil, refined	4,461	3	27	4,490
	Cottonseed oil, refined	2,242	1,283	432	3,957
	Groundnut oil, refined	6,681	405	442	7,529
	Maize oil	1,996	171	409	2,575
	Olive oil, refined	12,067	2,437	221	14,726
	Olive oil, virgin	11,826	2,388	217	14,431
	Palm oil, refined	4,787	1	182	4,971
	Palm kernel oil, refined	5,202	1	198	5,401
	Rapeseed oil, refined	3,226	438	636	4,301
	Sesame oil	19,674	1,183	936	21,793
	Soybean oil, refined	3,980	137	73	4,190
	Sunflower seed oil, refined	6,088	299	405	6,792
Edible sugar and starch	Cassava, starch	2,200	1	53	2,254
	Maize, starch	1,295	111	265	1,671
	Potato starch	1,005	173	333	1,512
	Rice, husked (brown)	1,488	443	242	2,172
	Rye flour	1,774	32	124	1,930
	Sorghum	2,857	103	87	3,048
	Sugarbeet	82	26	25	132
	Sugarcane, molasses	350	144	33	527
	Wheat, starch	1,004	269	163	1,436

required to assimilate pollutant load as per the existing ambient water quality standards (Mekonnen and Hoekstra, 2011). These measures of water footprints provide indicators of the direct and indirect uses of freshwater resources.

Globally, the average water footprint of crop production was 7404 Gm^3 per year for the 1996–2005 period, with a breakdown of green water (78%), blue water (12%), and gray water (10%). For key biojet fuel feedstocks, the total water footprint proportions globally are wheat (15%), maize (10%), soybean (5%), sugarcane (4%), oil palm (2%), coconut (2%), cassava (1%), and rapeseed (1%). Table 6.2 summarizes the global average water footprint of primary crop products and derived crop products for potential biojet fuel feedstocks. Derived crop products rather than primary crop products are selected as the components often have different water footprint compositions from the primary crop products.

Large variation was found for vegetable oil in terms of water footprint. Total water footprint ranges from 2575 m^3 per ton for maize oil to 24,740 m^3 per ton for castor oil. Edible sugar and starch feedstocks have a smaller absolute water footprint range from 132 m^3 per ton for sugarbeet to 3048 m^3 per ton for sorghum. From a water footprint point of view for the energy—water or even the food—water nexuses, the use of edible oil as feedstock will pile stronger pressure on water resource than edible sugar and starch. However, note that these values are just average, and they vary between regions. The general rule of thumb points to the average water footprint in Asia being lower than that of the Americas due to the greater yield.

Among plausible oil-based biojet fuel feedstock, maize and cottonseed will do well as biojet fuels when looking at water usage. In regions with heavy rainfall, coconut oil, palm oil, and palm kernel oil will be the choice to reduce water footprint as the plant relies almost exclusively on green water or rainwater. Castor oil, olive oil, and sesame oil are poor choices as feedstocks for biojet fuel production as they require huge water footprints.

For edible sugars and starch, sugarcane, sugarbeet, and wheat may be suitable biojet fuel feedstocks for water conservation. In regions with high rainfall, cassava will be a good choice as the plant taps primarily green water and uses almost no surface- and groundwater.

The water withdrawal and consumption issues for biofuels, although inclusive of bioethanol and biodiesel, are expected to grow as shown in Fig. 6.10 (IEA, 2020). Water withdrawal as an indicator helps to evaluate demands from agricultural, domestic, and industrial users by quantifying the total amount of water withdrawn from blue water source. Water consumption refers to the amount of withdrawn water permanently lost from its source. The water is lost due to consumption by human and livestock, transpired by plants or evaporated. Water stress evaluation must incorporate both indicators.

Presently, water withdrawal from the energy sector is dominated by coal (150.3 bcm) and nuclear energy (99.9 bcm) in the power generation energy sector. Water withdrawal due to biofuels (31.3 bcm) stands in third place and is comparable with the much larger fossil fuel energy sector (21.1 bcm) and gas (26.1 bcm). The water withdrawal for the energy sector under the Sustainable Development Scenario projected for 2030 in a report by the International Energy Agency (IEA) showed that biofuels will withdraw the most water with 107.6 bcm, representing a 244% increase from 2016. Biofuels will also command a share of 39.5% of total water withdrawal for

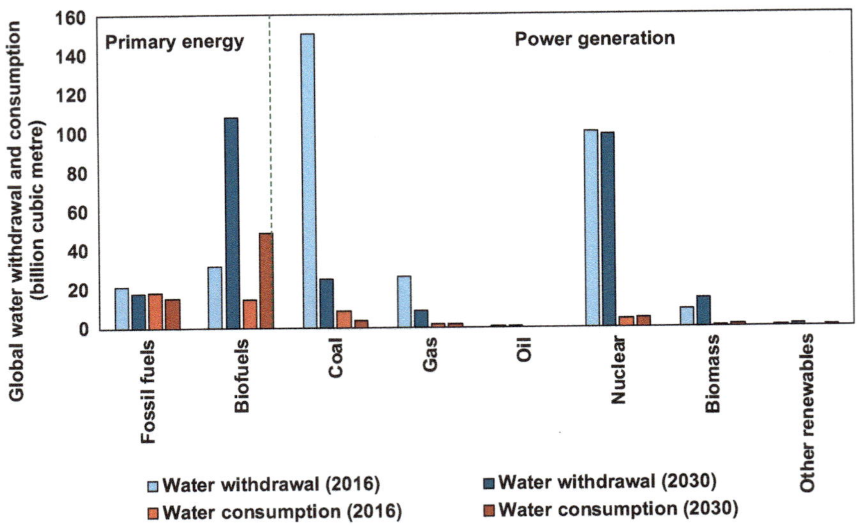

Figure 6.10 Global water withdrawal in the energy sector by fuel type in the Sustainable Development Scenario, 2016–30. *(Adapted from IEA, 2020. International Energy Agency, if the Energy Sector Is to Tackle Climate Change, it Must Also Think about Water. https://www.iea.org/commentaries/if-the-energy-sector-is-to-tackle-climate-change-it-must-also-think-about-water.)*

the energy sector in 2030 as compared with only 9.3% in 2016. Together with biomass, biofuels are the only fuel type with an expected increase in water withdrawal.

Similar trends were found for water consumption, where water consumption for biofuel is ranked second at 14.5 bcm behind fossil fuel at 18.0 bcm. This represents 30.8% global water consumption for the energy sector. Water consumption for biofuels is expected to increase by 34–48.5 bcm in 2030. This leads to biofuels (64.8%) consuming more water than the other fuels types in the entire energy sector. This also means that a drought will adversely affect biojet fuel production.

When conversion pathways are factored in, the oil-based feedstocks are comparable with the edible sugar and starch-based feedstocks in water consumption. This reverses the impression obtained when evaluating water footprint at crop planting level. Table 6.3 shows the water consumption for the different biojet fuel production pathways. A good way forward is to combine the use of feedstock with low water footprint at cultivation level and low water usage at process stage.

The cultivation of microalgae for energy has the potential to treat wastewater. Microalgae are efficient at removing nutrients from municipal

Table 6.3 Water consumption for the different biojet fuel production pathways (Wei et al., 2019).

Pathway	Feedstock	Water use (m^3/GJ)
Hydrogenated esters and fatty acids	Pongamia oil	5.50–11.80
	Microalgae oil	6.40–13.90
	Rapeseed oil	57.91–143.00
	Jatropha	66.45–75.03
	Soybean	63.65–106.79
Direct sugar-to-hydrocarbon	Sugarcane	15.60–147.00
	Switchgrass	92.39–104.74
	Corn grain	76.46–85.81

wastes. Municipal wastewater can provide the appropriate quantities of micronutrients, macronutrients, and dissolved salts. The relative proportions of carbohydrate, lipid, and protein can be altered through nutrient change (Dickinson et al., 2013). In fact, the effectiveness of microalgae in removing N and P will abate the effects of eutrophication and improving water quality. As such, the coupling of bioremediation and biojet fuel production using algae will be a possible win–win solution to satisfy the energy–water nexus. This also shows that the cultivation of microalgae could shift away from using clean water mixed with nutrients.

If energy production from biofuels is reliant on water as a resource, then water also relies on energy in its entire supply chain. The global water sector consumed 120 Mt of oil equivalent (mtoe) in 2014 (Popp et al., 2014). The majority of this quantity is used in the form of electricity, leading to a 4% electricity consumption globally. The major uses of electricity in the water sector are water extraction (40%), wastewater treatment (25%), and water distribution (20%). The interreliance of the energy–water nexus is shown when we factor in the 15% of global freshwater withdrawal in production of energy worldwide.

6.8.2 Energy–food nexus in biojet fuel production

First-generation biojet fuel feedstocks utilize edible oil, sugar, and starch. This meant that potential food is diverted to the energy sector. As the limiting factor for food production manifests itself in the form of cropland, the diversion of food to the energy sector should be calculated in terms of cropland area. The proportion of global cropland utilized for biofuels stands at about 2.5% or 40 million gross hectares (Popp et al., 2014). The net land requirement can be reduced to 1.5% globally by considering the by-products substituted for grains and oilseeds. The usage has a large variation among countries and regions. The reliant of biofuels on arable land will increase from 40 million gross hectares to 100 million gross hectares in 2050, representing 6% of total arable land today. Such an expansion would include existing cropland, pastures, and currently unused land. It is expected that the land requirement projection might be higher considering that biojet fuel was not yet prominent when the bioethanol and biodiesel-centric projection was calculated.

Conversely, the food sector (which edible oil, sugar, and starch are a subset of) is also dependent on energy resource. Energy is required along the entire food production value chain. From farm-to-table, 32% of the total global energy demand is used by the food sector (OFID, 2017). The number represents the primary production of food, storage, logistic, processing, and food preparation. It can be seen that the energy and food sectors have huge overlapping.

Brazil managed food and energy resources efficiently with their successful bioethanol and biodiesel industry. Sugarcane for the bioethanol industry used only 8.5 Mha or 1% of total land area in Brazil and was responsible for 15.7% of the country's 2014 domestic energy supply (Cortez et al., 2016). Soybean for the biodiesel industry used approximately 7.6 Mha or 32% of soybean production for 80% of biodiesel production in Brazil. This in turn contributed to 4.2% of overall diesel consumption in 2014. Firewood and charcoal from the planted forests of eucalyptus and pinus contributed to 8.1% of domestic energy supply from just 2 Mha of land or 31% of wood plated area. The aforementioned feedstocks can also be used for biojet fuel production. This places Brazil in the driving seat to expand their success in biofuels production to also include biojet fuel production. Furthermore, Brazil has strong conservation laws, large areas of legally protected native vegetation, and significant swathe of land already cleared for agriculture. Biojet fuel industries around the world could use Brazil as an example to balance the energy—food nexus.

One of the keys to solving the energy and food puzzle is from productivity gains. Brazil's food sector saw a productivity gain of 175% for grains from 1992 to 2012, while only increasing cultivated land usage by merely 49% as shown in Fig. 6.11. This represents a food productivity gain to land usage ratio of 3.57. In the same vein, productivity gain for sugar of 266% came from 102% land usage increases for sugarcane cultivation (Cantarella et al., 2015). Among measures taken include the introduction of double-cropping system for two crops to be harvested in the same year alternately. In fact, it is estimated that only 1.9 Mha or roughly 1% of

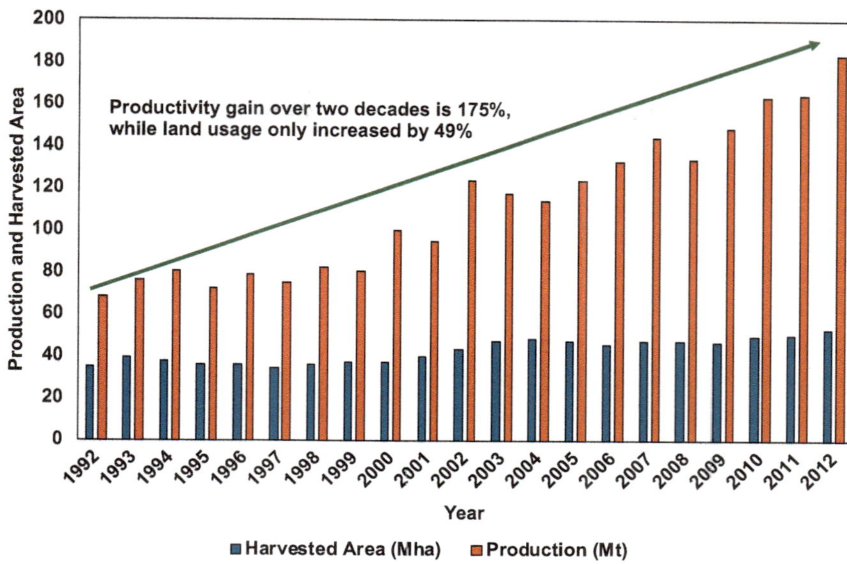

Figure 6.11 Production of grains and harvested area in Brazil. *(Adapted from Cortez, L.A.B., Cantarella, H., Moraes, M.A.F.D., Nogueira, L.A.H., Schuchardt, U., Franco, T.T., et al., 2016. Roadmap to a sustainable aviation biofuel: a Brazilian case study. In: Biofuels for Aviation, 339–350.)*

pasture land in Brazil is required to replace all of Brazil's jet fuel consumption if sugarcane ethanol is used as feedstock to produce biojet fuel. By comparison, the US ethanol productivity is lower than that of Brazil's, leading to the United States needing to deal with a significant agricultural limitation (Archer and Szklo, 2016). Brazil shifted the "food versus fuel" debate and epitomizes "food and fuel" instead.

The energy–food nexus is rarely straightforward. To illustrate the complexity of the energy–food relationship, crops meant for food will invariably be wasted, of which the food waste can be used as feedstock to be reconverted into biojet fuels. In general, 30%–50% of food produced globally is wasted prior to human consumption (Chuck et al., 2016). In the developing world, the mechanism of wastage is predominantly focused on poor transportation, harvesting, and storage practices. In the developed world, the main wastages are consumer behavior oriented, such as disposal due to arbitrary sell-by dates, not meeting esthetics standards. It is estimated that 15% of total food production can be collected as urban food waste (UFW) and be directed for biofuel production. Proportions of UFW as percentage of continent are Asia (51%), Americas (23%), Europe (16%), and Africa (10%). Food residues in Europe are expected to increase to 126 million tons by 2020, while in Asia, it is projected to rise to 416 million tons by 2025 (Tsiligiannis and Tsiliyannis, 2019). The ability of biojet fuel to be produced from various pathways means that almost the entirety of the UFW quantity can be used as feedstock, as opposed to just the oil portion for biodiesel and the sugar and starch portions for bioethanol. The practical drawbacks include high logistic and pretreatment costs.

Looking from an EWF nexus lens, the use of second-generation feedstocks such as jatropha oil, sugarcane bagasse, agricultural waste, and WCO will neither divert food nor will it use precious arable land for energy production. However, water requirement would have to be evaluated. Production of transportation fuel from jatropha requires more water than most other first-generation edible oil, sugar, and starch crops as shown in Fig. 6.12. This is at odds with some findings of jatropha needing only minimum water requirements as compared with other crops (Giwa et al., 2018).

China also faced a potential food versus fuel dilemma, of which the bioenergy targets threaten to derail the aims to attain self-sufficiency on grain supply. However, China has a ready-made solution in the form of

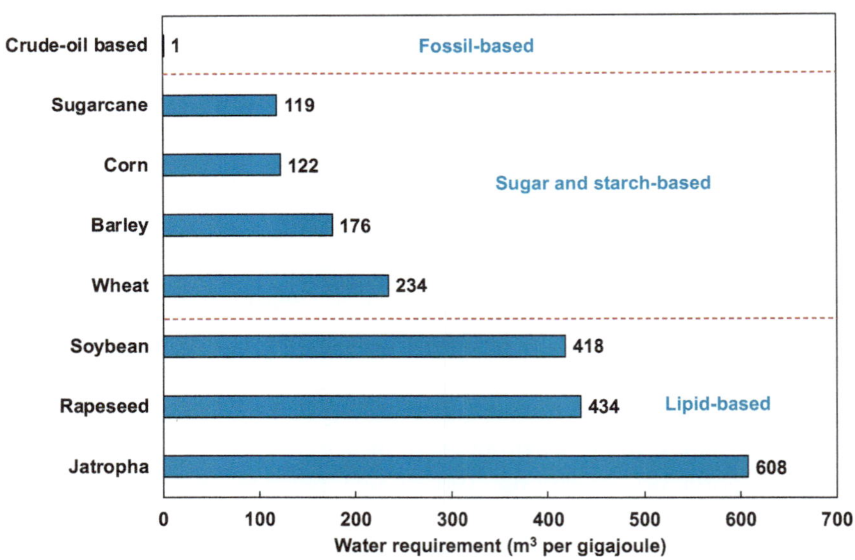

Figure 6.12 Water requirement of transportation fuel production from various feedstocks. *(Adapted from OFID, 2017. The OPEC Fund for International Development, the Energy—Water—Food Nexus: Managing Key Resources for Sustainable Development. https://opecfund.org/var/site/storage/original/application/80be162d98453051ded87e13032727cf.pdf.)*

Table 6.4 Potential generation and collectable potential of cellulosic biomass in China (Chang et al., 2012).

	Generation potential (Mt)	Collectable potential (Mt)		
Residues	**2010**	**2010**	**2020 (estimated)**	**Current utilization**
Agriculture	650.55	540	563.79	Traditional burning (35%), fodder (28%), returning to the field (15%), agricultural and industry use (7%)
Forest	854.56	460.79	520.55	Fiber and paper manufacturing (32%), returning to the forest (25%), traditional burning (16%)

cellulosic feedstock, which can also be used for biojet fuel productions. The vast quantities of agricultural and forest residues in China could unlock high quantities of feedstock without jeopardizing food supply. The potential generation and collectable potential of cellulosic biomass in China are summarized in Table 6.4. Another alternative available to China that will not cause potential food insecurity and still meet current land administration policies is the tapping into marginal land for oil-bearing trees. China's vast landmass meant that the combined forest barren lands, barren mountains, barren sand areas, and unutilized land resources could provide more than 36 Mha for oil-bearing crops. This is roughly the total landmass size of Malaysia (32.98 Mha) or the average arable land of the G7 group (37.12 Mha). It is posited that growing cellulosic crops on degraded lands could sequester carbon, improve soil health, and improve habitat for wildlife (Costello and Ayoub, 2019).

Similar to China, the moderately sized Spain in Europe also has low-cost and readily harvestable biomass from agriculture and forest. Table 6.5 tabulates the biomass quantity and the estimated bioenergy that can be unlocked from the biomass (Paredes-Sánchez et al., 2019). The potential biomass and bioenergy were estimated for the Spanish Renewable Energy Pan 2011–20. Spain alone has 0.726 EJ of bioenergy awaiting to be unlocked. These biomasses can be used for biojet fuel production without adversely affecting the energy-food nexus.

Table 6.5 Potential biomass and bioenergy in Spain.

Source	Biomass (Mt/y)	Bioenergy (Mtoe/y)	Bioenergy (PJ)
Lumber industry residue	2.98	0.64	26.88
Exploitation of entire tree	15.73	3.41	143.22
Herbaceous agricultural residue	14.44	6.39	268.38
Woody agricultural residue	16.12		
Herbaceous mass from agricultural terrain	17.74	3.59	150.78
Woody mass collected from agricultural terrain	6.60	1.47	61.74
Woody mass collected from forest terrain	15.07	1.78	74.76
Total	88.68	17.29	726.18

Adapted from Paredes-Sánchez, J.P., López-Ochoa, L.M., López-González, L.M., Las-Heras-Casas, J., Xiberta-Bernat, J., 2019. Evolution and perspectives of the bioenergy applications in Spain. J. Clean. Prod. 213, 553–568.

More broadly in the European Union, it is estimated that biomass production for the years 2005, 2030, and 2050 are 90, 212, and 356 Mtoe, respectively (Benito and Alonso, 2018). The largest growth of biomass production is expected to come from second-generation crops from none in 2005 to 127 Mtoe in 2050. Biomass production for second-generation crops is expected to dwarf first-generation crops by a factor of 18.1. Biomass from waste will also increase by 248% to 87 Mtoe in 2050. The European Union is also expected to increase waste import by 1200% over the 2005–50 period, although in absolute terms, it amounts to 24 Mtoe. This is close to the expected increase of agricultural residues from 17 Mtoe to 49 Mtoe for the same period.

Seaweed or macroalgae are consumed as food, but its potential for energy is even greater. From an EWF standpoint, seaweed has advantages over terrestrial cultured crops and microalgae. The cultivation of seaweed does not require freshwater. Additional production of seaweed will add to food supply without reducing available arable land. From an energy viewpoint, it has high solar efficiency of up to 8%, making it four times better than any terrestrial crops (Chuck et al., 2016). Seaweed farms located in coastal regions, offshore, and open sea colonies can produce up to 35, 110, and 6000 EJ, respectively. It could also democratize energy as seaweed is sufficiently robust to be inhabiting every marine ecosystem, allowing it to be cultivated in large scale by any country. In short, seaweed satisfies all of the EWF nexus considerations. The question remains is the logistics involved in ramping up production of seaweed to support the biojet fuel industry. The other considerations not covered by the EWF nexus include ecosystems and biodiversity. The danger of one feedstock dominating the landscape is the displacement of native species, habitat fragmentation, disruption to food chain, and possibility of reducing the resilience of an area to natural disaster.

6.8.3 Energy—water—food nexus and holistic considerations for biojet fuel production

6.8.3.1 Limiting factors

An analysis was conducted to evaluate the EWF nexus in biojet fuel production around the world. Limiting factors were determined using an unpublished data-driven model by the authors. In addition to the EWF nexus, economic factors were also considered. As biojet fuel itself represents the energy dimension, the limiting factor is not in the quantity, but instead the energy diversity afforded through the introduction of sustainable biojet fuel. For this, the Herfindahl—Hirschman Index (HHI) was used to determined energy type concentration. Countries that ended up with a poorer HHI for energy type will have an energy diversity limiting factor. The limiting factor of the water dimension is defined by water stress level. Food limiting factor is defined by the inadequacy of food triggered by the development of a biojet fuel industry in the country. Lastly, economic factors are defined by the prevailing jet fuel price and refinery costs. Economic factors are important as there will be no biojet fuel industry without profitability. The strongest of the limiting factors will be identified as the limiting factor for the biojet fuel industry of a country. Fig. 6.13 shows the global limiting factor map based on the EWF nexus and economic factors.

The model only considers exported feedstocks to avoid food versus fuel conflict. It also uses three conversion pathways of oil-to-jet (OTJ), ethanol-to-jet (ETJ), and gas-to-jet (GTJ) representing the three diverse branches of production methods. It is apparent that the bulk of countries with biojet fuel potentials are limited by economic factors first and foremost, rather than EWF resource factors. Nearly all of Africa, the Middle East, the Iberian Peninsula, and Central Asia are limited by economic factors. They are joined prominently by Mexico in North America; Peru and Bolivia in South America; Malaysia and Cambodia in South-East Asia; and New Zealand. This clearly shows that before sustainability and resource concerns are factored in, the biojet fuel industry must first resolve the profitability aspects of producing biojet fuel.

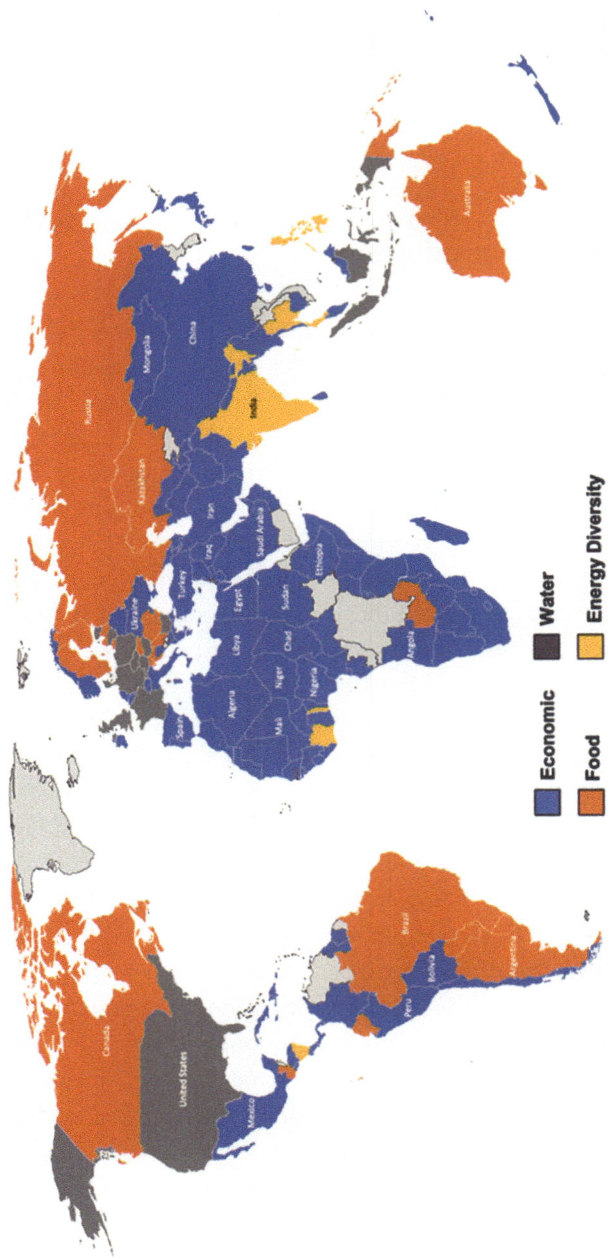

Figure 6.13 Global limiting factors of the biojet fuel industry for each countries based on economic and EWF nexus. *EWF*, energy–water–food.

There are a handful of countries, such as India, Myanmar, Thailand, the Philippines, Ghana, Togo, and Nicaragua, which are limited by energy diversity concerns if they introduce large-scale biojet fuel industry in their countries. It is likely that these countries share the characteristics of already being either heavily vested in biofuels or already have a balanced mixed of energy type, hence not benefitting greatly from adding biojet fuel to the mix.

Food as a limiting factor affects Canada, Russia, Kazakhstan, Australia, Brazil, Argentina, and a host of other nations. It is striking that most of the countries mentioned have large landmass, but not necessarily having high proportion of arable land. Most of these countries are located in either the upper end of the Northern Hemisphere or southern end of the Southern Hemisphere. This makes them located away from the fertile tropical region, so food production in these countries would not be able to withstand a huge diversion for the purpose of producing fuel.

Water stress is a limiting factor for the United States, Indonesia, and most of Europe. The existing thriving economic sectors in the United States and Europe are already tapping into the finite water resources. The addition of the biojet fuel industry for these wealthy nations will generally improve energy diversity in general but will put too much undue pressure on water supply. Unlike food and energy which can be imported, water is primarily a localized issue. Large quantity transfer of water over a large distance is a logistic impossibility.

Evaluating the limiting factors as a whole, economic factors will trump EWF concerns in the immediate term. Following it, food diversion and water concerns are equally pressing as limiting factors. Decreasing energy diversity is a minor concern globally.

6.8.3.2 Energy diversity

It is not disputed that the introduction of biojet fuel will help countries to achieve their decarbonization targets to battle climate change. However, the process of introducing biojet fuel might represent a shift in the energy mix. The monthly crude oil price in the entire of year 2020 hovered in the range of USD 17.28–71.68 per barrel, with the price on an upward trajectory and closed the year around the USD 50 per barrel range. Considering the previous high of 2020 and upward trajectory, the EWF nexus model projects the energy diversity change based on jet fuel price of USD 80 per barrel and production cost of biojet fuel are USD 0.355 per liter (OTJ), USD 1.00 per liter (ETJ), and USD 0.840 per liter (GTJ). Fig. 6.14 shows the global change in energy diversity as represented by

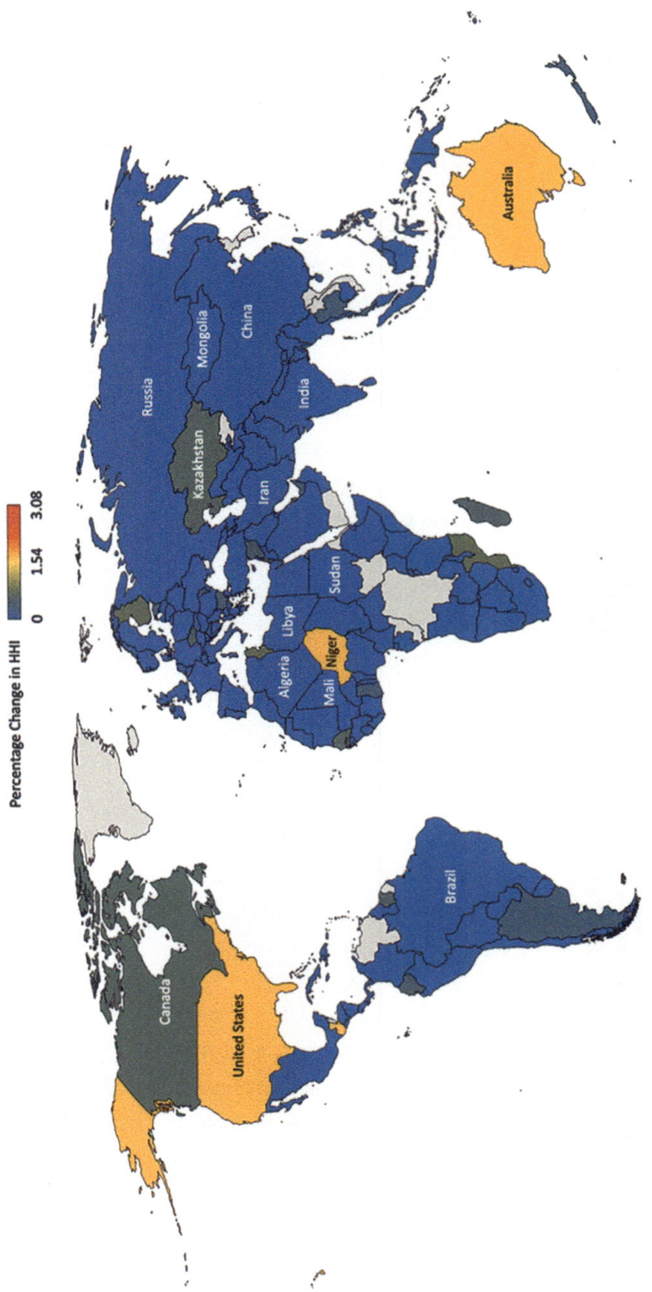

Figure 6.14 Global change in energy diversity as represented by percentage change in HHI in when conventional jet fuel price is USD 80 per barrel and baseline production cost of biojet fuel. *HHI*, Herfindahl–Hirschman Index.

percentage change in HHI in when conventional jet fuel price is USD 80 per barrel and production cost of biojet fuel is normal.

Using a realistic projection, biojet fuel is unlikely to change the energy mix of most countries. Countries that will see a notable positive change in energy diversity include the United States, Australia, Canada, Niger, Kazakhstan, Finland, Tunisia, Ecuador, Madagascar, and Argentina. For this scenario to improve, conversion technologies have to be more mature to bring production cost down. In this model, only the HEFA pathway contributed to biojet fuel into the energy mix in any meaningful quantity as it is the only production method sufficiently mature to take advantage of the higher jet fuel price. Thus, United States and Australia have the chance to build their biojet fuel industry if they strategically harness the potential of their feedstock. This can be done with present-day technology and without the governmental subsidies.

Several macroeconomics analysts have projected a postpandemic supernormal bull run for the oil and gas industry, where oil prices are expected to test the USD 200 per barrel psychological barrier. Considering that jet fuel prices mimic crude oil price, jet fuel price by extension is expected to also test the same price barrier. It is unclear if the recovery phase of the aviation industry will actually lead to the supernormal bull run, but a projection using the EWF model is conducted for jet fuel price of USD 180 per barrel and production cost of biojet fuel are USD 0.20 per liter (OTJ), USD 0.80 per liter (ETJ), and USD 0.84 per liter (GTJ). The lowered biojet fuel production cost reflects possible assistance provided by governments through subsidies. Such conditions represent the most optimistic of scenarios plausible, pending another "black swan" event akin to the COVID-19 global pandemic, which structurally changed all markets.

Fig. 6.15 shows the global change in energy diversity as represented by percentage change in HHI when conventional jet fuel price is USD 180 per barrel with favorable production cost of biojet fuel. Under this optimistic projection, Brazil, Angola, Poland, Iraq, New Zealand, Cambodia, Ukraine, and Colombia stand to improve their energy mix substantially. Australia's energy mix will not be improved further from this change in circumstances, while the United States and Niger will not diverse their energy sources. This further cements the idea that the EWF nexus is complex and dynamic where a single factor like cost will affect penetration of biojet fuel in a less than predictable manner.

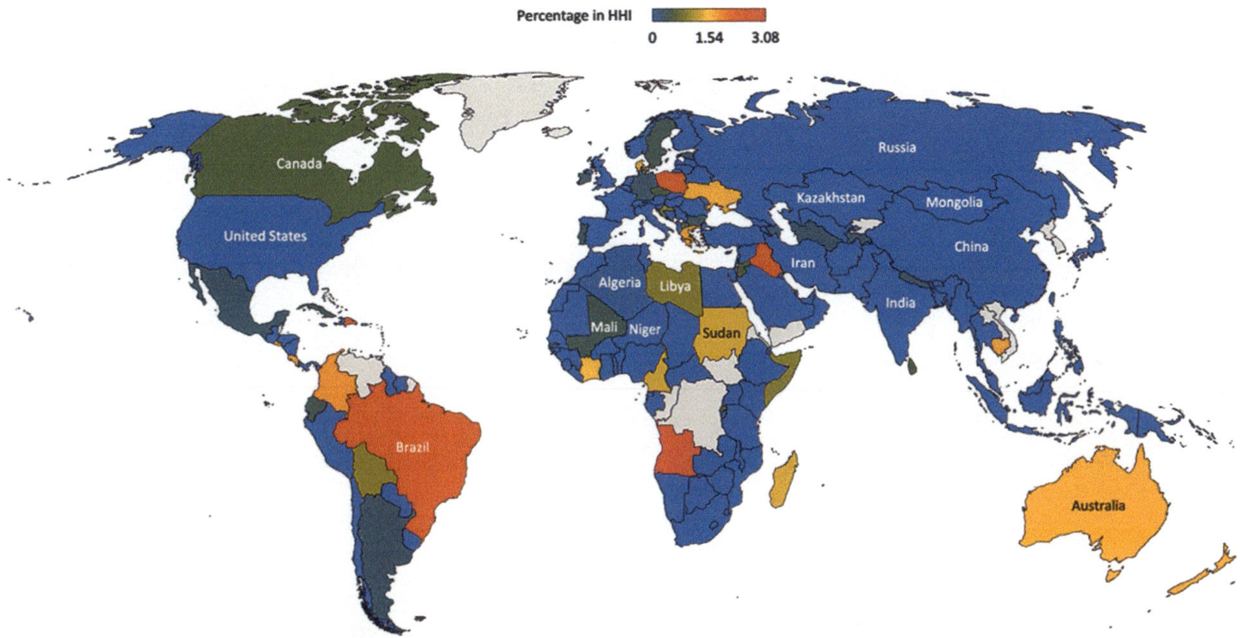

Figure 6.15 Global change in energy diversity as represented by percentage change in HHI in when conventional jet fuel price is USD 180 per barrel and production cost of biojet fuel are USD 0.20 per liter (OTJ), USD 0.80 per liter (ETJ), and USD 0.84 per liter (GTJ). *ETJ*, ethanol-to-jet; *GTJ*, gas-to-jet; *HHI*, Herfindahl–Hirschman Index; *OTJ*, oil-to-jet.

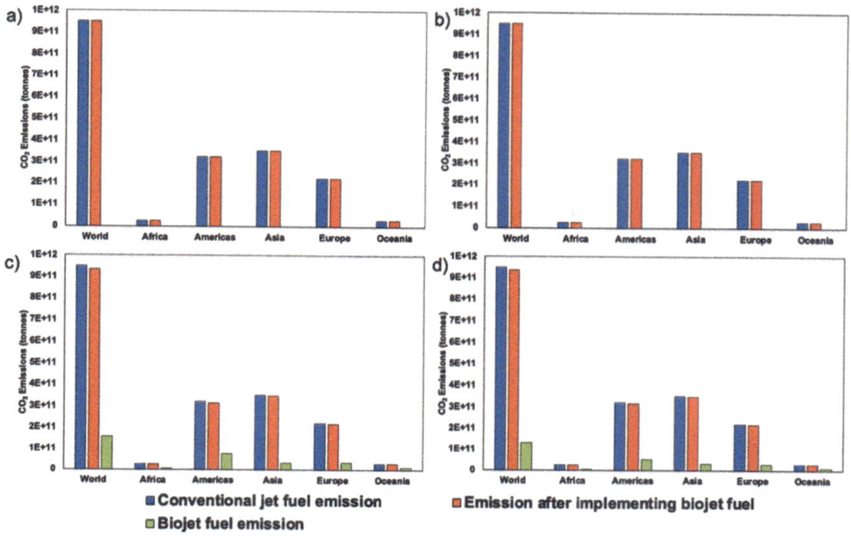

Figure 6.16 Aviation sector emissions scenarios for various jet fuel price (JFP) in USD per barrel and production cost per liter, where (A) JFP = 80, OTJ = −0.2, ETJ = 0.8, GTJ = 0.84, (B) JFP = 80, OTJ = 0.355, ETJ = 1.0, GTJ = 0.84, (C) JFP = 180, OTJ = −0.2, ETJ = 0.8, GTJ = 0.84, (D) JFP = 180, OTJ = 0.355, ETJ = 1.0, GTJ = 0.84. *ETJ*, ethanol-to-jet; *GTJ*, gas-to-jet; *OTJ*, oil-to-jet.

6.8.3.3 Emissions

The introduction of biojet fuel will displace conventional jet fuel and also the associated CO_2 emissions. Fig. 6.16 uses the EWF model by the authors to predict aviation sector emissions for four scenarios of various jet fuel prices and biojet fuel production costs. The four scenarios are as follows:

- Scenario 1 (optimistic O&G cycle with favorable production cost): Jet fuel price of USD 80 per barrel, production costs in USD per liter for OTJ = −0.2, ETJ = 0.8, and GTJ = 0.84
- Scenario 2 (optimistic O&G cycle with baseline production cost): Jet fuel price of USD 80 per barrel, production costs in USD per liter for OTJ = 0.355, ETJ = 1.0, and GTJ = 0.84
- Scenario 3 (supernormal O&G cycle with favourable production cost): Jet fuel price of USD 180 per barrel, production costs in USD per liter for OTJ = −0.2, ETJ = 0.8, and GTJ = 0.84
- Scenario 4 (supernormal O&G cycle with baseline production cost): Jet fuel price of USD 180 per barrel, production costs in USD per liter for OTJ = 0.355, ETJ = 1.0, and GTJ = 0.84

In scenarios 1 and 2, the emissions for the aviation sector are generally unchanged as biojet fuel is unlikely to have made any inroads to displace conventional jet fuel in any considerable quantity. As such, the emissions will be unchanged. This projection should form the lower bounds as governments around the world would either be subsidizing the biojet fuel industry or taxing the conventional jet fuel producers to meet the various decarbonization targets. Nonetheless, the free market prognosis is poor until production costs can be lowered. Two factors that are often understated include the rate at which production price tumbles as economies of scale are achieved, and pioneering nations figure out the best way to build the biojet fuel industry. Good examples include the solar energy industry pioneered in Germany and EV industries of the United States and China.

Scenarios 3 and 4 showed that biojet fuel has attained a market foothold, hence contributing the biojet fuel emissions. This should not be confused with LCA values. The prebiojet fuel and postbiojet fuel CO_2 emissions quantities do not vary much as biojet fuel and conventional aviation fuel have very similar fuel properties and are used in the same jet engines. Nonetheless, biojet fuel CO_2 emissions in the Americas, Asia and Europe are expected to be substantial but are unlikely to dominate over conventional jet fuel. It will be more prudent to expect emissions reduction from a life cycle perspective than combustion emissions perspective.

6.8.3.4 Energy–water–food nexus by biojet fuel generations

Table 6.6 summarizes the EWF nexus and other holistic concerns for the biojet fuels produced from different generations. The main strength of

Table 6.6 EWF and other holistic considerations of biojet fuel production by feedstock generations.

Generation	First-generation biojet fuel	Second-generation biojet fuel	Third-generation biojet fuel
Key feedstock	Edible oil, edible animal fat, edible sugar, edible starch	Nonedible oil, cellulosic materials, wastes	Microalgae, macroalgae, yeasts.
Energy dimension	Low energy sustainability, Net energy ratio (NER) ~0.8–1.5:1	High energy sustainability, NER as high as 5.4:1	Potentially high energy sustainable, NER ~0.2–3.0:1

Table 6.6 EWF and other holistic considerations of biojet fuel production by feedstock generations.—cont'd

Generation	First-generation biojet fuel	Second-generation biojet fuel	Third-generation biojet fuel
Water dimension	High water requirement	Potentially lower water requirement, as feedstock is by-product of process	High water requirement for conventional microalgae farming
Food dimension	Competes with food usage	Does not compete with food usage	Does not compete with food usage, and may even be used as food
Land usage	High land usage	High land usage, but can use marginal land	Low land usage as dominant feedstock of microalgae and macroalgae are not terrestrial based
Scalability	Large-scale production already in existence	Large-scale production is ramping, although yield for oil-based crops is inconsistent around the world. Logistics poses a huge barrier for cellulosic and waste feedstocks	Large-scale production requires more R&D, scalability studies and larger-scale pilot projects
Emissions	Significant pollution in CO_2 and GHG emissions	Significant reduction in CO_2 and other GHG emissions as compared with first-generation feedstocks	Potentially even lower CO_2 and other GHG emissions than second-generation feedstocks, although real-world large-scale project measurements are not yet available
Economic impacts	May cause food price increase, but improve livelihood of farmers	Improve rural economies	Generate modern agricultural or "high-technology" jobs

first-generation biojet fuel lies in the inherent scalability as it is directly a function of agricultural outcomes. Scaling of agricultural activities through crop intensification will occur with or without considering the biojet fuel industry. As agriculture modernizes, the energy sustainability might be improved, and water requirement could be reduced. As crop yield increases, food production might even lead to overproduction, which solves the food versus fuel debate. The possible food price increase from rising demand of first-generation feedstock is a double-edge sword. On one hand, it improves the livelihood of farmers. On the other hand, food might be priced beyond affordability of the poor. In spite of the relatively poor EWF nexus outlook, it will be sensible to first use the current generation feedstock in the interest of scaling up biojet fuel production to achieve economies of scale. This will also benefit the implementation of future generation biojet fuels.

The key strengths of second-generation biojet fuels are the twin factors of land usage reduction and noncompetition with food. The latter can be solved with increased yield, but the former is especially beneficial as land is a finite resource, which is ever dwindling. The scaling of second-generation feedstock looks promising as supply chains to collect wastes and residues are forming around the world, albeit still on scales which are orders of magnitude smaller than agricultural products. This is expected to be reversed in the next decade. Water usage is debatable as presently the feedstocks are considered as a by-product and not counted within the boundaries of water cycle calculations. Assuming that wastes and residues are considered as coproducts, then water stress imparted by these feedstocks will be factored in. This paradigm shift means that water requirement should be shared with first-generation crops and their agricultural residues. The EWF nexus considerations for second-generation biojet fuel are outstanding. As marginal lands could be utilized, rural economies can be stimulated. Thus, second-generation feedstocks should be the focus of the industry as the associated technologies are already available; the only requirement is the governmental will power and industry building know-how.

The pool of third-generation feedstock consists primarily of microalgae, macroalgae, and yeast-derived feedstocks. However, third generation is almost synonymous with microalgae due to its dominance in the research field. Third-generation feedstocks hold extreme promises in yield improvement and life cycle carbon reduction. Looking at the EWF nexus considerations, the food dimension will improve, while the water dimension will depend on the ability to use wastewater instead of fresh water. Energy dimension still requires more research to be conducted as the energy balance from a life cycle perspective is still unclear. When implemented correctly, third-generation feedstocks have the best EWF nexus balance. Unlike second-generation feedstocks where it is a matter of "when, not if," third-generation feedstock is a question of "if and when."

6.9 Summary

There exist different technologies and feedstocks that can be used to produce alternative jet fuel. Thorough assessment on the environmental and climatic impacts from the use of these alternative fuels is important to ensure the targets of negative carbon emissions are achieved. Life cycle assessment has become an important tool to assess the climatic impact from GHG emissions. The ICAO has set up a task force that assesses the LCA GHG emissions of alternative jet fuel and establishes the sustainability criteria of different jet fuel conversion pathways and feedstocks. From several LCA studies, it has been shown that alternative jet fuels derived from FT, HEFA, and ATJ pathways can lead to reduced GHG emissions up to 80% compared with conventional jet fuels. Among the conversion pathways, the carbon intensities for biojet fuels produced from lignocellulosic feedstocks, waste gas, and sugar-based feedstock via the ATJ routes tend to be lower. Lipid conversion technology via the HEFA pathway also shows promising reduced carbon footprint. Oil-bearing crops such as rapeseed, palm oil, soybean, jatropha, and camelina have shown potential GHG emissions reduction, but the GHG values can vary significantly when LUC effects are taken into considerations. The release of carbon into the atmosphere as a result of land conversion will lead to the spike in GHG emission intensity. The use of agrochemicals during the cultivation stage and the oil yield in the feedstock are among the factors that contribute to the GHG emissions. Gasification of biomass for production of FT fuel and combined pyrolysis and hydrotreatment process are viable options for low carbon jet fuel production. The use of renewable energies from solar and wind for production of hydrogen are strategies that can be adopted to further lower GHG emissions intensities. In addition to LCAs of carbon and energy, resource management in the form of the interdependent EWF nexus is equally important. Using an EWF nexus approach, the limiting factors in resources can be determined. Water stress issue is a concern for the United States and most of Europe, while food diversion has the potential to affect countries with large landmass such as Canada, Russia, Australia, Brazil, and Argentina. When EWF nexus is coupled with economic factors, sustainable development of the biojet fuel industry is curtailed by financial viability. Energy diversity can also be improved for most countries if a viable biojet fuel industry is successfully developed. The risks and opportunities in carbon emissions associated with alternative jet fuel production depend on the feedstock choice and production pathways. The promotion of alternative jet fuels with low carbon footprint and environmentally sustainable is essential to help achieve the decarbonization targets for the sector.

References

Archer, M., Szklo, A., 2016. Can increasing gasoline supply in the United States affect ethanol production in Brazil? Renew. Energy 95, 586–596.

ASTM D7566-19b, 2019. ASTM International, Standard Specification for Aviation Turbine Fuel Containing Synthesized Hydrocarbons.

Barbera, E., Naurzaliyev, R., Asiedu, A., Bertucco, A., Resurreccion, E.P., Kumar, S., 2020. Techno-economic analysis and life-cycle assessment of jet fuels production from waste cooking oil via in situ catalytic transfer hydrogenation. Renew. Energy 160, 428–449.

Benito, A., Alonso, G., 2018. Energy Efficiency and the Environment in Energy Efficiency in Air Transportation, pp. 153–199.

Budsberg, E., Crawford, J.T., Morgan, H., Chin, W.S., Bura, R., Gustafson, R., 2016. Hydrocarbon bio-jet fuel from bioconversion of poplar biomass: life cycle assessment. Biotechnol. Biofuels 9, 170.

Cantarella, H., Nassar, A.M., Cortez, L.A.B., Baldassin, R., 2015. Potential feedstock for renewable aviation fuel in Brazil. Environ. Develop. 15, 52–63.

Capaz, R., Posada, J., Seabra, J., Osseweijer, P., 2018. Life cycle assessment of renewable jet fuel from ethanol: an analysis from consequential and attributional approaches. In: 26th European Biomass Conference and Exhibition, 14–17 May 2018, Copenhagen, Denmark.

Carter, N.A., Stratton, R.W., Bredehoeft, M.K., Hileman, J.I., 2011. Energy and environmental viability of select alternative jet fuel pathways. In: 47th AIAA/ASME/SAE/ASEE Joint Propulsion Conference & Exhibit. AIAA, San Diego, California, 2011–5968.

Chang, S., Zhao, L., Timilsina, G.R., Zhang, X., 2012. Biofuels development in China: technology options and policies needed to meet the 2020 target. Energy Pol. 51, 64–79.

Chiaramonti, D., Goumas, T., 2019. Impacts on industrial-scale market deployment of advanced biofuels and recycled carbon fuels from the EU renewable energy directive II. Appl. Energy 251, 113351.

Chuck, C.J., McManus, M., Allen, M.J., Singh, S., 2016. Feedstocks for aviation biofuels. In: Biofuels for Aviation, pp. 17–34.

CORSIA, 2019. CORSIA Supporting Document - Life Cycle Assessment Methodology. https://www.icao.int/environmental-protection/CORSIA/Pages/default.aspx.

Cortez, L.A.B., Cantarella, H., Moraes, M.A.F.D., Nogueira, L.A.H., Schuchardt, U., Franco, T.T., et al., 2016. Roadmap to a sustainable aviation biofuel: a Brazilian case study. In: Biofuels for Aviation, pp. 339–350.

Costello, C., Ayoub, N., 2019. Exploring the potential for riparian marginal lands to enhance ecosystem services and bioenergy production. In: Biofuels, Bioenergy and Food Security, pp. 101–123.
Crossin, E., 2017. Life cycle assessment of a mallee eucalypt jet fuel. Biomass Bioenergy 96, 162–171.
de Jong, S., Antonissen, K., Hoefnagels, R., Lonza, L., Wang, M., Faaij, A., Junginger, M., 2017. Life-cycle analysis of greenhouse gas emissions from renewable jet fuel production. Biotechnol. Biofuels 10, 64.
Dickinson, K.E., Whitney, C.G., McGinn, P.J., 2013. Nutrient remediation rates in municipal wastewater and their effect on biochemical composition of the microalga Scenedesmus sp. AMDD. Algal Res. 2 (2), 127–134.
EU, 2018. Directive (EU) 2018/2001 of the European Parliament and of the Council of 11 December 2018 on the promotion of the use of energy from renewable sources. Off. J. Eur. Union 5, 82–209.
Fortier, M.-O.P., Roberts, G.W., Stagg-Williams, S.M., Sturm, B.S.M., 2014. Life cycle assessment of bio-jet fuel from hydrothermal liquefaction of microalgae. Appl. Energy 122, 73–82.
Ganguly, I., Pierobon, F., Bowers, T.C., Huisenga, M., Johnston, G., Eastin, I.L., 2018. 'Woods-to-Wake' life cycle assessment of residual woody biomass based jet fuel using mild bisulfite pretreatment. Biomass Bioenergy 108, 207–216.
Giwa, A., Adeyemi, I., Dindi, A., Lopez, C.G.-B., Lopresto, C.G., Curcio, S., Chakraborty, S., 2018. Techno-economic assessment of the sustainability of an integrated biorefinery from microalgae and Jatropha: a review and case study. Renew. Sustain. Energy Rev. 88, 239–257.
Han, J., Elgowainy, A., Cai, H., Wang, M.Q., 2013. Life-cycle analysis of bio-based aviation fuels. Bioresour. Technol. 150, 447–456.
Handler, R.M., Shonnard, D.R., Griffing, E.M., Lai, A., Palou-Rivera, I., 2016. Life cycle assessments of ethanol production via gas fermentation: anticipated greenhouse gas emissions for cellulosic and waste gas feedstocks. Ind. Eng. Chem. Res. 55, 3253–3261.
Hileman, J.I., Stratton, R.W., 2014. Alternative jet fuel feasibility. Transport Pol. 34, 52–62.
IATA, 2020. IATA, Jet Fuel Price Monitor. Available at: https://www.iata.org/en/publications/economics/fuel-monitor/. (Accessed 20 October 2020).
ICAO, 2016. IATA 2015 Report on Alternative Fuels, tenth ed. Retrieved from: https://www.iata.org/publications/Documents/2015-reportalternative-fuels.pdf.
ICAO, 2019a. 2019 Environmental Report - Aviation and Environment.
ICAO, 2019b. Annex 16 - Environmental Protection, CORSIA Default Life Cycle Emissions Values for CORSIA Eligible Fuels. https://www.icao.int/environmental-protection/CORSIA/Pages/default.aspx.
IEA, 2020. International Energy Agency, if the Energy Sector Is to Tackle Climate Change, it Must Also Think about Water. https://www.iea.org/commentaries/if-the-energy-sector-is-to-tackle-climate-change-it-must-also-think-about-water.
ISO 14040, 2006. ISO 14040:2006, Environmental Management - Life Cycle Assessment - Principles and Framework.
ISO 14044, 2006. ISO 14044:2006, Environmental Management - Life Cycle Assessment - Requirements and Guidelines.
Klein, B.C., Chagas, M.F., Junqueira, T.L., Rezende, M.C.A.F., Cardoso, T.d.F., Cavalett, O., Bonomi, A., 2018. Techno-economic and environmental assessment of renewable jet fuel production in integrated Brazilian sugarcane biorefineries. Appl. Energy 209, 290–305.

Li, X., Mupondwa, E., 2014. Life cycle assessment of camelina oil derived biodiesel and jet fuel in the Canadian Prairies. Sci. Total Environ. 481, 17–26.
Liu, H., Qiu, T., 2019. Life cycle assessment of Jatropha jet biodiesel production in China conditions. In: Proceedings of the 29th European Symposium on Computer Aided Process Engineering, June 16th to 19th, 2019, Eindhoven, The Netherlands.
Lokesh, K., Sethi, V., Nikolaidis, T., Goodger, E., Nalianda, D., 2015. Life cycle greenhouse gas analysis of biojet fuels with a technical investigation into their impact on jet engine performance. Biomass Bioenergy 77, 26–44.
Malins, C., Searle, S., Baral, A., 2014. A Guide for the Perplexed to the Indirect Effects of Biofuels Production. The International Council on Clean Transportation. Retrieved from: http://www.theicct.org/guide-perplexed-indirecteffects-biofuels-production.
Mekonnen, M.M., Hoekstra, A.Y., 2011. The green, blue and grey water footprint of crops and derived crop products. Hydrol. Earth Syst. Sci. 15, 1577–1600.
Neuling, U., Kaltschmitt, M., 2018. Techno-economic and environmental analysis of aviation biofuels. Fuel Process. Technol. 171, 54–69.
O'Connell, A., Kousoulidou, M., Lonza, L., Weindorf, W., 2019. Considerations on GHG emissions and energy balances of promising aviation biofuel pathways. Renew. Sustain. Energy Rev. 101, 504–515.
OFID, 2017. The OPEC Fund for International Development, the Energy–Water–Food Nexus: Managing Key Resources for Sustainable Development. https://opecfund.org/var/site/storage/original/application/80be162d98453051ded87e13032727cf.pdf.
Page, S.E., Morrison, R., Malins, C., Hooijer, A., Rieley, J.O., Jauhiainen, J., 2011. Review of peat surface greenhouse gas emissions from oil palm plantations in southeast Asia. In: White Paper No. 15. Indirect Effects of Biofuel Production Series. International Council on Clean Transportation. Available online at: http://www.theicct.org/sites/default/files/publications/ICCT_Peat-Emissions_Sept2011.pdf.
Paredes-Sánchez, J.P., López-Ochoa, L.M., López-González, L.M., Las-Heras-Casas, J., Xiberta-Bernat, J., 2019. Evolution and perspectives of the bioenergy applications in Spain. J. Clean. Prod. 213, 553–568.
Popp, J., Lakner, Z., Harangi-Rákos, M., Fári, M., 2014. The effect of bioenergy expansion: food, energy, and environment. Renew. Sustain. Energy Rev. 32, 559–578.
Stratton, R.W., Wong, H.M., Hileman, J.I., 2010. Life Cycle Green House Gas Emissions from Alternative Jet Fuels. Partnership of Air Transportation Noise and Emissions Reduction. Massachusetts Institute of Technology, Cambridge, MA. http://web.mit.edu/aeroastro/partner/reports/proj28/partner-proj28-2010-001.pdf.
Takriti, S.E., Pavlenko, N., Searle, S., 2017. Mitigating International Aviation Emissions. The International Council on Clean Transportation.
Tsiligiannis, A., Tsiliyannis, C., 2019. Renewable energy in cement manufacturing: a quantitative assessment of energy and environmental efficiency of food residue biofuels. Renew. Sustain. Energy Rev. 107, 568–586.
UNFCCC, 1998. Kyoto Protocol to the United Nations Framework Convention on Climate Change. Retrieved from: http://unfccc.int/kyoto_protocol/items/2830.php.
US DOE, 2012. Amyris Integrated Biorefinery. U.S. Department of Energy. https://www1.eere.energy.gov/bioenergy/pdfs/ibr_arra_amyris.pdf.
Valin, H., Peters, D., den Berg, M., Frank, S., Havlik, P., Forsell, N., et al., 2015. The Land Use Change Impact of Biofuels Consumed in the EU. Quantification of Area and Greenhouse Gas Impacts. Retrieved from: http://www.globiom-iluc.eu/iluc-study-now-available-online/. (Accessed 20 October 2020).

Vasquez, M.C., Martínez, A., Castillo, E.F., Silva, E.E., 2019. Holistic approach for sustainability enhancing of hydrotreated aviation biofuels, through life cycle assessment: a Brazilian case study. J. Clean. Prod. 237, 117796.
Vela-García, N., Bolonio, D., Mosquera, A.M., Ortega, M.F., García-Martínez, M.-J., Canoira, L., 2020. Techno-economic and life cycle assessment of triisobutane production and its suitability as biojet fuel. Appl. Energy 268, 114897.
Warshay, B., Pan, J., Sgouridis, S., 2011. Aviation industry's quest for a sustainable fuel: considerations of scale and modal opportunity carbon benefit. Biofuels 2, 33–58.
Wei, H., Liu, W., Chen, X., Yang, Q., Li, J., Chen, H., 2019. Renewable bio-jet fuel production for aviation: a review. Fuel 254, 115599.

Index

Note: 'Page numbers followed by "f" indicate figures and "t" indicate tables.'

C

FSC
www.fsc.org
MIX
Papier | Fördert
gute Waldnutzung
FSC® C083411

Kolibri 360 GmbH
Ferdinand-Jühlke-Straße 7
99095 Erfurt, Deutschland
produktsicherheit@kolibri360.de